AA007344

22nd International Summer School On Vacuum, Electron and Ion Technologies (VEIT 2021)

Journal of Physics: Conference Series Volume 2240

Sozopol, Bulgaria
20 - 24 September 2021

ISBN: 978-1-7138-5495-1
ISSN: 1742-6588

Printed from e-media with permission by:

Curran Associates, Inc.
57 Morehouse Lane
Red Hook, NY 12571

Some format issues inherent in the e-media version may also appear in this print version.

This work is licensed under a Creative Commons Attribution 3.0 International Licence.
Licence details: http://creativecommons.org/licenses/by/3.0/.

No changes have been made to the content of these proceedings. There may be changes to pagination and minor adjustments for aesthetics.

Printed with permission by Curran Associates, Inc. (2023)

For permission requests, please contact the Institute of Physics at the address below.

Institute of Physics
Dirac House, Temple Back
Bristol BS1 6BE UK

Phone: 44 1 17 929 7481
Fax: 44 1 17 920 0979

techtracking@iop.org

Additional copies of this publication are available from:

Curran Associates, Inc.
57 Morehouse Lane
Red Hook, NY 12571 USA
Phone: 845-758-0400
Fax: 845-758-2633
Email: curran@proceedings.com
Web: www.proceedings.com

TABLE OF CONTENTS

Preface VEIT 2021

Peer Review Statement

Phosphate Sealing Improvement of the Corrosion Resistance of Thin Cerium Oxides Coatings Electroless Deposited on Aluminium 1
R Andreeva, D Stoychev

Optimization of Atomic Layer Deposition of Al2O3 Films as Possible Template for Graphene Transfer........ 7
B S Blagoev, D A Delibatov, V B Mehandzhiev, P Sveshtarov, P Terziyska, I Avramova, P M Rafailov

Formation of Oxide Films on a Curved Aluminium Surface Upon Anodic Oxidation........ 12
B Tzaneva, V Videkov, A Bankova, M Mitov, D Dobrev

Atomic Layer Deposition of AlN on Different SiC Surfaces 18
M Beshkova, P Deminskyi, C-W Hsu, I Shtepliuk, I Avramova, R Yakimova, H Pedersen

Morphological Evolution of Thin AlN Films Grown by Atomic Layer Deposition 23
M Beshkova, B S Blagoev, V Mehandzhiev, R Yakimova, B Georgieva, I Avramova, P Terziyska, V Strijkova

Study of the Structure and Mechanical Properties of Electron-Beam-Welded Dissimilar Joints of Copper and Stainless Steel with and Without Offset 28
D Kaisheva, A Anchev, V Dunchev, B Stoyanov, S Valkov, M Ormanova, G Kotlarski, V Todorov, M Atanasova, S Rabadzhiyska, P Petrov

Optical Properties of Nanostructured Bimetallic Films from the Ag-In and Ag-Sb Systems and Their Surface-Enhanced Fluorescence Application........ 34
V Katrova, T Hristova-Vasileva, A Atanasova, V Strijkova, R Todorov

Effect of Laser Annealing on the Properties of Ag/ZnO Nanostructures........ 42
M E Koleva, A O Dikovska, N N Nedyalkov, D Karashanova

Structure and Phase Composition Study of Thin TiO2:C Films Deposited by R.f. Magnetron Sputtering 47
T Milenov, D Karaivanova, O Angelov, P Terziyska, G Avdeev, D Karashanova, B Georgieva, K Genkov, D Dimov, K Ivanov, S Kolev, E Valcheva

Formation and Characterization of CuO Coatings Deposited by Reactive Magnetron Sputtering........ 54
M Ormanova, G Kotlarski, S Valkov, D Dechev, N Ivanov, P Petrov

Study of the Structure, Roughness and Optical Properties of HfO2 Coatings Deposited on Microscopic Glass Substrates 60
S Rabadzhiyska, M Ormanova, S Valkov, D Dechev, P Terziyska, P Petrov

Comparative Analysis of Cavitation Erosion of Cr-C Carbide Coatings Depending on the Carbon Content 66
V Safonov, A Zykova, J Steller, T Seremak, N Donkov

Surface Treatment of Magnetron-Sputtered Tantalum Pentoxide Coatings and Its Effect on the Phenotype and Function of Immune Cells in Vitro 71
N Donkov, A Zykova, V Safonov, S Dudin, S Yakovin, A Goltsev, T Dubrava

Low Temperature Quaternary Hard Coatings with Cr- To Ti-Based Transition Composition........ 76
V Chitanov, St Kolchev, T Cholakova, L Kolaklieva, R Kakanakov, Ch Pashinski, N Hristeva

Spectral Readout of SPR Excited in Diffraction Grating 82
E Belina, V Mankov, H Kisov, T L Dimitrova, G Dyankov

Improvement in the Gas-Sensing Properties of a ZnO:Zn2TiO4 Composite Sensor by Light Irradiation 87
G Atanasova, T Dilova, A Dikovska, Ru Nikov, N Nedyalkov

Quartz Crystal Microbalance Sensor for NO2 Detection Based on Electrospun Amorphous Titanium Oxide Fibers 92
K Starbova, V Georgieva, N Starbov, P Stefanov, B Georgieva, J Lazarov

Study of Biogenic Iron Oxyhydroxide for Application in Electronics and Biotechnology 98
R Angelova, L Slavov, B Blagoev, Ch Ghelev, D Kovacheva, M Iliev, V Groudeva, I Nedkov

Silver Nanoparticles Synthesis and Their Effect on the SOPC Lipid Structure 103
Z Slavkova, J Genova, H Chamati, V Boev, D Yancheva

Double-Discharge Plasma System for Deposition of Carbon Nanostructures........ 108
S Marinov, V Vachkov, I Ivanov, Zh Kiss'ovski

Extraction of Accompanying Metals Through Electron Beam Refining of Copper 113
V Vassileva, K Vutova, M Markov

Coherent Optical Spectroscopy Characterization of the Magnetic Properties of Oriented Fe3O4 Nanoparticles........ 118
E Taskova, E Alipieva, S Kolev, T Koutzarova, D Brazhnikov

Effect of Cation Substitutions in Y-Type Ba0.5Sr1.5Me2Fe12O22 Hexaferrites on the Magnetic Phase Transitions 123
B Georgieva, S Kolev, Ch Ghelev, K Krezhov, D Kovacheva, B Vertruyen, R Closset, L-M Tran, M Babij, A Zaleski, T Koutzarova

Design of a Plasmonic Sensor Based on a Nanosized Structure for Biochemical Application 128
Hocine Ben Salah, Hocine Bahri, Abdesselam Hocini, Imane Zegaar, Sven Ingebrandt, Vivek Pachauri

Plasmonic Stop-Band Filter Based on an MIM Waveguide Coupled with Cavity Resonators........ 133
Imane Zegaar, Abdesselam Hocini, Djamel Khedrouche, Hocine Ben salah, Hocine bahri

Cirrus-Cloud Conditioned Ratio of Multiple-To-Single Backscatter Atmospheric Phase Functions 138
Ts Evgenieva, V Anguelov, L Gurdev

Aerosol Types Identification During Different Aerosol Events Over Sofia, Bulgaria, Using Sun-Photometer and Satellite Data on the Aerosol Optical Depth and Ångström Exponent 143
Ts Evgenieva, L Gurdev, E Toncheva, T Dreischuh

Characterization of the Aerosol Field Over Sofia, Bulgaria, Using Contact and Passive-Remote-Sensing Methods and Instruments 149
Ts Evgenieva, N Kolev, P Savov, L Gurdev, I Grigorov, T Dreischuh

Conversion of CO2 in Stabilized Low-Current Arc Discharge at Atmospheric Pressure 156
V Ivanov, S Lazarova, S Iordanova, Ts Paunska, N Georgiev, St Kolev

Organic Bulk Heterojunction Solar Cells Spin-Coated in Ambient Environment on Flexible and Glass Substrates 161
R Gergova, M Sendova-Vassileva, G Popkirov, Hr Dikov, G Grancharov, P Terziyska

Conversion of CO2 in a Pulsed Arc Discharge 166
V Vasilev, S Iordanova, E Vladkov, St Kolev

Vertical and Near-Horizontal Scanning Lidar Observations of Aerosol Loads Over Sofia, Bulgaria, in the Presence of Saharan Dust 171
Z Peshev, A Deleva, A Chaikovsky, V Pescherenkov, L Vulkova, T Dreischuh

Development of Lidar for Remote Methane Sensing Using an Optimal Configuration of High-Power Laser Diodes 180
V Pencheva, S Penchev, T Dreischuh

Thermodynamic Analysis of the Processes During Electron Beam Melting and Refining of Copper 185
V Stefanova, K Vutova, V Vassileva

Turbulent Flow Influence on the Discharge Parameters of a Magnetically-Stabilized Gliding Arc Discharge 191
P Tsonev, V Ivanov, St Kolev, Kh Tarnev, Ts Paunska

Simulations of a Capacitor with Graphene Electrodes 196
V Vachkov, Zh Kiss'ovski

Model-Based Multicriterial Optimization of Electron Beam Melting and Refining of Copper 200
E Koleva, L Koleva, V Vassileva, K Vutova, Ts Tsonevska

Multicriterial Optimization Strategies for Electron Beam Welding Processes 206
E Koleva, L Koleva, Dm Trushnikov, G Kolev, Z Petrova

Accuracy Comparison of Intraoral Versus Laboratory Scanners Used in the Contemporary Dental Practice 212
H Galeva, T Uzunov, Y Sofronov, G Todorov

Optimization of Titanium and Titanium Alloy Surface Properties by Ultra-Short Laser Processing for Improved Antibacterial Characteristics 217
D M Aceti, A Daskalova, L Angelova, E Filipov, L Sotelo, A Andreeva, A Trifonov, I Buchvarov

Altering the Surface Morphology and Wettability of Chitosan/graphene Coatings by Femtosecond and Nanosecond Laser Processing 222
E Filipov, L Angelova, D Aceti, V Marinova, D Karashanova, A Trifonov, I Buchvarov, A Daskalova

Ultra-Fast Laser Modification of Poly-Lactic Acid (PLA) -Towards Enhanced Biocompatibility 227
L Angelova, A Daskalova, R Mincheva, X Carette, A Trifonov, E Filipov, D Aceti, I Buchvarov

Duplex Surface Modification of Tool Steel - Deposition of a WN Film and Subsequent Electron-Beam Modification 233
S Valkov, M Ormanova, N Ivanov, D Dechev, R Bezdushnyi, P Petrov

Laser Processing of Nitride Ceramics 238
N Nedyalkov, T Dilova, A Dikovska, Ru Nikov, Ro Nikov, M Koleva, N Stankova, Al Daskalova, L Angelova

Laser Annealing of Bimetal Porous Structures Produced by PLD in Open Air 244
Ru Nikov, Ro Nikov, N Nedyalkov, A Dikovska, K Grochowska

Electrical Characterization of Memory Capacitors for Nonvolatile Memory Applications Based on Nanolaminated HfO2/Al2O3 and Al-Doped HfO2 Stacks 249
D Spassov, A Paskaleva, T Stanchev, Tz Ivanov

Effect of Growth Temperature on Nitrogen Incorporation into GaAsN During Liquid-Phase Epitaxy 255
M Milanova, V Donchev, S Georgiev, K Kirilov, P Terziyska

Nanosecond Laser Ablation of Nitride Ceramics in Liquid 260
Ro Nikov, N Nedyalkov, T Nikova, L Aleksandrov

Depth Profiling of Very Thin HfO2/Al2O3 Stacks by Ellipsometry 266
Y Karmakov, A Paskaleva, D Spassov

Comparative Study of the Sidewall Shape and Proximity Effect of the AR-N 7520 Negative Electron Beam Resist 271
I Kostic, K Vutova, E Koleva, A Bencurova, A Konecnikova, R Andok

Modification and Activation of the Surface of Medical-Grade PDMS After Irradiation by Ultrashort Laser Pulses 277
E Iordanova, G Yankov, N Stankova, N Nedyalkov

Author Index

Preface VEIT 2021

The International Summer School on Vacuum, Electron and Ion Technologies (VEIT) has been organized biennially since 1978, when the series of VEIT Schools was launched by the Institute of Electronics, Bulgarian Academy of Sciences, with the aim to act as a forum for exchange and dissemination of knowledge and ideas on the latest developments in electron-, ion-, and plasma-assisted technologies. The organizers of the 2021 edition of the event were the Institute of Electronics, Bulgarian Academy of Sciences, Sofia, Bulgaria and the Dutch Institute for Fundamental Energy Research, Eindhoven, The Netherlands. The School is a major scientific event and a meeting place for young scientists involved in research and development associated with high-tech industries. Many former School participants went on to become successful scientists in their field of research in leading institutions and companies all over the world. The School Proceedings have been published in Special issues of the international journals Vacuum, Plasma Processes and Polymers, Journal of Physics: Conference Series.

The twentieth-second edition of VEIT took place virtually via Zoom because of the COVID-19 pandemic, from 20 to 24 September 2021. It was attended by 106 participants (24 students and young researchers) from 16 countries on three continents: Algeria, Bulgaria, Czech Republic, France, Germany, Italy, Japan, Latvia, Portugal, Serbia, Slovakia, Sweden, Switzerland, The Netherlands, Ukraine, USA.

The School Program covered a large spectrum of fundamental problems of the nanotechnologies, the interaction of charged particles with solid surfaces, the deposition and characterization of thin films, which were balanced out by reports on a wide range of significant high-technology applications – from equipment for deposition of hard coatings or thin optical/protective films to nanostructures prepared via evaporation, sputtering or irradiation. By their subjects, the lectures and reports were grouped for presentation and discussion in three sessions, each of them including oral presentations at plenary sessions and posters.
The topics for VEIT 2021 are:
- Thin films deposition
- Surfaces and thin films processing and analysis
- Coatings for advanced applications
- New materials
- Green technologies
- Plasma-surface interaction and plasma diagnostics
- Modelling and computer simulation

The program comprised five topic lectures of 45 minutes, 18 progress reports of 30 minutes, and eight 20-minute reports by young scientists. Also, 79 original contributions were presented for 30 minutes each in Zoom at the three poster sessions. Six posters were active in parallel Virtual Rooms during the poster presentations in Zoom. Everyone was able to visit them via the main Zoom link of the Conference.

The only technical difficulty was the time zone difference. The organizers had to remind regularly and in advance each speaker about the time of the country in which they are located and the Bulgarian time. Despite the busy scientific program, the participants were completely satisfied by the organization, because everyone had the opportunity to present and get acquainted with the recent breakthroughs in the field of scientific research of all the involved contributors. Unfortunately, the social events were missed by everyone, even the pleasant conversations during the coffee breaks.

Following the tradition of publishing the VEIT Proceedings, a selection of 54 manuscripts presented at the event will be published in a special issue of Journal of Physics: Conference

Content from this work may be used under the terms of the Creative Commons Attribution 3.0 licence. Any further distribution of this work must maintain attribution to the author(s) and the title of the work, journal citation and DOI.
Published under licence by IOP Publishing Ltd

Series, under the originality and quality criteria of acceptance by the journal, including peer reviewing.

VEIT 2021 owes its success to many people. The International Advisory Committee shaped the scientific program and ensured high-quality plenary presentations by careful selection of the invited speakers. The Local Committee bore the brunt of the organization both during the virtual Zoom meetings and in handling correspondence, abstracts, and manuscripts for these proceedings. The members of the Local Committee are: Ch. Angelov, I. Balchev, M. Dimitrova (Chair), B. Georgieva, Ch. Ghelev, P. Ivanova, E. Taskova, E. Vasileva, all from the Bulgarian Academy of Sciences.

We are grateful to our sponsor the Ministry of Education and Science and the National Science Fund of Republic of Bulgaria for their steadfast support and generosity to manage the Zoom payment, printing conference materials, etc. We wish to express our appreciation to the members of the International Advisory Committee – N. Nedyalkov (Bulgarian Academy of Sciences, Sofia, Bulgaria), M. Dimitrova (Czech Academy of Sciences, Prague, The Czech Republic), R. Panek (Czech Academy of Sciences, Prague, The Czech Republic), Th. Czerwiec (Institut Jean Lamour, Ecole des Mines de Nancy, France), B. Rauschenbach (IOM and Leipzig University, Germany), H. Kersten (IEAP University of Kiel, Germany), D. Mataras (University of Patras, Greece), V. Guerra (Instituto Superior Tecnico, Lisbon, Portugal), G. Dinescu (National Institute for Laser, Plasma and Radiation Physics, Magurele-Bucharest, Romania), Z. Petrovič (Institute of Physics, Belgrade, Serbia), M. Mozetič (Jozef Stefan Institute, Ljubljana, Slovenia), K. Larsson (Uppsala University, Sweden), I. Katardjiev (Uppsala University, Sweden), M.C.M. (Richard) van de Sanden (Dutch Institute for Fundamental Energy Research, Eindhoven, The Netherlands), G.J. van Rooij (Dutch Institute for Fundamental Energy Research, Eindhoven, The Netherlands), M. Ürgen (Istanbul Technical University, Turkey), A. Ehiasarian, (Sheffield Hallam University, UK), I. Petrov (University of Illinois, Urbana, IL USA) for their continuous concern about maintaining the high scientific level and the future of the event, as well as for their contribution to inspiring new generations of outstanding young scientists and experts in the field. And, last but not least, we would also like to thank all authors for their valuable contributions to these proceedings and to the School, and all reviewers for their hard and tedious, but very important work.

The next conference in the series will be held in September 2023.

M.C.M. van de Sanden[1], Miglena Dimitrova[2], Chavdar Ghelev[3]
Guest Editors

[1] Dutch Institute for Fundamental Energy Research, Eindhoven, The Netherlands, M.C.M.vandeSanden@differ.nl

[2] Institute of Electronics, Bulgarian Academy of Sciences, Sofia, Bulgaria, miglena.dimitrova@gmail.com

[3] Institute of Electronics, Bulgarian Academy of Sciences, Sofia, Bulgaria, chavdarghelev@yahoo.com

Peer Review Statement

All papers published in this volume have been reviewed through processes administered by the Editors. Reviews were conducted by expert referees to the professional and scientific standards expected of a proceedings journal published by IOP Publishing.

- **Type of peer review:** Single Anonymous
- **Conference submission management system:** Morressier
- **Number of submissions received:** 54
- **Number of submissions sent for review:** 54
- **Number of submissions accepted:** 54
- **Acceptance Rate (Submissions Accepted / Submissions Received × 100):** 100
- **Average number of reviews per paper:** 1
- **Total number of reviewers involved:** 20
- **Contact person for queries:**
  Name: Miglena Dimitrova
  Email: miglena.dimitrova@gmail.com
  Affiliation: Institute of Electronics, BAS

Content from this work may be used under the terms of the Creative Commons Attribution 3.0 licence. Any further distribution of this work must maintain attribution to the author(s) and the title of the work, journal citation and DOI.
Published under licence by IOP Publishing Ltd

Phosphate sealing improvement of the corrosion resistance of thin cerium oxides coatings electroless deposited on aluminium

R Andreeva and D Stoychev[1]

Rostislaw Kaischew Institute of Physical Chemistry, Bulgarian Academy of Sciences, Acad. G. Bonchev Str., bl. 11, Sofia 1113, Bulgaria

E-mail: stoychev@ipc.bas.bg

Abstract. The corrosion-protective ability was investigated of electroless deposited ceria conversion coatings on substrate(s) of technically pure aluminium. By applying different electrochemical methods, a comparative study of the investigated samples was conducted regarding the effect of pre-treatment of the Al substrate and the type of phosphate post-treatment on the change of: E_{OCP} values, monitored E_{OCP} vs. time plot; course of i_a transients at the E_{pit} of Al, i.e., i_a vs. time plot; *R*p vs. time of exposure of the samples in a corrosion medium and connected with this change in the *CR* of the systems studied. The juxtaposition of the obtained results showed the order of stabilities of the investigated systems towards the appearance and development of pitting corrosion. The presented results can be connected with and explain the data obtained earlier about the calculated stoichiometry of the elements in the so-formed compounds in the surface layers ascertaining that they are built of phosphate and oxide compounds of Ce^{3+}, Ce^{4+} and Al^{3+}, as well as with the influence of the concentrations of Cu and Fe around them.
Keywords: aluminium, corrosion, ceria and phosphate(s) conversion coatings.

1. Introduction

Some authors consider that the formation of phosphate layers (PhL(s)) on conversion Ce-oxide coatings (CCOC) deposited on aluminium (Al) can have a positive impact on the extent of protective ability of these "green" protective systems. However, such considerations do not always correspond to the experimental findings.

We have previously reported our results regarding the effects of pre-treatment of Al substrates and additional phosphate post-treatment of CCOC deposited on Al substrates on the surface chemical composition of the so-formed Al/CCOC/PhL(s) systems. Based on the XPS investigations carried out with the studied samples, it was ascertained that there is a substantial influence of these "pre-" and "post-treatment" operations on the chemical composition and chemical state of the elements on their surface. It is expressed in a strong decrease in the concentration of Al_2O_3 and Ce_2O_3+CeO_2 components in CCOC at the expense of the formation of AlPO4 and AlOOH, $CePO_4$ as well as PO^{3-}, P_2O_5 and P_4O_{10} compounds with Al and Ce [1].

[1] To whom any correspondence should be addressed.

Content from this work may be used under the terms of the Creative Commons Attribution 3.0 licence. Any further distribution of this work must maintain attribution to the author(s) and the title of the work, journal citation and DOI.
Published under licence by IOP Publishing Ltd

The aim of the present corrosion investigations of the same systems was to obtain additional quantitative information about the influence of "pre-treatment and "post-treatment" operations connected with the formation of CCOC and PhLs with an increased corrosion-protective ability of Al 1050.

2. Experimental

Layers of CCOC (prepared by chemical conversion treatment) were deposited on substrates of "technically pure" Al 1050. Their pre-treatment, as described in [2], involved degreasing in organic solvent, etching in 1.5 M NaOH and consecutive activation in 5 M HNO_3 (the abbreviator of these samples in the text is "$Al_{(NaOH\&HNO3)}$"). The CCOCs were formed in a solution containing 0.5 M $CeCl_3 \times 7H_2O$ + 1×10^{-5} M $CuCl_2 \times 2H_2O$, as described in [3]. The abbreviator of these samples covered with CCOC is"$Al_{(NaOH\&HNO3)}/CCOC_{(Ce+Cu)}$". The thickness of the deposited CCOCs was 212 nm for the $Al_{(NaOH\&HNO3)}/CCOC_{(Ce+Cu)}$ system [3].

The post-treatment of the thin CCOC as-deposited on Al substrates aiming at the additional formation of phosphate layers (PhL(s)) on them was realized in 0.08 M Na_3PO_4 or 0.22 M $NH_4H_2PO_4$, as described in [2]. The abbreviations in these cases are: "$Al_{(NaOH\&HNO3)}/CCOC_{(Ce+Cu)}/PhL_{(Na3PO4)}$" or "$Al_{(NaOH\&HNO3)}/CCOC_{(Ce+Cu)}/ PhL_{(NH4H2PO4)}$", respectively. Open circuit potential (E_{OCP}) transients (trs.) and corrosion current (i_a) transients vs. time were measured in a 0.1 M NaCl corrosion medium (CM). The investigations of the changes in the polarization resistance (Rp), as well as the calculations of the corrosion rate (CR) of the samples studied, were carried out on a Gamry Interface 1000 (Software Analyzer EIS300). The Rp of the coatings was calculated by the Stern-Geary equation, which is based on the fact that a higher Rp value corresponds to a higher corrosion resistance and to a lower corrosion current as well, i.e., $R\text{p} \sim 1/i_{cor}$.

3. Results and discussion

Our prior investigation [2] of the potentiodynamic polarization curves (PDPC) of the ongoing cathode and anodic processes in CM established that: **I.** The deposition of CCOC on $Al_{(NaOH\&HNO3)}$ and the sealing of this system in $NH_4H_2PO_4$ (formation of the mixed system $Al_{(NaOH\&HNO3)}/CCOC_{(Ce+Cu)}/PhL_{(NH4H2PO4)}$) determines the strong influence on the kinetics of the cathode reaction of reduction of oxygen, which enhances the protection degree of this system; **II.** The similar post-treatment of the system $Al_{(NaOH\&HNO3)}/CCOC_{(Ce+Cu)}$ in both types of phosphate solutions leads to a substantial shift of the anodic curves towards a more positive E_{cor} and a lower i_{cor}; **III.** A characteristic feature of the anodic PDPC is the absence of a zone of passive state, which justifies the consideration that the corrosion potential (E_{cor}) of the studied systems coincides with the potential of pitting formation (E_{pit}), which is a prerequisite for the local character of the corrosion process under the conditions of stationary corrosion [2]. These results showed that the phosphated CCOC are not only static barrier coatings, but they change the kinetics of the conjugated electrochemical reactions, i.e., they also determine the electrochemical protection. At the same time, these results do not characterize fully the related corrosion reactions, since the real corrosion processes occur at the open circuit potential (E_{OCP}). To gain the needed additional insights, we conducted investigations of self-occurring corrosion processes in the systems of interest at E_{OCP} and studied the change in: E_{OCP} vs. time plot; i_{cor} transients at E_{pit}; Rp as a function of exposure time in CM (under E_{OCP} conditions), and the related change in CR.

3.1. Open circuit potential (OCP) measurements.

As it is known, the change in OCP as a function of the immersion time can be used to monitor the chemical stability and corrosion process of the studied samples, i.e., to assess their corrosion susceptibility. Figure 1 shows the results obtained for the studied samples after 1 h (figure 1 a) and 168 h (figure 1 b) of exposure in the CM. It is seen in figure 1 a that during the immersion for 1 h, the course and the changes in E_{OCP} for the bare sample (tr.1) and the system $Al_{(NaOH\&HNO3)}/PhL_{(NH4H2PO4)}$ (tr.3) practically coincide at ~ –0.705 V. A small shift in the positive direction (to ~ –0.683 V) is observed for the system $Al_{(NaOH\&HNO3)}/PhL_{(Na3PO4)}$ (tr.2). A consecutive increase of the E_{OCP} values are observed for the $Al_{(NaOH\&HNO3)}/CCOC_{(Ce+Cu)}$ (tr.4) – to ~ –0.655 V and post-treated in Na_3PO_4 or the $NH_4H_2PO_4$ $Al_{(NaOH\&HNO3)}/CCOC_{(Ce+Cu)}$ system – the mixed systems $Al_{(NaOH\&HNO3)}/CCOC_{(Ce+Cu)}/PhL_{(Na3PO4)}$ (tr.5) – to ~–0.645 V and $Al_{(NaOH\&HNO3)}/CCOC_{(Ce+Cu)}/PhL_{(NH4H2PO4)}$ (tr.6) – to ~ –0.640 V, respectively.

The big fluctuations observed for practically all studied systems point to dissimilarities in the potential of different intermetallic inclusions, metastable pit formation and re-passivation in the CM. The fluctuation in the OCP could be due as well to the partial incorporation of the aqua ion and formation of Al oxides/hydroxides. When the samples are exposed to the aggressive environment, the corrosive ions penetrate through the active regions and lead to a more negative OCP.

In total, the average positive change of E_{OCP} under the action of phosphate post-treatment of the CCOC is ~ 0.055 V. This positive shift (in the range of E_{OCP} from ~ –0.705 to ~ –0.640 V vs. Ag/AgCl) compared to the bare $Al_{(NaOH\&HNO3)}$ indicate the effective incorporation and also the stability of the phosphate-modified systems.

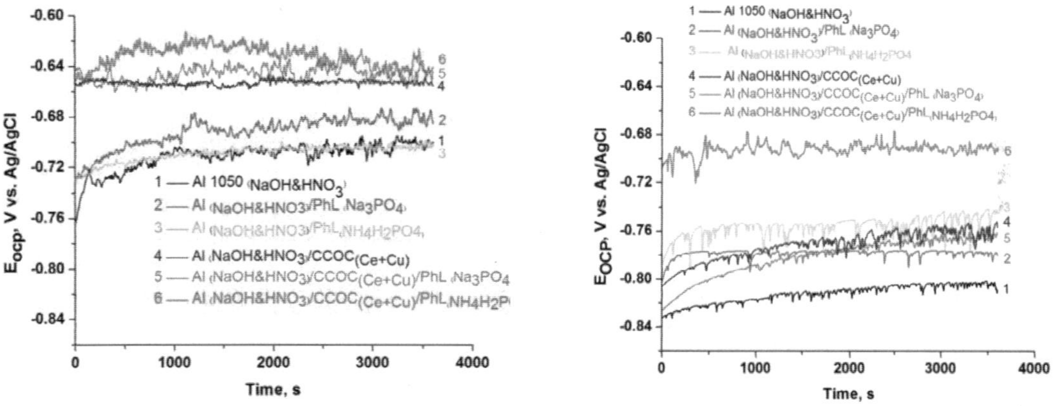

a) b)

Figure 1. OCP vs. time plots during immersion in a 0.1 M NaCl solution for as deposited samples (a) and after 168 h exposure in CM (b): (1) Bare $Al_{(NaOH\&HNO3)}$; (2) $Al_{(NaOH\&HNO3)}/PhL_{(Na3PO4)}$; (3) $Al_{(NaOH\&HNO3)}/PhL_{(NH4H2PO4)}$; (4) $Al_{(NaOH\&HNO3)}/CCOC_{(Ce+Cu)}$; (5) $Al_{(NaOH\&HNO3)}/CCOC_{(Ce+Cu)}/PhL_{(Na3PO4)}$; (6) $Al_{(NaOH\&HNO3)}/CCOC_{(Ce+Cu)}/PhL_{(NH4H2PO4)}$.

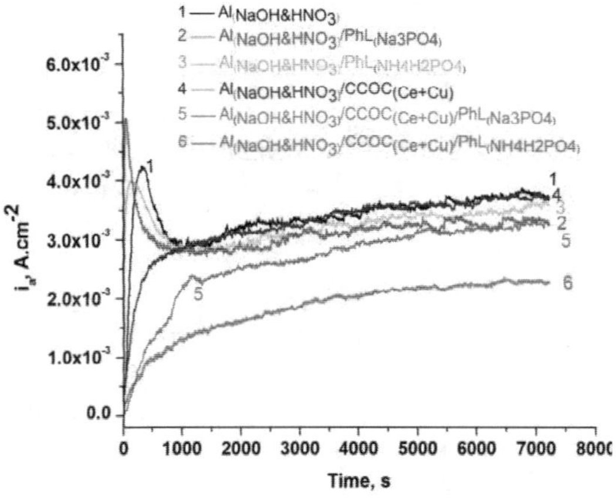

Figure 2. Chronoamperometric curves of the studied systems:
1 – $Al_{(NaOH\&HNO3)}$;
2 - $Al_{(NaOH\&HNO3)}/PhL_{(Na3PO4)}$;
3- $Al_{(NaOH\&HNO3)}/PhL_{(NH4H2PO4)}$;
4 - $Al_{(NaOH\&HNO3)}/CCOC_{(Ce+Cu)}$;
5 - $Al_{(NaOH\&HNO3)}/CCOC_{(Ce+Cu)}/PhL_{(Na3PO4)}$;
6 - $Al_{(NaOH\&HNO3)}/CCOC_{(Ce+Cu)}/PhL_{(NH4H2PO4)}$
in 0.1 M NaCl at the pitting potential of Al (E_{pit} = –0.500 V vs. SCE).

The analogous measurements after exposure of the same samples in the CM for 168 h are shown in figure 1 b. It is seen from the course of changes of the potential transients that the tendency of the E_{OCP}

change is similar to that after 1 h exposure in the CM. It is important to note that in this case (quite a longer exposure in the CM), the positive shift of the E_{OCP} for the system $Al_{(NaOH\&HNO3)}/CCOC_{(Ce+Cu)}/PhL_{(NH4H2PO4)}$ is more than twice as high compared to the bare $Al_{(NaOH\&HNO3)}$. However, the E_{OCP} transients are shifted strongly in the negative direction (in the E_{OCP} range from ~ –0.810 V to ~ –0.690 V vs. Ag/AgCl). This large decrease in the OCP can be attributed to a degradation of the formed protective coatings, as well as to a decrease of the concentration of CCOC and PhL (based on particularly omitted or dissolved parts of conversion layers). Figure 1 b shows also the stronger effect of the complex $NH_4H_2PO_4$ salt by direct phosphating of Al (trs. 2 and 3) after 168 h exposure in the CM. This effect is pronounced (more than 120 mV in positive direction) in the case when Al is covered consecutively by CCOC and $PhL_{(NH4H2PO4)}$ ($Al_{(NaOH\&HNO3)}/CCOC_{(Ce+Cu)}/PhL_{(NH4H2PO4)}$ system – tr. 6).

The results obtained and discussed above distinguish the $Al_{(NaOH)}/CCOC_{(Ce+Cu)}/PhL_{(NH4H2PO4)}$ system as the most stable one. This conclusion is in a very good quantitative agreement with the noted increase in the protective effect of CCOC following the additional deposition of PhL [2], including mixed $AlPO_4$ and AlOOH, $CePO_4$, as well as PO_3^-, P_2O_5 and P_4O_{10} compounds with Al and Ce [1].

3.2. Chronoamperometric investigations.

Based on the course of the registered curves (figure 2), we could deduce the character of the corrosion attack and the appearance of pitting damages. We observed that for the $Al_{(NaOH\&HNO3)}$, after its immersion in CM at E_{pit}, the i_a density increases sharply (until the ~65-th second of exposure) up to values ~ 4.22×10^{-3} A cm^{-2}, whereupon the surface film on the Al is disrupted (figure 2, curve 1), which is a prerequisite for the appearance and development of pitting corrosion during the interaction with the CM. After breaking through the passive film, there starts a process of local corrosion characterized by values of the anodic current (i_a) ~ 3.40×10^{-3} A cm^{-2} (at 3600 s) and the appearance of current oscillations specific for it owing to the unstable pits that are repassivated/activated. Similar behavior is also observed for the current transients of the $Al_{(NaOH\&HNO3)}/PhL_{(Na3PO4)}$ and $Al_{(NaOH\&HNO3)}/PhL_{(NH4H2PO4)}$ systems (figure 2, curves 2 and 3).

The course of the current transients is essentially different for the samples protected by CCOC and post-treated in phosphate solutions (figure 2, curves 4, 5 and 6). No current fluctuations characteristic for localized breakdown of the passive film are seen that would lead to the initiation and growth of corrosion pits in the CM. The transients have a process course characteristic of general corrosion; the i_a of the $Al_{(NaOH)}/CCOC_{(Ce+Cu)}$ system increases gradually, coinciding with the transient for the Al substrate ($Al_{(NaOH)}$) after about 1000 s. The transients characterizing the systems $Al_{(NaOH)}/CCOC_{(Ce+Cu)}/PhL_{(Na3PO4)}$ and $Al_{(NaOH)}/CCOC_{(Ce+Cu)}/ PhL_{(NH4H2PO4}$ (figure 2, curves 5 and 6) remain lower. This difference in the drop in i_a is enhanced substantially to 2.6×10^{-3} A cm^{-2} and 1.7×10^{-3} A cm^{-2}, respectively, vs. 3.4×10^{-3} A cm^{-2}, the i_a of the Al substrate.

The juxtaposition of the current transients in the case of a consecutive pre-treatment of the Al substrate in NaOH and HNO_3 shows that the order of stabilities of the systems towards the appearance and development of pitting corrosion is the following: $Al_{(NaOH\&HNO3)}$. < $Al_{(NaOH\&HNO3)}/CCOC_{(Ce+Cu)}$ < $Al_{(NaOH\&HNO3)}/PhL_{(NH4H2PO4)}$ < $Al_{(NaOH\&HNO3)}/PhL_{(Na3PO4)}$ < $Al_{(NaOH\&HNO3)}/CCOC_{(Ce+Cu)}/PhL_{(Na3PO4)}$ < $Al_{(NaOH\&HNO3)}/CCOC_{(Ce+Cu)}/PhL_{(NH4H2PO4)}$.

3.3 Investigations of Rp at E_{OCP}.and calculations of CR.

Table 1 shows the results reflecting the changes in *R*p and *CR* (E_{cor} and i_a, respectively) of the studied systems as a function of the exposure time in the CM. The analysis of the results obtained showed that in the direct immersion treatment of the Al substrate ($Al_{(NaOH\&HNO3)}$) by PhLs, better results from a corrosion point of view are obtained by post-treatment by Na_3PO_4. The effect of this phosphate post-treatment for the *R*p and *CR* for time of exposure of 48 h is highly impressive.

CCOC deposited on $Al_{(NaOH\&HNO3)}$ (the $Al_{(NaOH\&HNO3)}/CCOC_{(Ce+Cu)}$ system) is characterized by the lowest protective indicators compared to those of the systems $Al_{(NaOH\&HNO3)}$/PhLs, most probably because of the higher concentration of Cu and Fe [2].

The phosphate(s) processing of the $Al_{(NaOH\&HNO3)}$/$CCOC_{(Ce+Cu)}$ system, however, improves significantly the protective indicators of the mixed $Al_{(NaOH\&HNO3)}$/$CCOC_{(Ce+Cu)}$/PhLs systems. This improvement is connected with the noteworthy decrease in the deviations of *R*p and i_{cor} starting from the beginning of exposure in the CM. The maximum protective effect is reached for the system $Al_{(NaOH\&HNO3)}$/$CCOC_{(Ce+Cu)}$/$PhL_{(NH4H2PO4)}$. The deviations and oscillations in *R*p and *CR* (E_{cor} and i_a, respectively) for this system decrease and stabilize to a maximal extent.

Table 1. Changes in the *R*p, *CR*, E_{cor} and i_a of the studied systems vs. the exposure time in CM.

Sample	*R*p, kΩcm²	*CR*, μm/y	E_{cor}, mV	i_a, A/cm²
$Al_{(NaOH\&HNO3)}$ 1 h	9.49	29.97	-710.2	2.8E-06
$Al_{(NaOH\&HNO3)}$ 24 h	4.16	68.34	-684.3	6.3E-06
$Al_{(NaOH\&HNO3)}$ 48 h	0.56	505.72	-675.5	4.6E-05
$Al_{(NaOH\&HNO3)}$ 120 h	107.00	2.65	-821.3	2.4E-07
$Al_{(NaOH\&HNO3)}$ 144 h	95.50	2.98	-782.4	2.7E-07
$Al_{(NaOH\&HNO3)}$ 168 h	71.80	3.96	-808.2	3.6E-07
$Al_{(NaOH\&HNO3)}$/$PhL_{(Na3PO4)}$ 1 h	10.40	27.20	-737.2	2.5E-06
$Al_{(NaOH\&HNO3)}$/$PhL_{(Na3PO4)}$ 24 h	22.70	12.50	-757.3	1.2E-06
$Al_{(NaOH\&HNO3)}$/$PhL_{(Na3PO4)}$ 48 h	156.00	1.82	-770.2	1.7E-07
$Al_{(NaOH\&HNO3)}$/$PhL_{(Na3PO4)}$ 120 h	41.40	6.86	-792.2	6.3E-07
$Al_{(NaOH\&HNO3)}$/$PhL_{(Na3PO4)}$ 144 h	73.30	3.87	-808.5	3.6E-07
$Al_{(NaOH\&HNO3)}$/$PhL_{(Na3PO4)}$ 168 h	25.30	11.20	-802.0	1.0E-06
$Al_{(NaOH\&HNO3)}$/ $PhL_{(NH4H2PO4)}$ 1 h	3.05	93.16	-662.6	8.6E-06
$Al_{(NaOH\&HNO3)}$/ $PhL_{(NH4H2PO4)}$ 24 h	7.57	37.49	-651.2	3.4E-06
$Al_{(NaOH\&HNO3)}$/ $PhL_{(NH4H2PO4)}$ 48 h	91.50	3.11	-674.4	2.9E-07
$Al_{(NaOH\&HNO3)}$/ $PhL_{(NH4H2PO4)}$ 120 h	19.70	14.45	-753.7	1.3E-06
$Al_{(NaOH\&HNO3)}$/ $PhL_{(NH4H2PO4)}$ 144 h	18.00	15.80	-754.8	1.5E-06
$Al_{(NaOH\&HNO3)}$/ $PhL_{(NH4H2PO4)}$ 168 h	14.80	19.13	-768.9	1.8E-06
$Al_{(NaOH\&HNO3)}$ /$CCOC_{(Ce+Cu)}$ 1 h	4.12	68.99	-631.2	6.3E-06
$Al_{(NaOH\&HNO3)}$ /$CCOC_{(Ce+Cu)}$ 24 h	29.60	9.60	-720.7	8.8E-07
$Al_{(NaOH\&HNO3)}$ /$CCOC_{(Ce+Cu)}$ 48 h	5.79	49.05	-644.5	4.5E-06
$Al_{(NaOH\&HNO3)}$ /$CCOC_{(Ce+Cu)}$ 120 h	8.50	33.40	-631.6	3.1E-06
$Al_{(NaOH\&HNO3)}$ /$CCOC_{(Ce+Cu)}$ 144 h	8.74	32.51	-670.1	3.0E-06
$Al_{(NaOH\&HNO3)}$ /$CCOC_{(Ce+Cu)}$ 168 h	12.20	23.34	-665.2	2.1E-06
$Al_{(NaOH\&HNO3)}$ /$CCOC_{(Ce+Cu)}$/$PhL_{(Na3PO4)}$ 1 h	1.52	187.47	-702.4	1.7E-05
$Al_{(NaOH\&HNO3)}$ /$CCOC_{(Ce+Cu)}$/$PhL_{(Na3PO4)}$ 24 h	9.57	29.65	-711.8	2.7E-06
$Al_{(NaOH\&HNO3)}$ /$CCOC_{(Ce+Cu)}$/$PhL_{(Na3PO4)}$ 48 h	19.40	14.63	-649.7	1.3E-06
$Al_{(NaOH\&HNO3)}$ /$CCOC_{(Ce+Cu)}$/$PhL_{(Na3PO4)}$ 120 h	137.00	2.07	-795.1	1.9E-07
$Al_{(NaOH\&HNO3)}$ /$CCOC_{(Ce+Cu)}$/$PhL_{(Na3PO4)}$ 144 h	74.40	3.81	-792.7	3.5E-07
$Al_{(NaOH\&HNO3)}$ /$CCOC_{(Ce+Cu)}$ /$PhL_{(Na3PO4)}$ 168 h	71.00	4.00	-755.7	3.7E-07
$Al_{(NaOH\&HNO3)}$/$CCOC_{(Ce+Cu)}$/$PhL_{(NH4H2PO4)}$ 1 h	117.00	2.43	-628.5	2.2E-07
$Al_{(NaOH\&HNO3)}$/$CCOC_{(Ce+Cu)}$/$PhL_{(NH4H2PO4)}$ 24 h	80.50	3.53	-661.1	3.2E-07
$Al_{(NaOH\&HNO3)}$/$CCOC_{(Ce+Cu)}$/$PhL_{(NH4H2PO4)}$ 48 h	181.00	1.57	-670.2	1.4E-07
$Al_{(NaOH\&HNO3)}$/$CCOC_{(Ce+Cu)}$/$PhL_{(NH4H2PO4)}$ 120 h	48.60	5.84	-723.2	5.4E-07
$Al_{(NaOH\&HNO3)}$/$CCOC_{(Ce+Cu)}$/$PhL_{(NH4H2PO4)}$ 144 h	48.70	5.83	-731.8	5.4E-07
$Al_{(NaOH\&HNO3)}$/$CCOC_{(Ce+Cu)}$/$PhL_{(NH4H2PO4)}$ 168 h	34.60	8.21	-722.1	7.5E-07

4. Conclusions

By applying different electrochemical and analytical techniques, a comparative study of the investigated systems was conducted regarding the effect of pre-treatment of Al substrates and the type of phosphate post-treatment on the change in E_{OCP}, i_a at E_{pit}, Rp and *CR*. The juxtaposition of the obtained results shows that their order of stability to pitting corrosion is: $Al_{(NaOH\&HNO3)}$ $<Al_{(NaOH\&HNO3)}/CCOC_{(Ce+Cu)}$ $<$ $Al_{(NaOH\&HNO3)}/PhL_{(NH4H2PO4)}$$<Al_{(NaOH\&HNO3)}/PhL_{(Na3PO4)}$$<Al_{(NaOH\&HNO3)}/CCOC_{(Ce+Cu)}/PhL_{(Na3PO4)}$$<Al_{(NaOH\&HNO3)}/CCOC_{Ce+Cu)}/PhL_{(NH4H2PO4)}$. These results can be related to the data obtained earlier about the calculated stoichiometry of the elements in the so-formed compounds in the surface layers ascertaining that they are built of phosphate and oxide compounds of Ce^{3+}, Ce^{4+} and Al^{3+}, as well as to the influence of the presence of high concentrations of Cu and Fe around them.

References:

[1] Andreeva R, Stoyanova E, Tsanev A and Stoychev D 2019 *C. R. Acad. Bulg. Sci.* **72**(10) 1336
[2] Andreeva R, Stoyanova E, Tsanev A and Stoychev D 2020 *J. Phys.: Conf. Ser.* **1492** 012019
[3] Andreeva R, Stoyanova E, Tsanev A and Stoychev D 2018 *Int J Electrochem Sci* **13** 5333

Optimization of atomic layer deposition of Al_2O_3 films as possible template for graphene transfer

B S Blagoev[1,4], D A Delibatov[1,2], V B Mehandzhiev[1], P Sveshtarov[1], P Terziyska[1], I Avramova[3] and P M Rafailov[1]

[1]G. Nadjakov Institute of Solid State Physics, Bulgarian Academy of Sciences, 72 Tsarigradsko Chaussee, 1784 Sofia, Bulgaria
[2]Technical University of Sofia, 8 St. Kliment Ohridski Blvd., 1000 Sofia, Bulgaria
[3]Institute of General and Inorganic Chemistry, Bulgarian Academy of Sciences, Acad. G. Bonchev Street, bl. 10, 1113 Sofia, Bulgaria

E-mail: blago_sb@yahoo.com

Abstract. Nanofilms of aluminium oxide Al_2O_3 (ALO) were grown on Si, Cu and graphene/Cu substrates by atomic layer deposition (ALD). Trimethylaluminium (TMA) and deionized (DI) water were used as precursors for the ALD process. In order to obtain the ALD temperature window for the above-mentioned precursors, ALO nanofilms were deposited on Si substrates at temperatures ranging from 50 °C to 300 °C. Based on ellipsometry measurements of the films thickness, the growth rate of ALO films obtained was between 1.15 Å/cycle and 1.25 Å/cycle. Ellipsometry and XPS analyses showed that the optimal growth temperature for ALO films is between 150 °C and 200 °C. A graphene layer was grown on a Cu foil by chemical vapor deposition (CVD) and coated with an ALD ALO nanofilm. Raman spectroscopy was used to characterize the influence of ALO films deposition on the graphene layer. A Raman intensity enhancement for all graphene bands in the Al_2O_3/graphene/Cu sample was observed.

1. Introduction

Atomic layer deposition (ALD) is a process based on sequential alternating pulses of gaseous chemical precursors that react with the substrate in a self-limiting way. Based on this, ALD is characterized by large area thickness uniformity, atomically flat and smooth coatings, nanometer scale control of the thickness, perfect 3D conformability [1, 2]. Aluminium oxide (Al_2O_3) possesses a high dielectric constant, excellent stability, a favorable thermal conductivity, high hardness and low refractive index [3]. These properties make it very useful for various applications.

A good quality graphene is usually grown by CVD on copper foil [4]. After that, the graphene film must be transferred to a suitable insulating substrate. A standard graphene transfer is performed by a chemically deposited polymer layer [5]. Such a transfer is accompanied by many technological problems. Therefore, graphene researchers seek alternative transfer methods.

In this paper, we report on the optimization of the ALD process of Al_2O_3 amorphous films on Si and graphene/Cu. The temperature ALD window of Al_2O_3 films is investigated and analyzed.

[4] To whom any correspondence should be addressed.

Content from this work may be used under the terms of the Creative Commons Attribution 3.0 licence. Any further distribution of this work must maintain attribution to the author(s) and the title of the work, journal citation and DOI.
Published under licence by IOP Publishing Ltd

2. Experimental

Graphene was grown in an Oxford Instruments Plasmalab 100 PECVD (plasma enhanced chemical vapor deposition) reactor. Electropolished copper foil (99.8 %, Alfa Aesar) was used as a substrate. During the process, the working pressure and nitrogen flow were 4 Torr and 1500 sccm, respectively. The Cu foil was first heated up to 1065 °C and annealed in a hydrogen gas flow at 200 sccm. Subsequently, for the growth stage, a 10 sccm methane (CH4) flow was let through the reactor chamber for 30 min. The obtained graphene/Cu samples were then slowly cooled down and transferred to the ALD reactor.

Aluminium oxide (Al_2O_3, ALO) nanofilms were deposited in a Beneq TFS-200 atomic layer deposition (ALD) system. To optimize the ALD window of ALO, standard n- and p-Si substrates were used as a reference. Further, ALO was grown on Cu and graphene/Cu substrates for testing the coexistence of ALO and graphene. The ALD growth of Al_2O_3 films was performed by trimethylaluminium (TMA) and deionized (DI) water sequential precursor pulses with a duration of 180 ms. The nitrogen purging duration between pulses was 3 s. During the deposition, a flow of 300 sccm nitrogen gas was maintained through the reactor. The reference substrates were grown at temperatures from 50 °C to 300 °C, while the temperature of the Cu and Cu/graphene substrates was kept at 200 °C. The ALO deposition on Cu and graphene/Cu substrates was preceded by a water treatment of the substrates: 200 ms DI water / 2 s purging; 10 repetition cycles.

The ellipsometry measurements were performed using a Woollam M2000D ellipsometer in the spectral range from 193 nm to 1000 nm. The data acquisition and analysis software used was CompleteEASE 5.10 J. A. Woollam Co., Inc.

The thicknesses of the ALO layers deposited on Si substrates were determined by analyzing the spectroscopic ellipsometry data Ψ and Δ using a model consisting of a Si substrate with a 2.5-nm native silicon oxide, an ALO layer, represented by a Cauchy layer, and a roughness layer (represented by the effective medium approximation of 50% by 50% voids/Al_2O_3).

The XPS data of the different ALO thin films deposited on n-Si substrates at different temperatures were recorded by an ESCALAB MKII spectrometer using non-monochromatic Al Kα radiation (1486.6 eV).

The Raman spectra were measured on a Jobin Yvon Labram spectrometer with a CCD detector and entrance slits set at a 1.5-cm^{-1} spectral width. The 632.8 nm (1.92 eV) line of a He/Ne-laser was used for excitation, the absolute accuracy being 1 – 2 cm^{-1}. The laser beam was focused on a spot of about 2 μm in diameter on the sample surface using microscope optics with a long working-distance objective of 50× magnification.

3. Results and discussions

The as-obtained Al_2O_3 (ALO) films grown at low temperatures (50 °C and 75 °C) appear non-uniform, matte and non-transparent. These films have diffuse reflection, which precludes ellipsometry measurements. The other ALO films were characterized by ellipsometry and their thicknesses were calculated. Figure 1 presents the growth rate of ALO films. This result is close to the results presented by other authors [6]. For further deposition of ALO on Cu and graphene/Cu substrates, we chose the middle point – 200 °C.

As calculated from ellipsometry data, the growth rate of the ALO films deposited on Si, Cu and graphene/Cu at 200 °C (including H_2O pretreatment) is 1.21 Å/cycle (on Si) and 1.22 Å/cycle (on Cu and graphene/Cu). The value for ALO on Si is higher than that for ALO films without H_2O pretreatment (figure 1). Furthermore, the native copper oxide under the graphene layer is negligible (or does not exist), compared to the native copper oxide formed on the bare copper foil. Cu foil naturally undergoes a slow oxidation process when exposed to air. Before the graphene growth, the Cu foil was heated up to 1065 °C and annealed in a hydrogen flow for 30 min whereby the Cu oxide was reduced (or completely vanished).

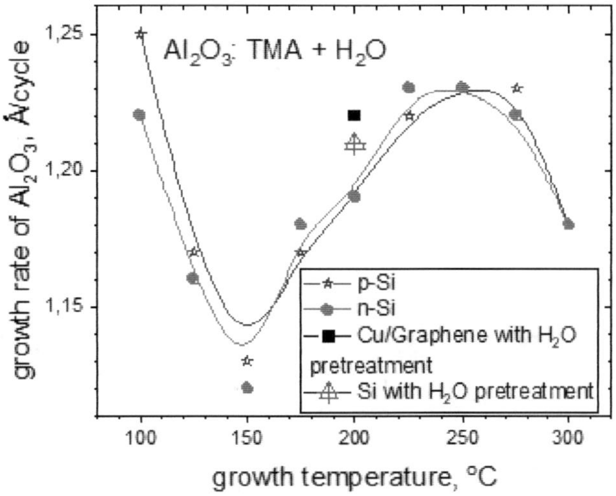

Figure 1. Growth rate of ALD Al_2O_3 films on different substrates.

XPS studies were carried out to obtain information about the differences in the surface compositions of the Al_2O_3 films deposited on n-Si substrates. For all films studied in the present work, the O1s spectra are broad and resolved into two-component species (figure 2). The peak at 531.4 ± 0.5 eV suggests the formation of oxygen (O_o) in oxides. The second peak observed at 532.4 ± 0.2 eV indicates the presence of chemisorbed oxygen (O_a). The ALO film deposited at the lowest temperature of 50 °C is characterized by the highest chemisorption, i.e., the largest O_a/O_o peak ratio, $O_a/O_o > 1$. This ratio decreases with the increase of the deposition temperature – at 150 °C, the O_o peak exceeds that of O_a, i.e., $O_a/O_o < 1$ and the O_a/O_o peak ratio reaches a minimum. The further increase of the deposition temperature increases again the O_a/O_o peak area, but up to 250 °C the O_o peak remains bigger ($O_a/O_o < 1$).

The variation of the O_a/O_o peak ratio could be correlated with the ALD temperature window. At low temperatures, the oxygen energy is not sufficient to incorporate it in the crystal lattice, so that oxygen chemisorption (O_a) occurs predominantly. As the deposition temperature is raised, the oxygen energy increases, which is probably the reason for the higher rate of oxygen lattice incorporation. Therefore, the O_a/O_o peak ratio decreases below 1. At high enough deposition temperatures, a decomposition process takes place and the O_a/O_o peak ratio starts slowly increasing.

The Al 2p peak is present at a binding energy of 74.5 ± 0.5 eV. This particular binding energy position actually corresponds to the formation of the Al_2O_3 phase (figure 2).

The oxygen to aluminium ratio (O/Al) in the deposited alumina thin films was obtained from the respective total amounts of oxygen and aluminium related to the formation of Al_2O_3. The composition of the thin film deposited at 100 °C is closer to the stoichiometric O/Al ratio. For the thin films deposited at 150 °C, 200 °C and 250 °C, the calculated O/Al ratio provides evidence for oxygen deficiency in the grown films.

Figure 3 shows microphotographs of an Al_2O_3/graphene/copper foil; a graphene/copper foil and a copper foil directly coated with Al_2O_3 for reference. It is seen that the ALD coating with Al_2O_3 creates a characteristic orange coloring of both bare and graphene-coated Cu foil.

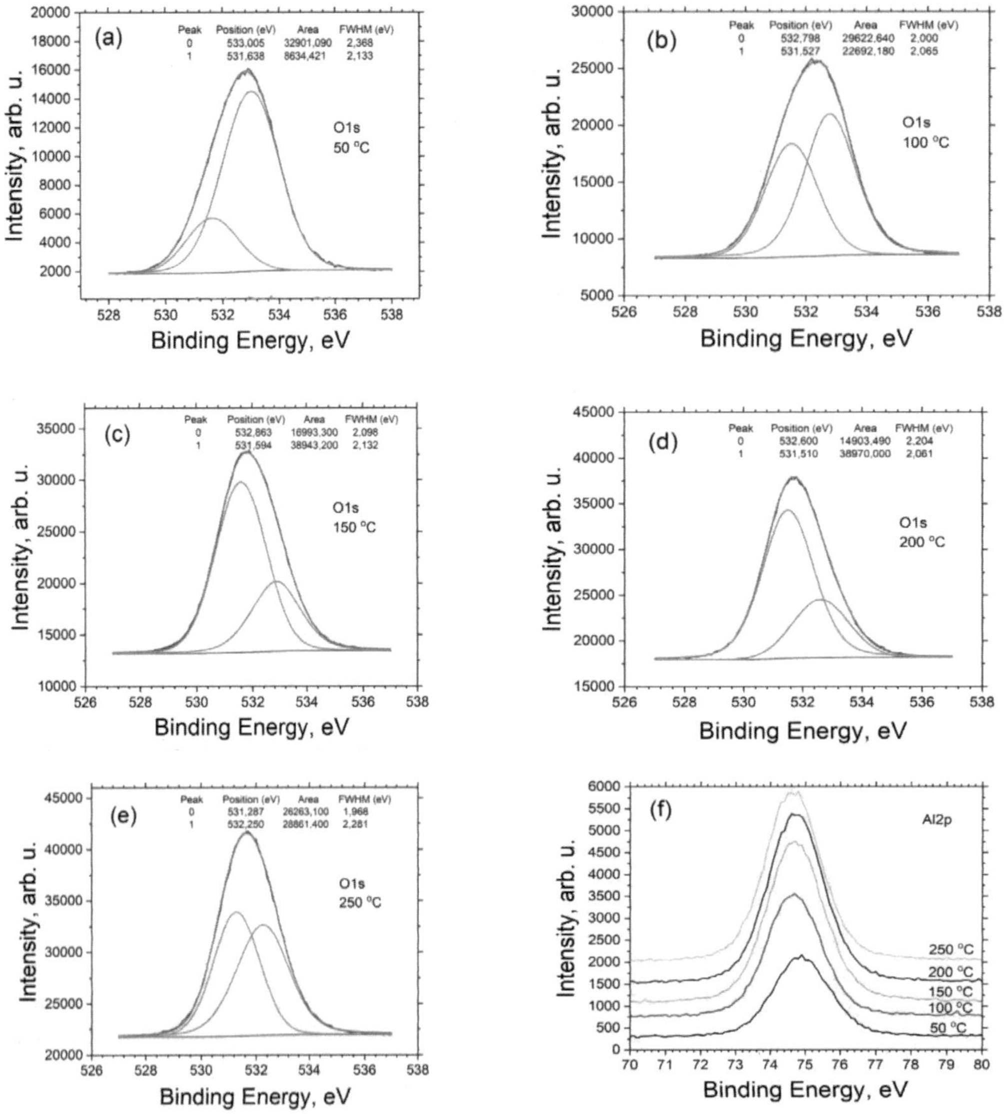

Figure 2. Binding energy of O1s and Al2p spectra of ALO films deposited on n-Si at different temperatures; O1s at (a) 50°C; (b) 100°C; (c) 150°C; (d) 200°C; (e) 250°C and f) Al2p.

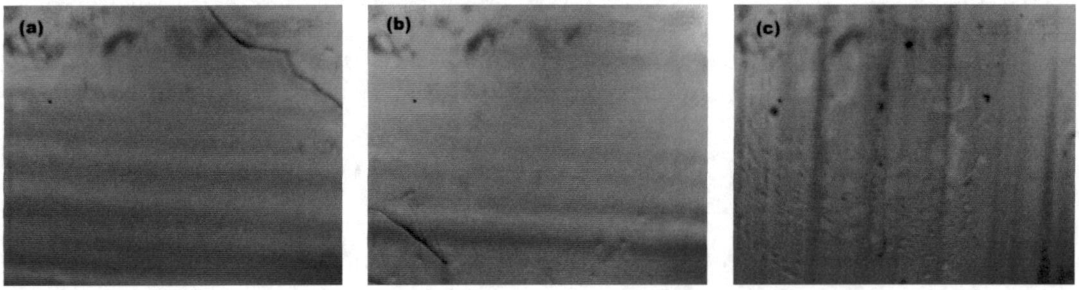

Figure 3. Microphotographs of Al_2O_3/graphene/Cu (a); graphene/Cu (b) and Al_2O_3/Cu (c).

Figure 4 shows Raman spectra of Al_2O_3/graphene/Cu, graphene/Cu and a reference spectrum of the copper foil coated directly with Al_2O_3. The high-intensity ratio of the 2D to the G band, the single-peak shape of the 2D band and the low intensity of the D band, which are representative for the multiple spatially resolved Raman measurements of these samples, provide evidence that the underlying graphene is predominantly single- and double-layer graphene with a good quality, which is retained after ALD ALO deposition on top of it. Although the Raman features of Al_2O_3 itself are not detected in these spectra, the presence of Al_2O_3 is manifested by a multiple enhancement of the Raman intensity of all graphene bands.

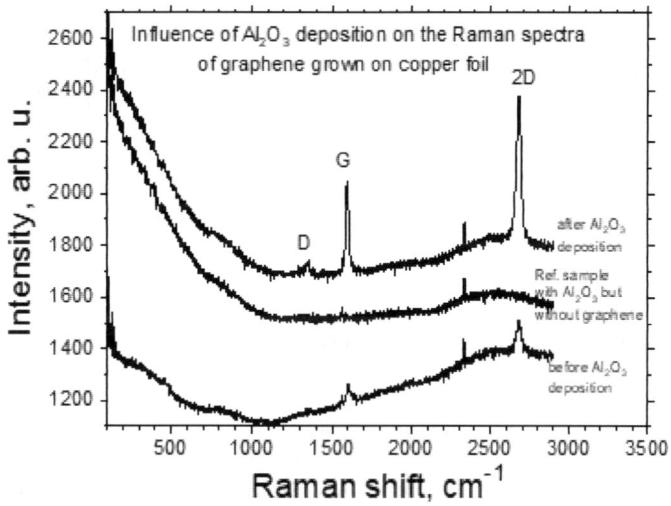

Figure 4. Raman shift of ALO/graphene/Cu, ALO/Cu and graphene/Cu samples.

4. Conclusions

ALO nanofilms were successfully deposited on Si substrates; the ALD temperature window for ALO films grown from TMA and DI water was obtained; good quality graphene on Cu foil was grown by CVD; an Al_2O_3 (ALO) film was deposited on a graphene/Cu substrate by ALD; the ALO films in the ALO/graphene/Cu samples are uniform and smooth as evidenced by spectroscopic ellipsometry and, furthermore, a Raman intensity enhancement for all graphene bands was observed. Graphene in the ALO/graphene/Cu samples retains its good quality; thus, the obtained heterostructure can be regarded as a prospective template candidate for graphene transfer.

Acknowledgments
This work was supported by the Bulgarian National Science Fund under Grant KP-06-N38/10.

References

[1] Johnson R W, Hultqvist A and Bent S F 2014 *Materials Today* **17** (5) 236
[2] Blagoev B S, Aleksandrova M, Terziyska P, Tzvetkov P, Kovacheva D, Kolev G, Mehandzhiev V, Denishev K and Dimitrov D 2018 *Journal of Physics: Conf. Series* **992** 012027
[3] Shi Sh, Qian Sh, Hou X, Mu J, He J and Chou X 2018 *Advances in Condensed Matter Physics* **2018** 7598978
[4] Muñoz R and Gómez-Aleixandre C 2013 *Chemical Vapor Deposition* **19** 297
[5] Ullah S, Yang X, Ta H Q, Hasan M, Bachmatiuk A, Tokarska K, Trzebicka B, Fu L and Rummeli M H 2021 *Nano Research* **14** 3756–3772
[6] Cheng Ch-Ch, Chien Ch-H, Luo G-L, Liu J-Ch, Kei Ch-Ch, Liu D-R, Hsiao Ch-N, Yang Ch-H and Chang Ch-Y 2008 *Journal of The Electrochemical Society* **155** (10) G203–G208

Formation of oxide films on a curved aluminium surface upon anodic oxidation

B Tzaneva[1,4], V Videkov[1], A Bankova[1], M Mitov[2] and D Dobrev[3]

[1]Technical University of Sofia, 8 Kliment Ohridski bld., 1000 Sofia, Bulgaria
[2]Visteon Electronics Bulgaria Ltd., Capital Fort Building, 90 Tsarigradsko Chaussee, 1784 Sofia, Bulgaria
[3]Institute of Mathematics and Informatics, Bulgarian Academy of Sciences, Acad. Georgi Bonchev Str., Bld. 8, 1113 Sofia, Bulgaria

E-mail: borianatz@tu-sofia.bg

Abstract. The aluminium oxide obtained by anodization in acidic solutions has a highly ordered nanoporous structure consisting of straight parallel pores. The oxide layer has good electrical insulating properties on planar surfaces. However, the electrochemical formation of nanoporous film on convex surface induces effects such as crack formation and deterioration of the insulating properties. The present work is focused on investigating the conditions of growth of a thick oxide layer on aluminium wires with a diameter of 0.2 mm to 0.5 mm. In order to prevent cracking, the oxide growth was carried out by reducing the anodization voltage according to a mathematical model in view of obtaining conical pores. In this study, the minimum thickness at which a crack-free oxide layer grows was determined.

1. Introduction

Anodic aluminium oxide (AAO) is a well-known material used for the synthesis of nanostructures, such as nanowires, nanotubes and nanorods, and for developing advanced sensing devices [1-7]. Anodic layers formed on aluminium under acidic electrolyte conditions are grown into nanoporous structures, producing regular arrays of hexagonal cells with cylindrical holes [8]. The resulting AAO structures usually resemble a honeycomb-like pore structure [9]. The oxide layers obtained are considered materials of interest due to the simplicity of implementation, the highly ordered arrangement of uniform nanopores, and the structures' repeatability and reproducibility [10, 11]. Thus, despite the decades of experience of using them, the AAO films continue to be an attractive object of research regarding their practical applications.

The majority of the research on pore formation has been performed on flat Al substrates. Anodization on a nonplanar surface is yet to be well studied or applied in nanofabrication. AAO growth on curved Al surfaces can be considered incidental, resulting in the impossibility of obtaining a continuous layer on convex surfaces with a radius of less than 7 mm [12]. The authors in [13] have shown that anodizing convex Al substrates affects the properties of the AAO layer and the morphology of the AAO pores. Due to the larger volume of porous alumina compared to that of aluminium, during anodizing the outer layers are subjected to high tensile stress, which causes the

[4] To whom any correspondence should be addressed.

Content from this work may be used under the terms of the Creative Commons Attribution 3.0 licence. Any further distribution of this work must maintain attribution to the author(s) and the title of the work, journal citation and DOI.
Published under licence by IOP Publishing Ltd

appearance of radial cracks. Such cracks do not occur when the oxide grows on inwardly curved surfaces [14]. In addition to disrupting the ordered nanoporous structure, the presence of cracks in the anodic layer significantly impairs its protective and insulating properties, eliminating the possibility of using it as a dielectric base in conductive coating applications.

The experiments published so far in the literature have been performed at a constant anodization voltage, determining a uniform pore diameter along the anodic layer.

The present study aims to examine the possibility of growing a defect-free AAO layer by forming pores with a conical profile. The minimum crack-free layer thickness obtainable on a thin aluminium wire was experimentally determined.

2. Experimental

In this study, nanoporous alumina was grown on high purity (>99.99%) 40-mm long Al wires with diameters of 0.2 mm, 0.25 mm, and 0.5 mm. Before anodization, all samples were degreased in acetone and stained in a 4% NaOH solution for 1 minute at room temperature. The anodization of Al was conducted in an electrochemical cell using 0.4-M oxalic acid at room temperature under constant (40 V) and decreasing (from 40 V to 20 V) cell voltage according to a mathematical model for the formation of conical pores. A VOLTCRAFT 4005 programmable laboratory power supply was used. The anodized wires were extracted from the oxalic acid solution, rinsed in distilled water, and dried in hot air. To determine the thickness of the formed structures and to detect the presence of cracks, the wires were cast in epoxy resin and cut crosswise. The cross-sectional samples were examined by optical and scanning electron microscopy.

3. Results and Discussion

3.1. Anodization at constant voltage

The first experimental studies of anodized wire were performed with a minimum diameter of 0.2 mm at a constant voltage of 40 V. The cross-section optical microscopy observation of the samples anodized for various times in 0.4-M oxalic acid at 40 V showed the appearance of uneven cracks. Figure 1 presents typical optical images of samples after anodization for 1.5 and 3.0 hours. The coefficient of volume expansion (volume of porous AAO formed/volume of Al consumed), determined by the cross-sectional views is 1.25.

In the microscopic views of the cross-section of samples anodized at a constant voltage to the complete consumption of the aluminium core, the following characteristic features are observed:

- The cracks originate at a layer thickness of more than 2 μm and their number remains unchanged until the core is fully depleted, increasing only the width of the surface gap;

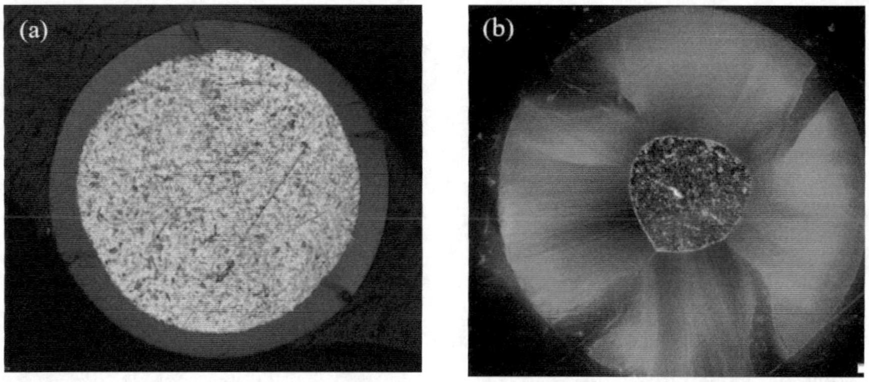

Figure 1. Optical microscope cross-sectional images of Al wires anodized at 40 V for (a) 1.5 h and (b) 3 h.

- In the course of the anodization process, the Al core shape changes (figure 1 b). The effect observed shows that the internal stresses during anodization occur at the aluminium/oxide interface. The induced stresses are unevenly distributed in the anodic layer; in some sections, this causes plastic deformation in the aluminium, while in others it contributes to the growth of the oxide layer.

Therefore, the formation of pores and AAO structures on outwardly curved surfaces at a constant anodization voltage leads to the generation of internal mechanical tensile forces. On the periphery (the oxide/electrolyte interface), the tensile stresses result from the higher volume expansion of the oxide compared to the aluminium base. The forces acting inside the oxide layer are considerably more complex. It seems likely that due to the presence of constant voltage in an electrochemical cell, a corresponding tendency exists for preserving the pores diameter. Meanwhile, the surface from which the AAO pores begin to grow decreases. This, combined with the increasing curvature, leads to a decrease in the number of pores at the Al/AAO interface and a tendency to form flat regions in the cross-section. The defects are concentrated at the crack line, which allows the pores in the remaining layer to maintain a relative parallelism.

Assuming that if the pores had conical shapes so that their number remained constant on the decreasing aluminium surface, the deformation conditions would change. This in turn would reduce the cracks width on the peripheral area and, by a suitable choice of pores geometry, cracking of the layer should be completely preventable. Since the pore diameter is proportional to the applied voltage, a relation has to be found between the radius of the aluminium wire and the rate reduction of the anodization voltage to form conical pores.

3.2. Mathematical model

The purpose of the model is to determine the mathematical relation between the voltage change over time, which applies to different diameters of the used aluminium wire and different starting voltages. The mathematical model is based on the following assumptions:

(I) The volume expansion of the oxide is 25% and its effect on the pores diameter can be neglected. The generated tensile stresses at the periphery are assumed to not affect the oxide structure.

(II) The main part of induced stresses in the oxide can be reduced by modeling the conical profile of the pore, as shown in the following diagram (figure 2).

In a layer of thickness $\boldsymbol{h}$, the reduction of the pore diameter from one initial diameter $\boldsymbol{d_1}$ to a new smaller value $\boldsymbol{d_2}$ can be expressed by a reduction coefficient $\boldsymbol{k_r}$:

$$d_2 = \frac{d_1}{k_r} \tag{1}$$

The reduction coefficient $\boldsymbol{k_r}$ of the pore diameter at depth $\boldsymbol{h}$ is expressed via the similarity of two triangles:

$$k_r = \frac{R_1}{R_1 - \frac{3}{4}h} \tag{2}$$

where $\boldsymbol{R_1}$ is radius of the wire before anodizing (initial radius); $\boldsymbol{h}$ is thickness of the oxide layer and $\frac{3}{4}\boldsymbol{h}$ is the aluminium layer consumed to form an oxide layer of thickness $\boldsymbol{h}$.

(III) Assumption of linear increase of AAO.

The voltage decrease will not have a significant effect on the growth rate of the oxide and it remains constant for the time of the experiment (the oxide thickness $\boldsymbol{h}$ is proportional to the anodization time $\boldsymbol{t}$).

Assuming that the AAO layer thickness is linearly increasing, the pore diameter should also decrease linearly:

$$d_2 = a.t + d_1 \tag{3}$$

where ***t*** is the time for obtaining a layer of thickness ***h***. Further, using equation (3) we can determine the slope of the function, which gives the rate of pore diameter change. Using eqs. (1), (2) and (3), we can find the relationship between the constant ***a***, the pore diameter and the radius of the wire:

$$a = \frac{\frac{d_1}{k_r} - d_1}{t} = \frac{d_1(1-k_r)}{k_r t} = -\frac{0.75 d_1 h}{R_1 t} \tag{4}$$

For nanoporous oxide thicknesses of up to 15 μm, it has been found experimentally that the growth rate of AAO (*h*/*t*) is 0.1 μm/min. Taking into account that oxalic acid anodizing is typically performed at $\boldsymbol{U_1}$ = 40 V ($\boldsymbol{d_1}$ ≈ 40 nm), the slope ***a*** reaches:

$$a = -\frac{3}{R_1}, \text{ nm/min} \tag{5}$$

(IV) The anodization voltage ***U*** is proportional to the pore diameter ***d***:

$$d = kU \tag{6}$$

where the coefficient of proportionality ***k*** is approximately 1 nm/V.

Considering the assumptions and calculations described above, dependence (3) takes the following final form:

$$U = U_1 - \left(\frac{3}{R_1}\right).t \tag{7}$$

where the starting voltage $\boldsymbol{U_1}$ = 40 V when using oxalic acid.

Based on this basic mathematical model, experiments were performed on annealed high-purity aluminium wires (99.99% Al) with diameters of 200 μm, 250 μm and 500 μm (initial radiuses of 100 μm, 125 μm and 250 μm). The voltage reduction steps calculated by dependence (7) are presented in table 1.

Table 1. Steps of anodization voltage decreasing from U_1 = 40 V on aluminium wires with different diameters.

$\boldsymbol{R_1}$ (μm)	***a*** (nm/min)	**Δ*U*** (V)	**Δ*t*** (min)
100	-0.030	0.1	3.3
125	-0.024	0.1	4.1
250	-0.012	0.1	8.3

3.3. Anodization according to the mathematical model

Figure 3 shows the anodization time dependency of the anodic current density (J_a) for a 200-μm diameter Al wire in 0.4-M oxalic acid and under stepwise reduction of the anodizing voltage from 40 V to 30 V according to the mathematical model. Each step of voltage reduction by 0.1 V clearly affects the curve in the way of a small decrease in J_a, after which the current returns to a value close to the previous step. However, over time, a gradual decrease in J_a is observed which is probably related to an increase in the layer thickness of the AAO and the associated voltage drop due to diffusion difficulties related to the thickness increase. For the duration of the experiment of 5 hours, the decrease is 0.2 mA/cm^2.

Microscopic observation made on the cross-sections of the samples anodized under decreasing voltage again showed cracks in the oxide layers at thicknesses above 2 μm (figure 4). No connection was found between the diameter of the Al wire and the number of cracks, as the number of cracks on samples from the same series varied widely (from 4 to 8 cracks for all wire diameters). In the case of decreasing anodizing voltage, some cracks end inside the oxide without reaching the aluminium core, which point to a partial reduction in the internal stresses (figure 4, b). Using the microscopic images, we determined the coefficient of thickness increase of the oxide $\boldsymbol{h_{ox}}$ with respect to the starting aluminium layer $\boldsymbol{h_{Al}}$, which is consumed for its production:

$$h_{ox} = 1.2 h_{Al} \tag{8}$$

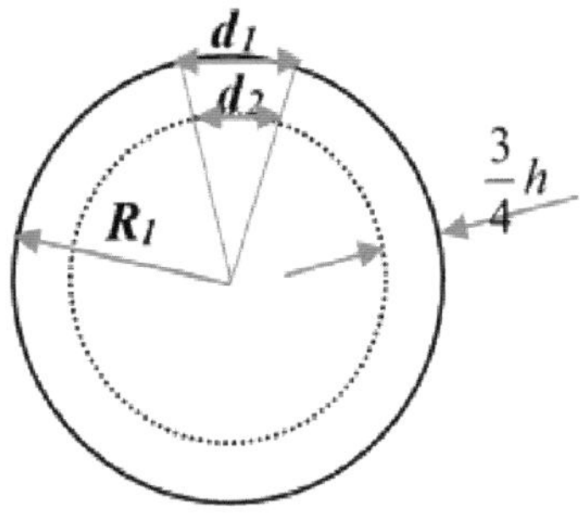

Figure 2. Schematic diagram of the wire cross section showing the pore parameters.

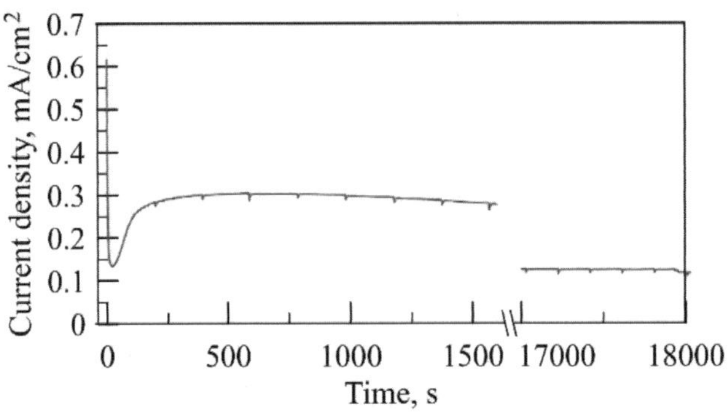

Figure 3. Time dependence of the current density during anodization of a 0.2-mm diameter Al wire in 0.4 M oxalic acid under decreasing voltage from 40 V to 30 V.

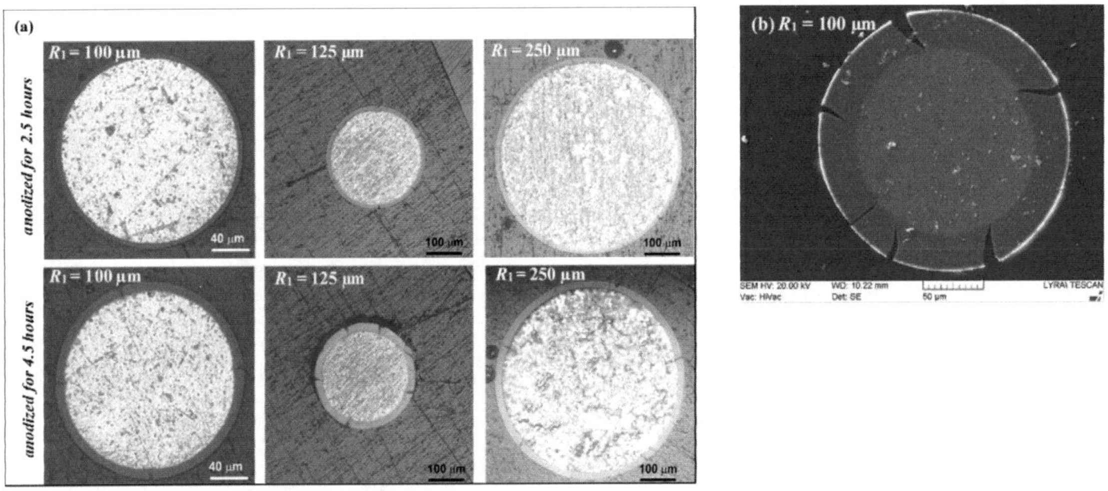

Figure 4. Optical (a) and SEM (b) images of Al wire cross-sections with initial diameters of 200 µm, 250 µm and 500 µm anodized under decreasing voltage according to the mathematical model for conical pores, Table 1.

4. Conclusions

AAO pores growth was studied on Al wires with initial diameters of 200 µm, 250 µm and 500 µm subjected to decreasing anodization voltages according to a mathematical model for obtaining conical pores. Through optical and SEM observation, the pore structure parameters and the volume expansion coefficient were obtained. The minimum thickness for crack-free AAO on a curved surface was determined to be 2 µm. Beyond this thickness, the number of cracks remains unchanged until the core is fully depleted. The results of this mathematical model were not satisfying. Therefore, we plan the design a new model to consider the volume expansion of AAO relative to the starting aluminium substrate.

Acknowledgments
The work was supported by research project KP-06-N29-1 funded by the Bulgarian National Science Fund.

References
[1] Larsson A, Abbondanza G, Linpé W, Carlà F, Mousley P, Hetherington C, Lundgren E and Harlow G S 2020 *J. Electrochem. Soc.* **167** 122514
[2] Baik J M, Schierhorn M and Moskovits M 2018 *J. Phys. Chem.* C **112** 2252
[3] Luo L, Wang X, Ma X, Zhang Y, Jin L and Yuan Z 2020 *Vacuum* **172** 109051
[4] Okamoto T, Shimizu T, Takase K, Ito T and Shingubara S 2020 *Micro Nano Eng.* **9** 100071
[5] Mansourian A, Saidpour H and Mannan S H 2020 *Appl. Surf. Sci.* **532** 147426
[6] Kumar S, Raina K K and Islam T 2021 *Sens. Actuators B Chem.* **329** 128908
[7] Podgolina S K, Petukhov D I, Dorofeev S G and Eliseev A A 2020 *Talanta* **219** 121248
[8] Lee W and Park S J 2014 *Chem. Rev* **114** 7487
[9] Masuda H and Fukuda K 1995 *Science* **268** 1466–1468
[10] Liu K S, Tian J and Zhang W 2021 *Nanotechnology* **32** 222001
[11] Domagalski J T, Xifre-Perez E and Marsal L F 2021 *Nanomaterials* **11** 430
[12] Kasi J K , Kasi A K, Bokhari M and Sohai M 2016 *Am. J. Condens. Matter Phys* **6** 36–40
[13] Tiong T, Ooi L, Dee C, Hamzah A A and Majlis B Y 2018 *Nanotechnology* **29** 435601
[14] Yoo B Y, Hendricks R K, Ozkan M and Myung N V 2006 *Electrochim. Acta* **51** 3543

Atomic layer deposition of AlN on different SiC surfaces

M Beshkova[1,5], P Deminskyi[2], C-W Hsu[3], I Shtepliuk[3], I Avramova[4], R Yakimova[3] and H Pedersen[2]

[1]Institute of Electronics, Bulgarian Academy of Sciences, 72 Tsarigradsko Chaussee, 1784, Sofia, Bulgaria
[2]Chemistry Division, Department of Physics, Chemistry and Biology-IFM, Linköping University, S-58183 Linköping, Sweden
[3]Division of Semiconductor Materials, Department of Physics, Chemistry and Biology-IFM, Linköping University, S-58183 Linköping, Sweden
[4]Institute of General and Inorganic Chemistry, Bulgarian Academy of Sciences, Acad. G. Bonchev Street, bl. 11, 1113 Sofia, Bulgaria

E-mail: mbeshkova@yahoo.com

Abstract. Thin AlN films were grown using a Picosun R-200 atomic layer deposition (ALD) reactor on SiC surfaces with different crystallographic orientation: on-axis 4H-SiC (0001) and 8° off-axis 4H-SiC. TMA (trimethylaluminium) and NH_3 were used as precursors while hydrogen and nitrogen plasma were applied for in-situ substrate cleaning. The substrate temperatures were 400 °C and 450 °C, with 20 ALD cycles. The surface morphology was investigated by scanning electron microscopy (SEM), which revealed nanometer-sized islands in all films. The AlN films deposited on on-axis 4H-SiC at 450 °C substrate temperature exhibited a relatively small roughness of about 0.255 nm. The chemical composition and bonding states were investigated by X-ray photoelectron spectroscopy. For all layers, high-resolution XPS showed Al 2p and N 1s spectra that are characteristic of AlN. These results are a good prerequisite of establishing the growth conditions of AlN films for surface acoustic wave (SAW) devices.

1. Introduction

Wurtzite aluminium nitride is a wide band gap material (6.2 eV at 300 °C) with high sound speed (up to 6000 m/s) and dielectric constant ($\varepsilon_0 = 8.5$), which make it a candidate material as a dielectric in SiC-based metal-insulator-semiconductor structures for high-power devices and high-temperature applications, as well as for surface acoustic wave (SAW) devices for harsh environment [1-3].

Because of the similar thermal expansion coefficients [4], AlN can be grown with a high crystalline quality onto SiC, usually using the chemical vapor deposition (CVD) technique at temperatures above 1000 °C [5-6].

The atomic layer deposition is a version of low-temperature CVD technique, which may provide a means of growing highly conformal thin films with a sub-nanometer thickness control, since the precursors are supplied in sequential pulses which make the material deposition closely dependent on the process cycles.

[5] To whom any correspondence should be addressed.

Content from this work may be used under the terms of the Creative Commons Attribution 3.0 licence. Any further distribution of this work must maintain attribution to the author(s) and the title of the work, journal citation and DOI.
Published under licence by IOP Publishing Ltd

In this work, we investigated the results of depositing thin AlN films on the (0001) face of on-axis 4H-SiC in comparison with depositing thin AlN films on 8° off-axis 4H-SiC(0001) under identical conditions at substrate temperatures of 400 °C and 450 °C.

2. Experimental

AlN thin layers were deposited on *in situ* cleaned by H_2 and N_2 plasma 10×10 mm on-axis 4H-SiC(0001) and, for comparison, on 8° vicinal cut toward $[11\bar{2}0]$ 4H-SiC pieces. The depositions were carried out in a Picosun R-200 ALD reactor without a load-lock chamber at a total pressure of 6 hPa. Trimethylaluminium (TMA) and ammonia (NH_3) were the Al and N sources, with one cycle consisting of 0.5-s TMA pulse/10-s N_2 purge, and 40-s Ar/NH_3 flow (100/300 sscm)/10-s N_2 purge. The ALD cycles were set to 20 in order to obtain the desired thickness. Figure 1 shows the experimental set up.

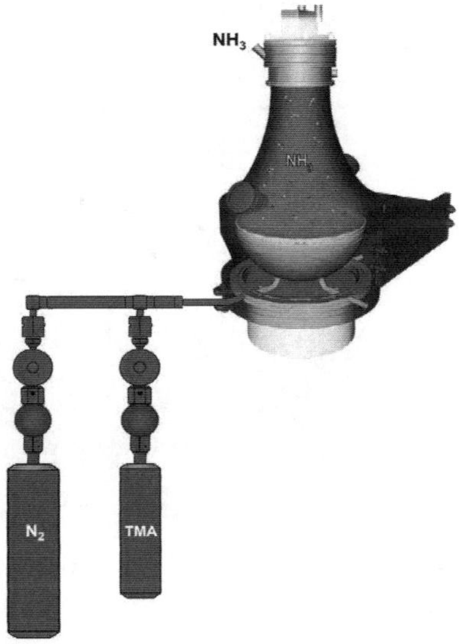

Figure 1. Experimental set up of Picosun R-200 ALD reactor without a load-lock chamber.

To investigate the effect of the initial stage of nucleation on the AlN film properties, AlN layers were deposited on the (0001) face of on-axis 4H-SiC and compared with AlN thin films deposited on 8° off-axis 4H-SiC(0001) at substrate temperatures of 400 °C and 450 °C. The relatively narrow substrate temperature range was chosen based on the saturation curve of aluminium precursor obtained at 400 –425°C and reported in our previous work [7].

The initial AlN growth on on-axis and vicinal 4H-SiC (0001) substrate was inspected by a high-resolution Leo 1550 Gemini field-emission scanning electron microscope (SEM). The surface coverage was determined from the SEM images by morphology analysis using the ImageJ software, while atomic force microscopy (AFM) operating in a contact mode was used to reveal the surface roughness of the thin AlN films.

The films' chemical nature was investigated by X-ray photoelectron spectroscopy (XPS) with Al Kα radiation (1486.69 eV) and a total instrument resolution of 1 eV. The binding energies (BEs) were determined utilizing the C 1s line at 285.0 eV as a reference.

3. Results and discussion

The SEM assessed morphology of AlN films grown on on-axis 4H-SiC(0001) at 400 °C and 450 °C, respectively, are shown in figure 2a,b. At both substrate temperatures, the AlN films exhibit a grain morphology with uniform islands of size of less than 10 nm and coverage close to 100% (table 1). The term "~100% coverage" can be interpreted as continuous coverage.

As shown in the AFM images (figure 2 c, d), the surface roughness of the AlN films obtained decreases as the substrate temperature is increased. This is related to an improvement of the lateral mobility of the adatoms and a growth with better surface control [8].

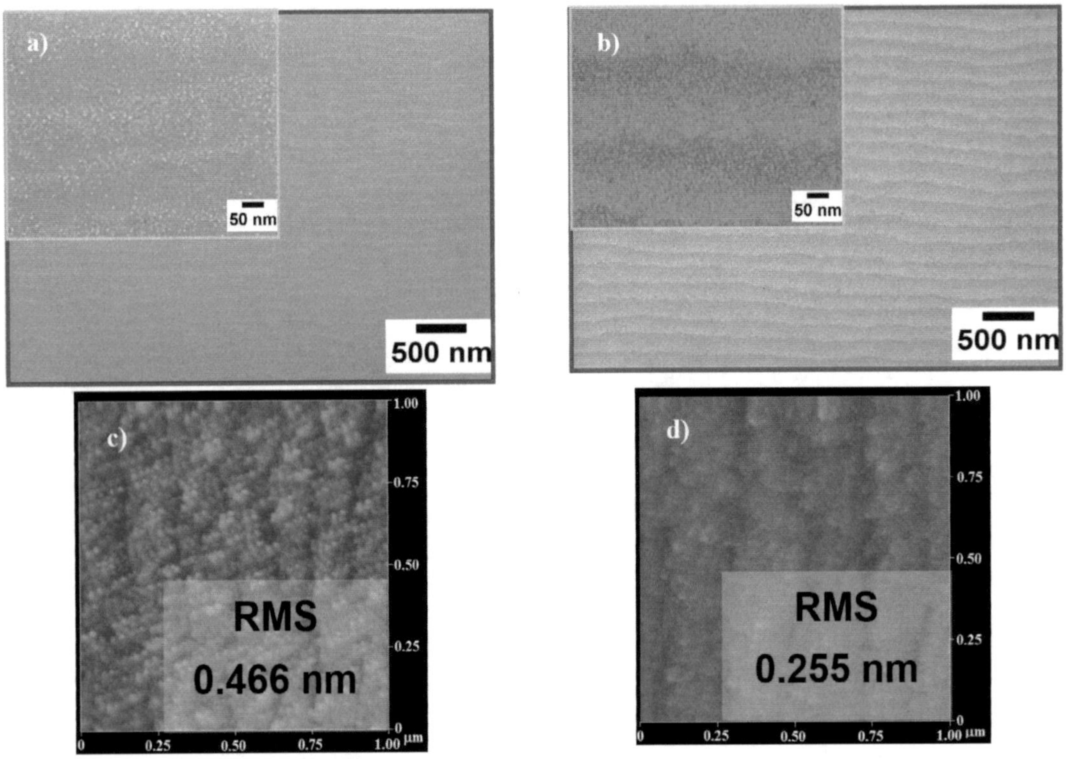

Figure 2. SEM morphology of films deposited on on-axis 4H-SiC. a, b) Top view of films deposited at substrate temperatures 400 °C and 450 °C, respectively [7]; c, d) AFM images of AlN films deposited at 400 °C and 450 °C, respectively.

AlN grown on off-axis 4H-SiC(0001) exhibits different morphology (figure 3 a, b). At both substrate temperatures, the AlN films have a scratch grain morphology with an average grain size of ~10 nm with area coverage close to 100% (table 1). The scratches are probably inherited from the vicinal cut substrate that has polishing defects. The higher temperature raises the surface roughness of the AlN films grown on an off-axis SiC substrate. This can be explained by the existence of uniformly spread steps acting as additional nucleation centers when the mobility of adatoms is improved at high substrate temperatures. Also, the higher temperature reduces the surface roughness of the AlN films grown on an on-axis SiC substrate due to lower memory effect of the substrate (table 1).

Table 1. Deposition conditions of AlN films by ALD and root-mean-square (RMS) roughness of the sample from 1×1 μm AFM measurements, coverage area and average grain size.

Sample	Substrate	Deposit. Temp. (°C)	Number of ALD cycles	RMS [nm]	AlN Coverage %	Average grain size [nm]
AlN-1	4H-SiC on axis	450	20	0.255	~100	<10 nm
AlN-2	4H-SiC on axis	400	20	0.466	~100	<10 nm
AlN-3	4H-SiC 8 ° off axis	400	20	1.762	~100	<10 nm
AlN-4	4H-SiC 8 ° off axis	450	20	2.116	~100	<10 nm

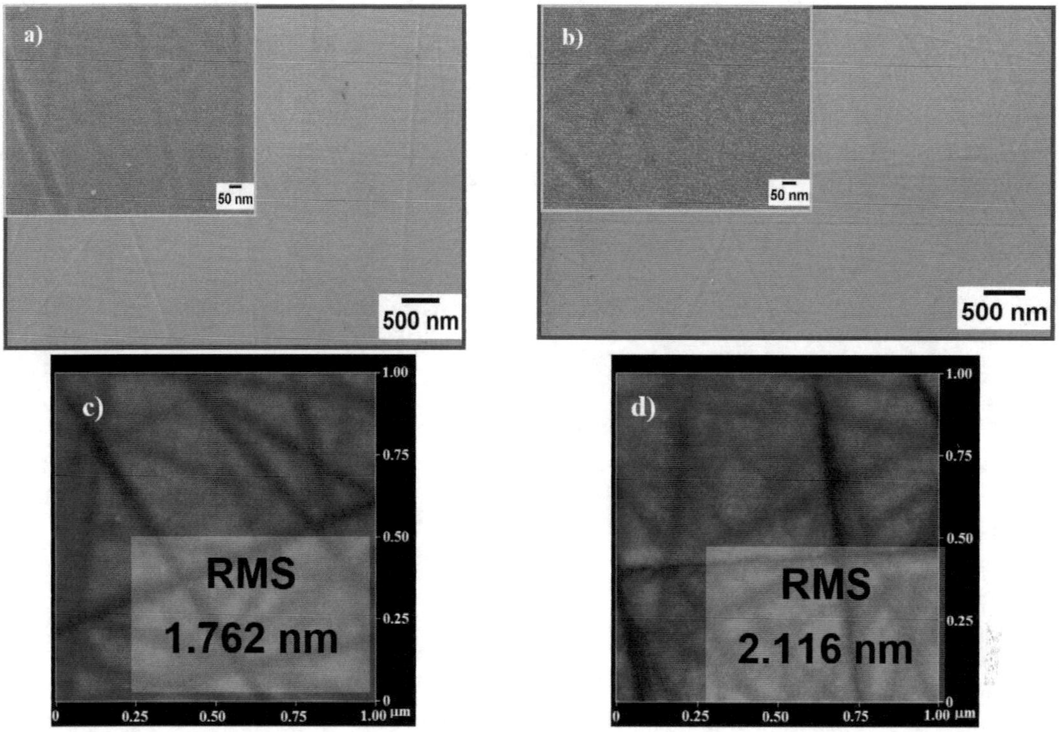

Figure 3. a, b) SEM morphology of films deposited on 8° off-axis 4H-SiC. Top view of films deposited at substrate temperatures 400 °C and 450 °C, respectively; c, d) AFM images of AlN films deposited at 400 °C and 450 °C, respectively.

The chemical nature of all samples was investigated by XPS. Survey scans detected peaks of aluminium, nitrogen, oxygen, and carbon in all samples. The Al 2p and N 1s peaks were fitted using two peaks, as presented in figure 4. The Al 2p photoelectron peaks at 73.3 eV and 74.3 eV are attributed to the Al-N and Al-O bond, respectively [9] (figure 4 a). The N 1s spectra in figure 4 b) were deconvoluted into two sub-peaks with binding energy of 396.6 eV and 398.0 eV that can be linked to the N-Al and N-C bond, respectively [10].

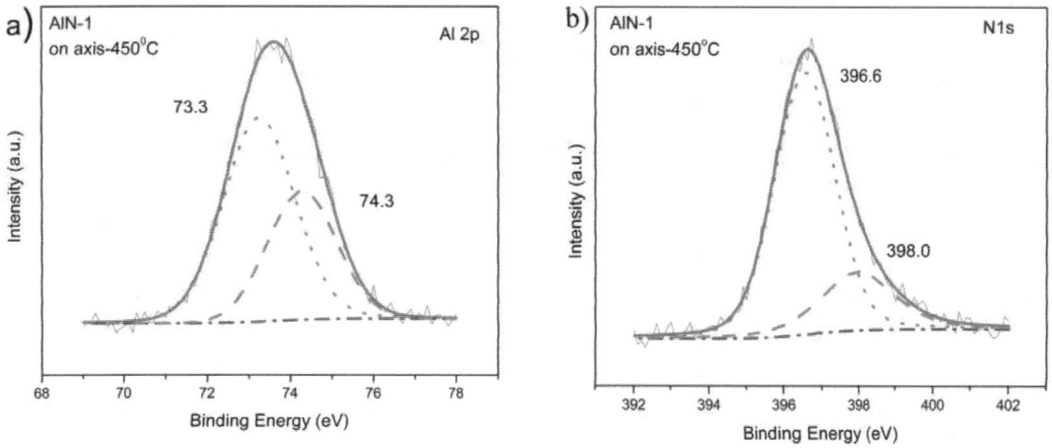

Figure 4. High-resolution XPS of a) Al 2p and b) N 1s photoelectron lines of the AlN film after deconvolution for AlN film deposited on on-axis 4H-SiC at a substrate temperature of 450 °C.

4. Conclusions

Here, we prepared thin AlN films on on-axis and on 8° off-axis 4H-SiC (0001) substrates at substrate temperatures of 400 °C and 450 °C by ALD from TMA and NH_3. SEM study shows grain morphology with a nanometer size of about ~10 nm for the AlN films deposited on on-axis 4H-SiC substrate, and a scratch grain morphology for the AlN films deposited on 8° off-axis 4H-SiC substrate, regardless of the substrate temperature. The RMS roughness of the thin AlN films deposited on on-axis SiC substrate is lower by a factor of eight compared to that of the AlN films deposited on off-axis SiC at a substrate temperature of 450 °C.

The XPS results reveal the presence of Al-N bonds in both the Al 2p and N 1s regions for all samples. The results obtained may be considered promising regarding applications in SAW devices and sensors.

Acknowledgements

The work was supported by the Bulgarian National Science Fund under contract DN 18/6 and by the Swedish Foundation for Strategic Research through project No SSF-RMA 15-0018 and by the Knut and Alice Wallenberg Foundation through project No KAW 2013.0049. P.D. acknowledges the Carl Trygger Foundation for a post-doctoral scholarship at the Linköping University and INFRAMAT (part of the Bulgarian National roadmap for research infrastructures) supported by the Bulgarian Ministry of Education and Science.

References

[1] Meinschen J, Behme G, Falk F and Stafast H 1999 *Appl. Phys.* A **69** 683

[2] Lamanna L, Rizzi F, Bhethanabotla V R and Vittorio M De 2020 *Sensors and Actuators A: Physical* **315** 112268

[3] Ng E J, Sharma J, Ching E W L, Wu G, Pohl D and Vancauwenberghe O 2021 *IEEE 34th International Conference on Micro Electro Mechanical Systems (MEMS)* 370

[4] Strite S and Morkoç H 1992 *J. Vac. Sci. and Technol.* B **10** 1237

[5] Zollner C J, Almogbel A, Yao Y, Burhan K. SaifAddin B K, Wu F, Iza M, DenBaars S P, Speck J S and Nakamura S 2019 *Appl. Phys. Lett.* **115** 161101

[6] Foronda H M, Wu F, Zollner C, Alif M E, Saifaddin B, Almogbel A, Iza M, Nakamura S, DenBaars S P and Speck J S 2018 *J Cryst. Growth* **483** 134

[7] Beshkova M, Deminskyi P, Hsu C-W, Shtepliuk I, Avramova I, Yakimova R and Pedersen H 2021 *Phys. Status Solidi* A 2000684

[8] George S M 2010 *Chem Rev.* **110** 111

[9] Liao J X, Xia L F, Sun M R, Liu W M, Xu T,Yang C R, Chen H W, Fu C L and Leng W J 2005 *Appl. Surf. Sci.* **240** 71

[10] Zhu J, Zhao D, Luo W B, Zhang Y and Li Y R 2008 *J. Cryst. Growth* **310** 731

Morphological evolution of thin AlN films grown by atomic layer deposition

M Beshkova[1,6], B S Blagoev[2], V Mehandzhiev[2], R Yakimova[3], B Georgieva[1], I Avramova[4], P Terziyska[2] and V Strijkova[5]

[1]Institute of Electronics, Bulgarian Academy of Sciences, 72 Tsarigradsko Chaussee, 1784 Sofia, Bulgaria
[2]Institute of Solid State Physics, Bulgarian Academy of Sciences, 72 Tsarigradsko Chaussee, 1784 Sofia, Bulgaria
[3]Department of Physics, Chemistry and Biology, Linköping University, 58183 Linköping, Sweden
[4]Institute of General and Inorganic Chemistry, Bulgarian Academy of Sciences, Acad. G. Bonchev Street, bl. 11, 1113 Sofia, Bulgaria
[5]Acad. J. Malinowski Institute of Optical Materials and Technologies, Bulgarian Academy of Sciences, Acad. G. Bonchev Str. bl. 109, 1113 Sofia, Bulgaria

E-mail: mbeshkova@yahoo.com

Abstract. Thin AlN films were grown using a Beneq TFS-200 ALD reactor. TMA (trimethylaluminium) and NH_3 were used as precursors. The substrate temperature was 330 °C, the ALD cycles, 550. The TMA and NH_3 doses (pulses) lasted 180 ms and 90 ms, followed by 2-s and 9-s nitrogen gas purge, respectively. In order to study the morphological evolution of the thin AlN films, substrates providing different surface kinetics were used: Si-face and C-face of 4° off axis and on-axis 4H-SiC, and graphene grown on 4H-SiC by sublimation. As revealed by atomic force microscopy (AFM), the lowest RMS surface roughness of about 0.8 nm was exhibited by the AlN film deposited on Si-face on-axis 4H-SiC due to its higher surface energy which provides for better film nucleation. The chemical composition and bonding states were investigated by X-ray photoelectron spectroscopy (XPS). The existence of AlN is justified by the presence of the XPS peaks of Al 2p and N 1s at about 73.3 eV and 396.6 eV, respectively. These results are promising in view of further studies of thin AlN films properties for application in surface acoustic wave devices (SAW).

1. Introduction

The low-temperature growth of thin AlN films with homogeneous and well-controlled film thickness down to the sub-nanometer scale, high chemical stability, and suitable step coverage is necessary to enable the integration of AlN in surface acoustic wave (SAW) devices [1-3]. Aluminium nitride (AlN), which is a III-V family compound, has outstanding properties, such as SAW speed up to 6000 m/s in comparison with GaN (3693 m/s) [4]. Additionally, AlN has a wider band gap (6.2 eV) in comparison with GaN, which often has a considerable background carrier density at room temperature.

[6] To whom any correspondence should be addressed.

Content from this work may be used under the terms of the Creative Commons Attribution 3.0 licence. Any further distribution of this work must maintain attribution to the author(s) and the title of the work, journal citation and DOI.
Published under licence by IOP Publishing Ltd

When applying thin AlN films in SAW devices, the structure is required to have a low surface roughness. Silicon carbide (SiC) is an excellent substrate material for SAW applications because of its large elastic constant [5], low propagation loss characteristics [6], a small lattice mismatch between the SiC and AlN (less than 1% [7]), and similar thermal expansion coefficients [8].

Therefore, it is of great significance to study the surface roughness of AlN films on different type of SiC substrates in terms of face polarity and orientation.

In this work, we investigated the surface roughness dependence of thin AlN films deposited under identical conditions by atomic layer deposition (ALD) on Si-face and C-face of 4°-off axis and on-axis 4H-SiC substrate, and on 4H-SiC(0001) with a seed layer of 1-2 ML graphene grown by sublimation.

2. Experimental

A Beneq TFS-200 thin-film deposition research system was used for deposition. Five different types of 4H-SiC substrates were used for AlN deposition. The first two types were of 4H-SiC with 4° misorientation toward $[11\bar{2}0]$, Si- and C-terminated; the next two were of on-axis 4H-SiC, also Si and C-terminated. The last one was of 4H-SiC with 1-2 ML graphene grown by sublimation [9, 10]. Before loading into the reactor, the 4H-SiC substrates were cleaned by sequentially boiling in organic solvents (acetone and ethanol) for 3 min and dipping in HF:H_2O (1:1) for 1 min to remove the oxide on the surfaces, rinsing with DI water, and drying with high-purity argon gas; the substrate of 4H-SiC(0001) with 1-2 ML graphene was used as received. During deposition, the substrate temperature was 330 °C. 550 ALD cycles were used in order to obtain the desired thickness of ~25 nm. The precursors for the films were trimethylaluminium $Al_2(CH_3)_6$, TMA, and ammonia (NH_3). One ALD cycle consisted of four stages – 180-ms TMA dose, 2-s nitrogen purge, 90-ms NH_3, followed by another 9-s nitrogen purge. The deposition conditions are summarized in table 1.

Table 1. Deposition condition of AlN films by ALD, root-mean-square (RMS) roughness of the sample from 5×5 μm and the highest peak in the *z*-axis using AFM.

Sample	Substrate	Deposit. Temp. (°C)	Number of ALD cycles	RMS [nm]	z-axis max. [nm]
AlN-1	Si face 4H-SiC 4° off axis	330	550	1.576	65.116
AlN-2	C face 4H-SiC 4° off axis	330	550	2.374	8.239
AlN-3	Si face 4H-SiC on axis	330	550	0.826	4.215
AlN-4	C face 4H-SiC on axis	330	550	3.541	79.985
AlN-5	Si face 4H-SiC with seed layer 1-2 ML graphene	330	550	3.101	33.909

The thickness of the AlN films were assessed using spectroscopic ellipsometry over the wavelength range of 193 – 1000 nm on a J.A. Woolman ellipsometer.

An MFP-3D Asylum Research AFM system was used to map the sample surface and measure the surface roughness. Silicon tips were used in a non-contact AC mode.

The chemical composition of the films was studied by X-ray photoelectron spectroscopy (XPS) using a VG Escalab MKII apparatus equipped with an Al (1486.69 eV) X-ray source.

3. Results and discussion

The AFM surface images of the AlN films (figure 1 a), c) display brighter spots indicating columnar structure of the films for both off-axis wafers. The RMS surface roughness of the AlN films deposited on Si- and C-face is 1.576 nm and 2.374 nm (table 1), respectively. The higher value can be explained by the scratch morphology of AlN deposited on C-terminated 4H-SiC. These scratches are inherited from the substrate. The C-face 4H-SiC$(000\bar{1})$ has a lower surface energy (0.75 J/m^2) compared to that of Si-face 4H-SiC(0001) (1.8 J/m^2), which can explain the better film nucleation on the Si-face 4H-SiC than on the C-face 4H-SiC [11].

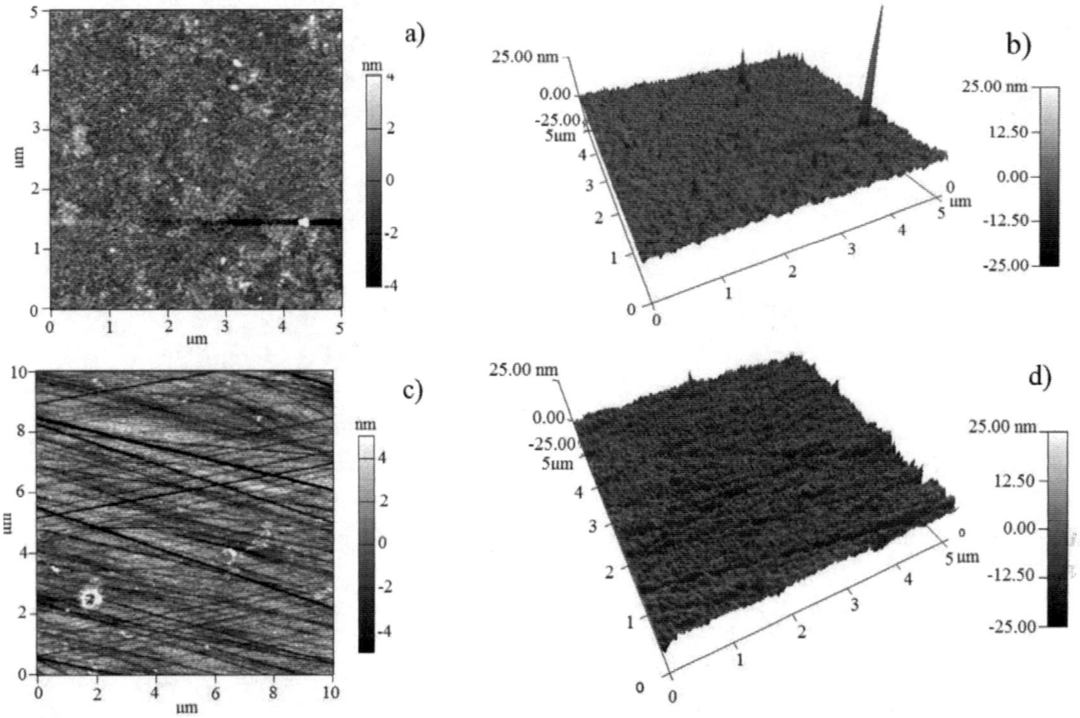

Figure 1. AFM surface images of AlN films deposited on 4° off-axis 4H-SiC a, b) Si-face, c, d) C-face.

The AFM surface images of the AlN films (figure 2 a), c) reveal brighter spots indicating columnar structure for the AlN film deposited on C terminated SiC. Furthermore, the AFM image of a 5×5-μm area quantifies a higher surface roughness of 3.541 nm (figure 2 d) for the AlN film deposited on on-axis C-face 4H-SiC(000$\bar{1}$) compared to the surface roughness of AlN deposited on Si-face 4H-SiC. The irregular nucleation can be explained by the larger initial terraces on the on-axis wafer, as well as the lower surface energy of the C-face 4H-SiC substrate.

The AFM image of the AlN film grown on 4H-SiC(0001) with a seed layer of 1-2 ML graphene (figure 3 a) is notably different. It consists of three regions, one mixed and a higher and a lower region. These can be related with the inhomogeneity of the graphene seed layer [9]. Moreover, the AFM image reveals a region without coverage, as shown in figure 3 b) (4). This is related to the chemical inertness of graphene due to lack of dangling bonds. Additionally, the lower surface energy of graphene (0.119 J/m^2 [12]) compared to that of Si-face and C-face 4H-SiC can explain the better film nucleation on SiC [13-15]. This film's RMS surface is 3.101 nm.

The high resolution XPS analyses show presence of aluminum, nitrogen, oxygen, and carbon. Al 2p data was fitted with two sub-peaks at 73.5 eV and 74.6 eV, which are assigned to Al-N and Al-O bond, respectively [16]. The N 1s data was also fitted with two peaks at 396.8 eV assigned to the N-Al bond and confirming the presence of AlN. A peak at 398.1 eV on the other hand corresponds to the N-C bond [17].

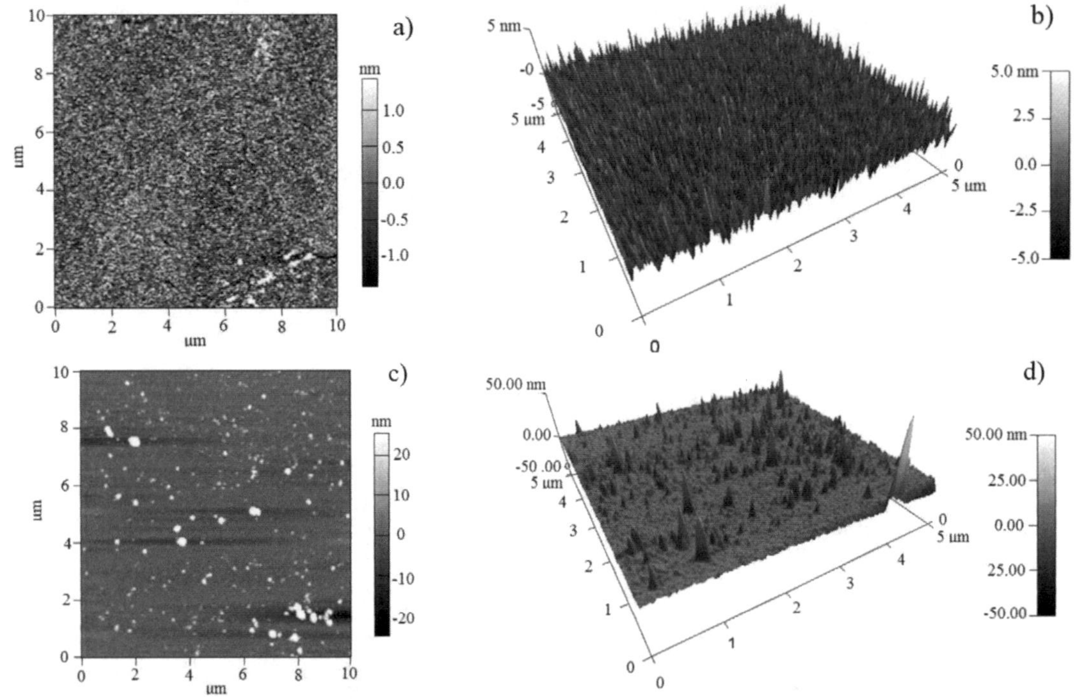

Figure 2. AFM surface images of AlN films deposited on on-axis 4H-SiC a,b) Si-face, c,d) C-face.

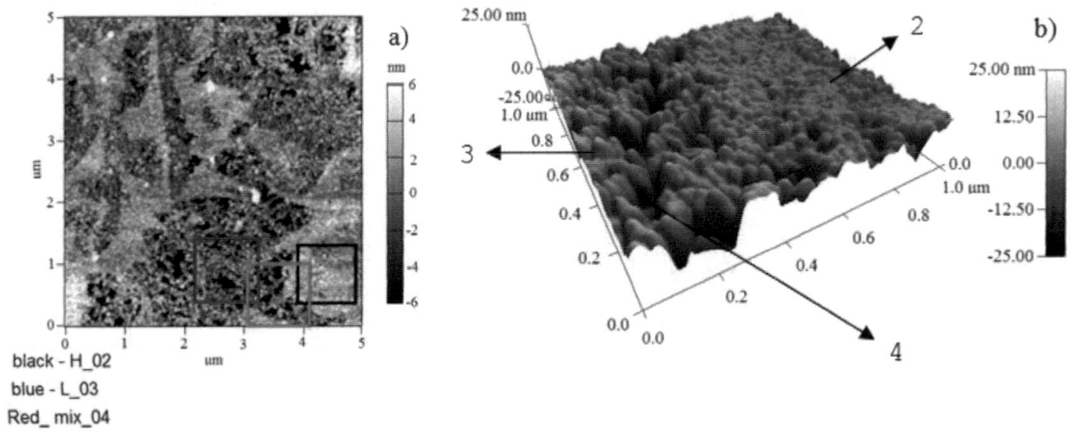

Figure 3. AFM surface images of AlN films deposited on on-axis 4H-SiC(0001) with a seed layer of 1-2 ML graphene a) 5×5 μm scale, b) 1×1 μm scale.

4. Conclusions

In this study, using ALD with TMA and NH_3 we deposited thin AlN films on different types of 4H-SiC substrates in regard to their face polarity and orientation, as well on 4H-SiC with a seed layer of 1-2 ML graphene. The AFM analysis revealed that the AlN film deposited on Si-face on-axis 4H-SiC substrate had the lowest RMS surface roughness of about 0.8 nm, as well as the highest peak in the z-axis direction of about 4.215 nm, which conforms with the requirements of SAW devices for the surface roughness of AlN films. The XPS results show the presence of Al-N bonds in both the Al 2p and N 1s regions for all samples.

Acknowledgements
The work was supported by the Bulgarian National Science Fund under contract DN 18/6, and INFRAMAT (part of the Bulgarian national roadmap for research infrastructures) supported by the Bulgarian Ministry of Education and Science.

References
[1] Meinschen J, Behme G, Falk F and Stafast H 1999 *Appl. Phys.* A **69** 683
[2] Lamanna L, Rizzi F, Bhethanabotla V R and Vittorio M De 2020 *Sensors and Actuators A: Physical* **315** 112268
[3] Ng E J, Sharma J, Ching E W L, Wu G, Pohl D and Vancauwenberghe O 2021 *IEEE 34th International Conference on Micro Electro Mechanical Systems (MEMS)* 370
[4] Takagaki Y, Santos P V, Wiebicke E, Brandt O, Schönherr H -P and Ploog K H 2002 *Appl. Phys. Lett.* **81**(14) 2538
[5] Mehregany M, Zorman C A, Roy S, Fleischman A J, Wu C-H and Rajan N 2000 *Int. Mater. Rev.* **45** 85
[6] Didenko I S, Hickernell F S and Naumenko N F 2000 *IEEE Trans. Ultrason. Ferroelect. Freq. Control* **47** 179
[7] Liu L and Edgar J H 2002 *Mater. Sci.Eng.* **R37** 61
[8] Strite S and Morkoç H 1992 *J.Vac.Sci. and Technol.* B **10** 1237
[9] Yakimova R, Iakimov T and Syvajarvi M, Process for growth of graphene 2005 Patent No.: US 9, 150, 417 B2
[10] Beshkova M, Deminskyi P, Hsu C-W, Shtepliuk I, Avramova I, Yakimova R and Pedersen H 2021 *Phys. Status Solidi* A 2000684
[11] Syväjärvi M, Yakimova R and Janzén E 1999 *J. Electrochemical Soc.* **146** 1565
[12] van Engers C D, Cousens N E A, Babenko V, Britton J, Zappone B, Grobert N and Perkin S 2017 *Nano Lett.* **17** 3815
[13] Cui T, Yip K, Hassan A, Wang G, Liu X, Sun Y and Filleter T 2020 *Sci. Adv.* **6** eabb1335
[14] Kim E S, Hwang J Y, Lee K H, Ohta H, Lee Y H and Kim S W 2017 *Adv. Mater.* **29** 1604899
[15] Tang H, Tao J, Ruzsinszky A and Perdew J P 2019 *J. Phys. Chem.* C **123** 13748
[16] Liao J X, Xia L F, Sun M R, Liu W M, Xu T,Yang C R, Chen H W, Fu C L and Leng W J 2005 *Appl. Surf. Sci.* **240** 71
[17] Zhu J, Zhao D, Luo W B, Zhang Y and Li Y R 2008 *J. Cryst. Growth* **310** 731

Study of the structure and mechanical properties of electron-beam-welded dissimilar joints of copper and stainless steel with and without offset

D Kaisheva[1,2,4], A Anchev[3], V Dunchev[3], B Stoyanov[3], S Valkov[1,3], M Ormanova[1], G Kotlarski[1], V Todorov[3], M Atanasova[3], S Rabadzhiyska[1] and P Petrov[1]

[1]Institute of Electronics, Bulgarian Academy of Sciences, 72 Tsarigradsko Chaussee, 1784 Sofia, Bulgaria
[2]Neofit Rilski South-Western University, 66 Ivan Michailov Str., 2700 Blagoevgrad, Bulgaria
[3]Technical University of Gabrovo, 4 H. Dimitar Str., 5300 Gabrovo, Bulgaria

E-mail: darinakaisheva@ie.bas.bg

Abstract. We present the results of an investigation of the structure and mechanical properties of dissimilar joints of copper and 304L stainless steel formed by electron-beam welding. The samples studied were welded without a beam offset and with a beam offset towards either welded materials to a distance of 0.3 mm. The phase composition was determined via X-ray analysis. The structure was investigated using scanning electron microscopy. The mechanical properties, including hardness and tensile strength, were measured. The phase composition of the considered specimens consisted of a double-phase structure of face-centered cubic (fcc) and body-centered cubic (bcc) phases. The sample welded without offset exhibited the highest yield strength and tensile strength values. The microhardness of all samples increased in the fusion zone on the steel side and decreased in the fusion zone on the copper side compared to the initial hardness of the two materials.

1. Introduction

The need to combine two dissimilar materials is dictated by the purpose of using, in practice, the different useful properties possessed by each material separately. Thus, joints of copper and stainless steel combine the very good thermal and electrical conductivity of copper with the significant strength, wear resistance, and corrosion resistance of steel. Copper and stainless steel joints are used in heat transfer devices, nuclear devices, energy sources (batteries, accumulators), shipbuilding, aircraft, space, and other industries [1].

Durable welds made of dissimilar materials are difficult to obtain due to the materials' different physical-chemical properties. Their metallurgical compatibility and the welding technology are of major importance for their production. The metallurgical compatibility is determined by the mutual solubility of the joined metals in the liquid and solid states, as well as by the formation of intermetallic phases. The mutual solubility of the elements is determined by the similarity of the crystal lattices of the solutes, by the difference in the atomic radii and by the energy of the bond between the two elements [2].

---

[4] To whom any correspondence should be addressed.

Content from this work may be used under the terms of the Creative Commons Attribution 3.0 licence. Any further distribution of this work must maintain attribution to the author(s) and the title of the work, journal citation and DOI.
Published under licence by IOP Publishing Ltd

Electron-beam welding (EBW) is a technology that allows the successful welding of dissimilar materials; for some pairs of materials, this is one of the most widely applicable technologies. EBW is characterized by a high energy density, precise positioning and focusing of the electron beam, and a narrow heat-affected zone. All types of welding are accompanied by the appearance of residual stresses and micro-deformations, as well as defects (micro-cracks, etc.). In our previous works, we have studied the residual stresses that occur during EBW of both similar welds [3] and dissimilar welds [4]. As we found, one advantage of the EBW is that the level of residual stresses in EBW is lower than in other welding technologies [5]. Several papers have summarized various methods and techniques for welding dissimilar materials [6-8], focusing on the EBW and the outlook for its development. Experiments have been reported with the displacement of the laser beam during laser welding and the electron beam during EBW to the side of one metal, which leads to reduction of defects and improvement of the mechanical properties of the joint [9, 10]. The authors of [10] have investigated the effect of a beam offset on the weldability and structure of electron-beam welded Cu and 304 stainless steel (SS). The evaluated residual stresses in welds of the same materials and the influence of the technological conditions on it were presented in our work [4]. However, studies based on detailed investigations of the phase composition and mechanical properties of Cu - 304L SS are currently lacking in the scientific literature.

Therefore, the aim of the present work is to study the influence of the electron beam offset on the structure and mechanical properties of electron-beam welded samples of copper and 304L stainless steel.

2. Experiment

Three kinds of welded specimens of copper and AISI 304L stainless steel were examined – without an offset, with an offset towards the copper side (denoted by "+") and with an offset towards the steel side (denoted by "–"). The beam offset is shown in figure 1.

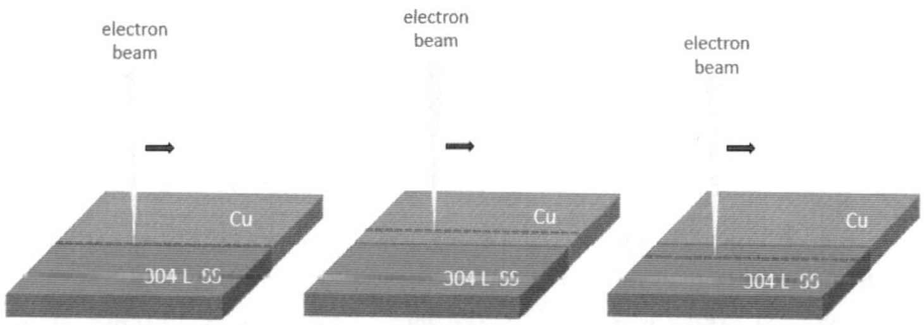

Figure 1. Schematic of the experimental design for EBW of copper and stainless steel AISI 304L.

The samples were flat plates with sizes of 100×50×8 mm. The stainless steel 304L used has the following chemical composition (wt %): 0.03% C; 2.0% Mn; 0.75% Si; 0.045% P; 0.030% S; 17.5 – 19.5% Cr; 8.0 – 12.0% Ni; 0.10% N; bal. Fe.

The electron-beam welding was carried out on an Evobeam Cube 400 welding unit manufactured by Evobeam. The technological parameters of the EBW process are presented in table 1. Due to the higher thermal conductivity of copper, we increased the beam power in the case of offset to copper and decreased it in case of offset to stainless steel.

Table 1. Technological parameters of EBW.

Sample	Accelerating voltage U, kV	Beam current I, mA	Focusing current I_f mA	Welding speed v, mm/s	Offset mm
1	60	50	1543	5	0
2	60	60	1543	5	+ 0.3
3	60	40	1547	5	– 0.3

The phase analysis was conducted by a Bruker D8 Advance X-ray diffractometer. The method applied was "Coupled Two Theta" in the range from 40° to 120° using Co Kα radiation with a wavelength of 1.78897 Å and line focus orientation. The X-ray generator current was 40 mA and the used voltage was 35 kV. Other characteristics of the test were a 0.05° step size and 0.25 s step time.

Secondary-electrons scanning electron microscopy (SEM) on an EVO MA10 Carl Zeiss instrument was used to study the microstructure of the welded specimens. The accelerating voltage was 20 kV. The chemical elements distribution in the fusion zone was analyzed by energy-dispersive X-ray spectroscopy (EDX) on a spot and on a line.

The tensile experiment was performed on a ZWICK Vibrophore 100 machine for static and dynamic tests in accordance with the requirements of ISO 6892-1, Method B. The experiment was conducted in two stages. In the first stage, the specimens were made only from the basic materials that are the subject of this study. In the second stage, specimens from both materials welded under different technological conditions were tested.

The microhardness experiment was performed on a ZHVμ-S ZWICK/Indentec semi-automatic microhardness tester. The metallographic cross-sections were made of welded materials in a direction transverse to the weld. The line along which the microhardness was measured was located at the middle depth of the weld seam. A load force of 0.49 N was used for all experimental points.

3. Results and discussion

The X-ray diffraction patterns of the welded joints are shown in figure 2. The phase composition of the considered specimens is a double phase structure of face-centered cubic (fcc) and body-centered cubic (bcc) phases. It should be noted that in both welded materials the fcc structure is the main phase, and the joint solidifies in the form of a substitutional solid solution. The diffractograms do not exhibit changes in the phase composition, meaning that the offset of the electron beam does not lead to a change in this crystallographic parameter. The phase composition of the untreated SS is a single phase gamma structure. It is known that the bcc phase is metastable and appears at temperature of about 1200 °C, and above. During the welding of the specimens in the present particular case, the temperature significantly exceeds this value, meaning that this phase has been formed. At the same time, it is well known that the cooling rate during the electron-beam welding process is quite high, and the solidifying molten material is not in equilibrium. In this case, the non-equilibrium solidification is the dominant process leading to the preservation of metastable phases, which is the formation mechanism of the bcc structure [11]. Considering sample 1 (i.e. without offset) it is obvious that the peak intensity of the bcc phase is the highest, meaning that the amount of this compound is the largest in comparison with the other specimens. In this case, copper melts first when heated by the electron beam due to its lower melting point and begins to form a vapor-gas channel known as a "keyhole". The molten pool initially consists only of molten copper. As the temperature rises, the steel begins to melt as well. The main mechanism

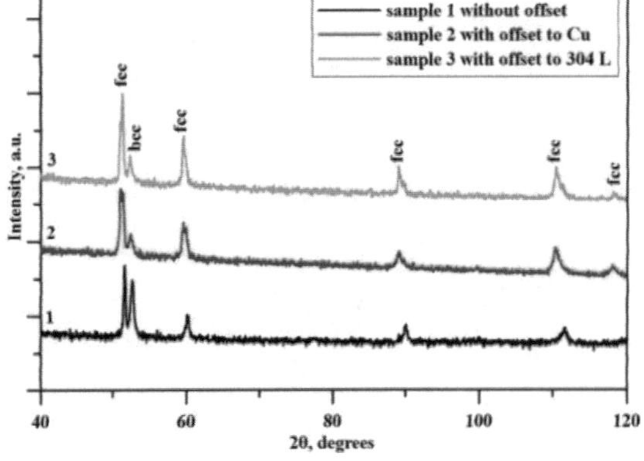

Figure 2. X-ray diffraction patterns of the welded joints.

for homogenizing the molten pool is the movement of the steam flow [12]. In the case of sample 1 (i.e. without beam offset), the thermal cycling gradient in the steel is the highest, leading to the largest amount of the bcc phase. In the case of sample 2, the cooling rate in the steel is lower due to the offset of the electron beam to the copper. Similarly, the lower cooling rate in the steel in the case of sample 3 (i.e. beam offset to the steel) is due to the lower value of the beam current.

The cross-sectional SEM images of the welded samples are shown in figure 3. Considering the specimen welded without offset (sample 1), the structure of the fusion zone (FZ) is significantly more homogeneous in comparison with the other specimens (samples 2 and 3), and no or a very small amount of cracks, pores and other structural defects can be observed. The FZ of the specimen welded by an offset towards the copper side consists of two clearly distinguished areas. The EDX results show that the Fe element in the area near the stainless steel is significantly higher than Cu, while in the area near the copper side the amount of Cu is larger than that of Fe. Considering sample 3 (offset towards SS) (figure 3 (c)), the FZ is not well visible, and the boundary between the FZ and SS plate is also indistinct. A large number of cracks, pores, and other structural defects can be seen. The boundary between the FZ and Cu is very sharp and well defined. The concentration of copper in the FZ is very low.

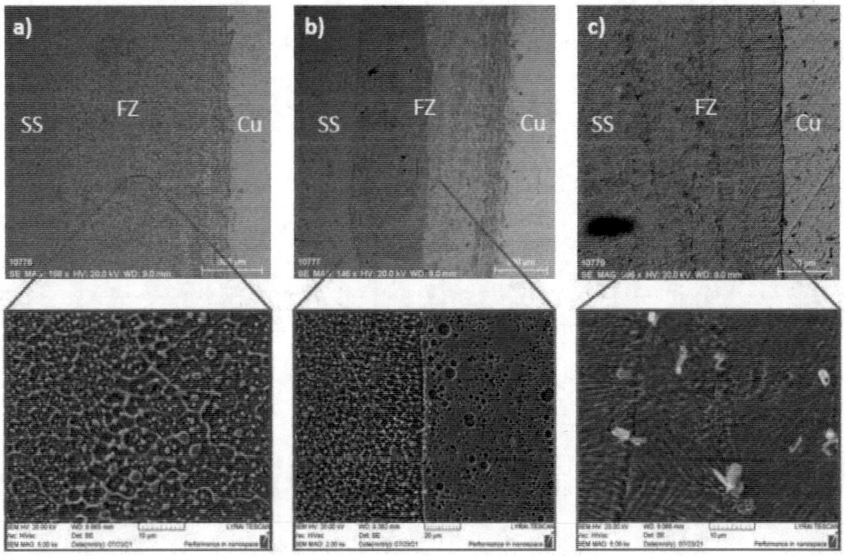

Figure 3. SEM images of sample 1 a), sample 2 b) and sample 3 c).

The results from the tensile test are presented in table 2. The best result of the mechanical tests was obtained for the sample with zero offset (sample 1), as it is clearly seen. Its yield strength is only 20.7% lower than the yield strength of pure copper. An offset of 0.3 mm to either material leads to a significant deterioration of the strength characteristics of the weld. In the case of offset to stainless steel and the above-mentioned technological conditions, the welding quality is unsatisfactory, as the value of only 2.9% elongation proves this fact. These results are consistent with those obtained by the XRD experiments. As already mentioned, the amount of the bcc phase is the largest in the case of sample 1 (i.e. without beam offset), leading to the enhancement of the mechanical properties.

Table 2. Results from the tensile tests.

Sample	Offset, mm	Yield strength $R_{p0,2}$, MPa	Tensile strength R_m, MPa	Elongation E, %
1	0	93	218	14.6
2	+ 0,3	52	204	18,7
3	- 0,3	54	59	2.9

These results are in agreement with those of the microstructure and phase composition of the welds considered. The mechanism of plastic deformation is based on the movement of dislocations, where the main mechanism is slipping. The slipping planes strongly depend on the crystallographic structure, with the bcc lattice having the largest number of slip systems. As already mentioned, the amount of bcc phase is the largest in the phase composition of sample 1. Therefore, the contribution of the discussed phase is the highest. Furthermore, the microstructure of sample 1 exhibits a significantly more homogeneous structure, without or with a very small amount of structural defects (such as cracks, pores etc.), which could be another reason for the best mechanical properties of the considered specimen.

The results of the measured cross-sectional microhardness in the middle of the width of the weld are shown in figure 4. In the weld area on the stainless steel side, an increase in the microhardness of approximately 22% is observed compared to the microhardness of the thermally unaffected zone of the steel. At the weld area of the copper side, a reduction in the microhardness of approximately 25% is observed compared to the thermally unaffected copper zone. The reduced value of the microhardness in the copper zone explains the subsequent destruction of the specimen in the mechanical test in this cross section.

For samples 1 and 2, the transition zone in which the microhardness is reduced from a maximum to a minimum is significantly wider than for sample 3. In this specimen, the transition is instantaneous, which corresponds to the result of the mechanical test and is proof of the lack of quality welding.

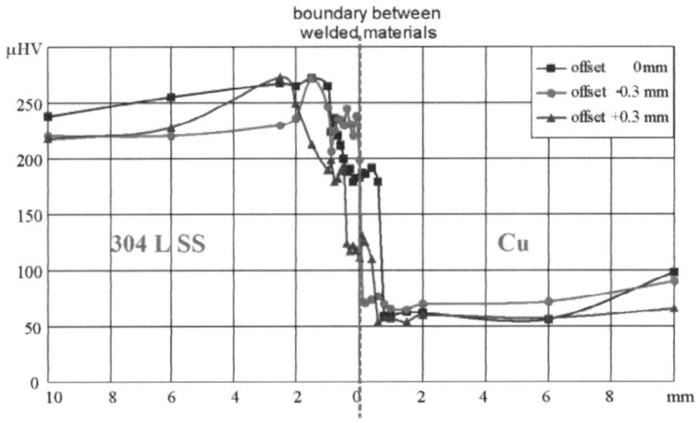

Figure 4. Distribution of the cross-sectional microhardness in the middle of the width of the welds.

4. Conclusions

Joints of copper and 304L stainless steel were successfully made via electron-beam welding without a beam offset and with a beam offset towards either welded material to a distance of 0.3 mm. The phase composition of the considered specimens is in the form of a double phase structure of face-centered cubic (fcc) and body-centered cubic (bcc) phases, with the amount of bcc being maximal in the weld formed without an offset. The specimen welded without an offset exhibits the highest values of yield strength and tensile strength. The change in the microhardness for all three samples has the same trend – it increases in the fusion zone on the steel side and decreases in the fusion zone to the copper side, compared to the initial hardness of the two materials. The area of this microhardness change is the narrowest for the sample with the beam offset to the steel side.

Acknowledgements

This work was supported by the Bulgarian National Scientific Fund under Grant KP 06-N47/6.

References

[1] Dilthey U and Stein L 2006 *Sci. Technol. Weld. Join.* **11** 135–142.
[2] Ryabov V 1984 *Welding of dissimilar materials* (In Russian) (Moskow: Mashinostroenie)

[3] Kaisheva D, Bokuchava G, Papushkin I and Petrov P 2020 *C. R. Acad. Bulg. Sci.* **73(4)** 475–484
[4] Petrov P, Kaisheva D, Bokuchava G, Papushkin I and Valkov S 2019 *Welding and welded structures (Zavarivanje i zavarene konstrukcije)* **64(4)** 149–156
[5] Bokuchava G and Petrov P 2020 *Metals*. **10(5)** 632
[6] Sun Z and Karppi R 1996 *J. Mater. Proc. Technol.* **59(3)** 257–267
[7] Kah P, Shrestha M and Martikainen J 2014 *Appl. Mech. Mater.* **440** 269–276.
[8] Martinsen K, Hu S and Carlson B 2015 *CIRP Ann. Manuf. Technol.* **64(2)** 679–699
[9] Sahul M, Tomcíková E, Sahul M, Pašák M, Ludrovcová B and Hodúlová E 2020 *Metals* **10** 1294
[10] Guo R, Zhou Q, Kong J, Peng Y, Xiang Y, Luo T, Wang K and Zhu J 2016 *Vacuum* **128** 205–212
[11] Fu J W, Yang Y S, Guo J J, Ma J C and Tong W H 2009 *J. Mater. Res.* **24** (7) 2385
[12] Tosto S, Nenci F, Jiandong Hu, Corniani G and Pierdominici F 2003 *Mater. Sci. Technol.* **19** 519

Optical properties of nanostructured bimetallic films from the Ag-In and Ag-Sb systems and their surface-enhanced fluorescence application

V Katrova[1], T Hristova-Vasileva[1,2], A Atanasova[1], V Strijkova[1] and R Todorov[1,3]

[1]Acad. J. Malinowski Institute of Optical Materials and Technologies, Bulgarian Academy of Sciences, Acad. G. Bonchev Str., bl. 109, 1113 Sofia, Bulgaria
[2]Institute of Solid State Physics, Bulgarian Academy of Sciences, 72 Tsarigradsko Chaussee, 1784 Sofia, Bulgaria

E-mail: rossen@iomt.bas.bg

Abstract. The possibility for tuned excitation of surface plasmon resonance in Ag-In and Ag-Sb films with different compositions and thicknesses was studied in terms of preparation and microstructural and optical properties. The analyses show that plasmon excitation can be achieved both by changing the thickness of the deposited bimetallic films and by varying their composition. The imaginary ε'' part of the complex permittivity of the thin films has a maximum due to the transverse oscillations of free electrons in the range of 1 eV to 3.5 eV. The films' applicability as amplifying substrates in surface-enhanced fluorescence was tested. Tryptophan and Cu (II)-phthalocyanine (CuPc) dye were used to analyze the efficiency of the localized surface plasmon resonance excitation in the ultraviolet spectral region. Amplification enhancing coefficient of 4.17 times was obtained in the case of CuPc dye.

1. Introduction

Plasmonics is based on studying the surface plasmon resonance (SPR) phenomenon as the quality of the plasmon resonance is closely related to the structure and properties of the materials [1]. Its implementation in analytical technologies, as surface-enhanced Raman spectroscopy (SERS), plasmon-enhanced fluorescence (PEF) and surface-enhanced infrared spectroscopy (SEIRA), for detection of chemicals and biological species, is important for various fields of life sciences, such as medical diagnostics, food control and security [2].

Localized surface plasmon resonance (LSPR) is a physical phenomenon caused by the interaction of light with metal nanoparticles (NPs) that are smaller than the wavelength of the incident light. The electromagnetic field causes a collective excitation of electrons from the conduction band, as a result of which coherent localized plasmon oscillations occur with a resonant frequency that strongly depends on the material's composition, the particles' size and shape, the distances between them, and the dielectric medium in which they are placed [3]. The optical properties of the materials are determined by the complex dielectric function $\varepsilon(\omega)$, as excitation of a surface plasmon (SP) is observed in that part of the electromagnetic spectrum in which the real part of the permittivity is less than zero ($\varepsilon' < 0$). A strong plasmon resonance is related to the relatively low and positive value of ε'' ($\varepsilon'' << |\varepsilon'|$) [4]. The most

[3] To whom any correspondence should be addressed.

Content from this work may be used under the terms of the Creative Commons Attribution 3.0 licence. Any further distribution of this work must maintain attribution to the author(s) and the title of the work, journal citation and DOI.
Published under licence by IOP Publishing Ltd

common materials for the preparation of nanoparticles are the noble metals Ag and Au, due to their strong LSPR in the visible spectral range caused by $d - d$ transitions of electrons [5]. Ag exhibits the sharpest and most intense absorption peaks of all metals, while Au is preferred for biological use due to its inert nature and biocompatibility [6]. Au and Ag, however, do not maintain good surface plasmon (SP) in the UV part of the spectrum because their imaginary dielectric constant increases in the energy range above 4 eV. The metals from the *5p* - block of the Periodic Table, such as Al, Ga, In, Sn, Tl, Pb and Bi, show the opposite trend and are promising candidates for use in UV plasmonics. The organics and biomolecules have strong absorption in the near UV range (3.1 – 12 eV), which is why the study of LSPR in this spectral region is important for amplifying techniques, such as SERS and PEF [7]. In [8], for example, the LSPR of In, Sn, Pb, Bi, and Sb was studied and an enhancing factor comparable to that of gold nanoparticles was shown.

An adjustment of the plasmon resonance can be achieved by using bimetallic alloys of the two metals [9]. Special attention to the plasmonic activity of silver-rich Ag-In alloys (with indium content up to 12 at %) was paid in earlier works, while the Ag-In alloys with a higher indium content based on mixture of intermetallic compounds and the In-rich materials have remained less studied. The authors reported a shift of the plasmonic frequency in the UV spectral range from 3.8 eV for pure silver to 4.2 eV. The increase of the structural mishmash in the crystalline structure of silver when the indium content is higher than 5 at % decreases the efficiency of excitation of surface plasmons [10]. The optical properties of thin Ag-Sb films are even less investigated [11]. The antimony (Sb), a semi-metal of the 5th group of the Periodic Table, has been rarely studied in terms of its plasmonic behavior, both experimentally and theoretically. However, the authors of [12] present an enhancement of the LSPR in antimony nanopolyhedrons with different types of morphologies.

The goal of the present work is to study the thickness dependence of the optical properties of thin Ag-In layers with selected compositions chosen according to their phase diagrams [7, 14], to evaluate the plasmonic properties on the basis of calculated spectra of complex permittivity and to check their applicability as amplifying substrates for surface-enhanced fluorometry tests.

2. Experimental details

Thin films of both Ag-In and Ag-Sb systems with different ratios of the chemical elements were deposited on Si and quartz substrates by thermal co-evaporation at a residual pressure of $\sim 1\times10^{-3}$ Pa. The composition of the deposited layers was chosen according to the specified regions in their phase diagrams [13, 14], and were realized through variation of the deposition rates of the two elements. In the process of evaporation, the thickness of the layers and the deposition rate were controlled by quartz monitoring. Thin films of investigated compositions with different thicknesses were obtained in one vacuum cycle.

The thin films' chemical composition was determined by energy dispersive X-ray microanalysis (EDAX) using a Philips 505 scanning electron microscope (SEM) with an EDAX 9100 microanalyzer. X-ray diffraction (XRD) patterns obtained using a Philips 1710 X-ray diffractometer with CuK$_\alpha$ emission (λ = 1.54056 Å) were used to analyze the level of alloying and the layers' microstructure.

An MFP-3D Asylum Research (Oxford Instruments) atomic force microscope (AFM) was used to observe the samples' surface morphology and to determinate their surface roughness. Standard AC 160 TS silicon probes with a frequency of 300 kHz and a spring constant of 26 N/m were used for these measurements, which were carried out in a non-contact AC mode.

The optical properties of the thin layers were investigated by spectrophotometric and ellipsometric measurements. The reflectance and transmittance coefficients were measured using a CARY 05E UV-VIS-NIR spectrophotometer in the spectral range 200 – 800 nm. The complex dielectric function $\varepsilon(\omega)$ was calculated from measurements made by a phase-modulation spectral ellipsometric UVISEL2 device (HORIBA Jobin Yvon) in the range 0.6 – 6.5 eV at angles of incidence of 50°, 60° and 70°. The Ag-Sb layers were tested for fluorescence signal amplification of the Cu (II)-phthalocyanine (CuPc) dye using a FluoroLog3-22 spectrofluorometer, (HORIBA JobinYvon). Fluorescence measurements of water solution of tryptophan on Ag-In thin films were performed on a USB4000 Ocean Optics spectrometer with an optical waveguide and a Hg lamp with emission line of 254 nm.

3. Results and discussion

The thin layers of both Ag-In and Ag-Sb systems were deposited by thermal co-evaporation. The thin Ag-In films were deposited at Ag:In deposition rates' ratios of 20:1, 1.8:1, 1.5:1, 1:2 and 1:30. The chemical compositions, obtained by X-ray microanalysis, for the above Ag:In ratios were Ag_{95}:In_5, Ag_{76}:In_{24}, Ag_{66}:In_{34}, Ag_{36}:In_{64} and Ag_4:In_{96}. The compositions of the thin films from the Ag-Sb system, deposited with expected Ag:Sb ratios of 19:1, 3:1 and 1:19, were Ag_{98}:Sb_2, Ag_{83}:Sb_{17}, Ag_6:Sb_{94}.

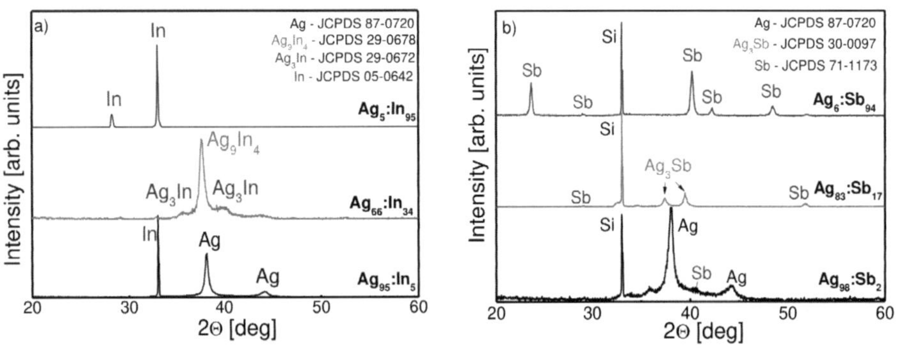

Figure 1. XRD patterns of thin films from the Ag-In and Ag-Sb systems.

The thickness (d) of all layers, derived from ellipsometric measurements, was in the range 10 – 40 nm. In view of comparing the samples' properties of interest, 300-nm thick layers, as well as pure silver, indium and antimony layers were also deposited.

Figure 1 presents the XRD patterns of the investigated thin films. The XRD patterns of the thin Ag-In films showed that the phase composition of the thin layers corresponds well to the phase diagram (figure 1 a). The presence of a silver-based solid solution, defined by a shift of the Ag reflexes towards the indium, and of unreacted indium was seen in the thin Ag_{95}:In_5 film. The thin Ag_{66}:In_{34} film is a mixture of Ag_3In and Ag_9In_4 crystals, and only indium appears in the diffraction pattern of the thin Ag_4:In_{96} layer. The diffraction peaks of silver, antimony and Ag_3Sb were registered in the Ag-Sb layers (figure 1 b). The Ag-rich sample Ag_{98}:Sb_2 exhibits the pattern of a silver-based solid solution. Formation of the Ag_3Sb compound is observed in the Ag_{83}:Sb_{17} layer. A residual elemental antimony is present in both samples. No significant amount of silver is registered in the Sb-rich sample.

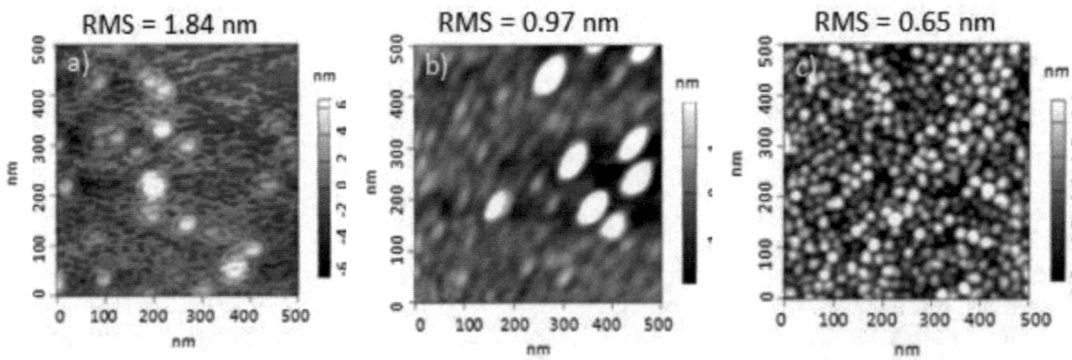

a) b) c)

Figure 2. AFM images of Ag-Sb bimetallic films with thickness of 10 nm: a) Ag_{98}:Sb_2 (d_g = 34.4 nm), b) Ag_{83}:Sb_{17} (d_g = 45.1 nm) and c) Ag_6:Sb_{94} (d_g = 17.1 nm).

The AFM images showed that the thin Ag-In films are built by grains with dimensions in the range of 20 – 70 nm. The root-mean-square (RMS) surface roughness increases slightly from 0.63 nm to 2.78 nm with the increasing In content. The AFM images of the thin Ag-In films with the smaller thickness of ~ 10 nm suggest an island microstructure and demonstrate an increase of the RMS values in the range of 2.13 nm to 6.53 nm. Similarly, the AFM images of the Ag-Sb coatings with a thickness of 10 nm are characterized by an island structure (figure 2). The grains in the Ag_{98}:Sb_2 layer are with an average size of d_g = 34.4 nm. The Ag_{83}:Sb_{17} layer, in which formation of the Ag_3Sb compound was observed, reveals ellipsoid-shaped particles with the largest average size of d_g = 45.1 nm. The Ag_6:Sb_{94} layer is characterized by closely spaced particles with d_g = 17.1 nm. The RMS surface roughness decreases from 1.84 nm to 0.65 nm with the increasing Sb content in the Ag-Sb layers.

The reflectance (*R*) and transmittance (*T*) spectra of the samples from the Ag-Sb system are shown in figure 3. Due to the strong influence of the substrate on the reflectance of the 10-nm films and in order to trace the compositional dependence, the spectra of *R* for 300-nm thick layers are shown (figure 3 a). The observed minimum is related to the plasma frequency ω_p of the material's lattice fluctuations, which is a longitudinal oscillation with respect to the direction of propagation of the electromagnetic wave, and represents a maximum in the spectrum of *T* (figure 3 b). For a thin silver layer, the minimum in reflection is observed at 315 nm (3.9 eV). The addition of antimony shifts the minimum towards the shorter wavelengths – from 308 nm to 247 nm for the Ag_{98}:Sb_2 and for the Sb-rich layer, respectively.

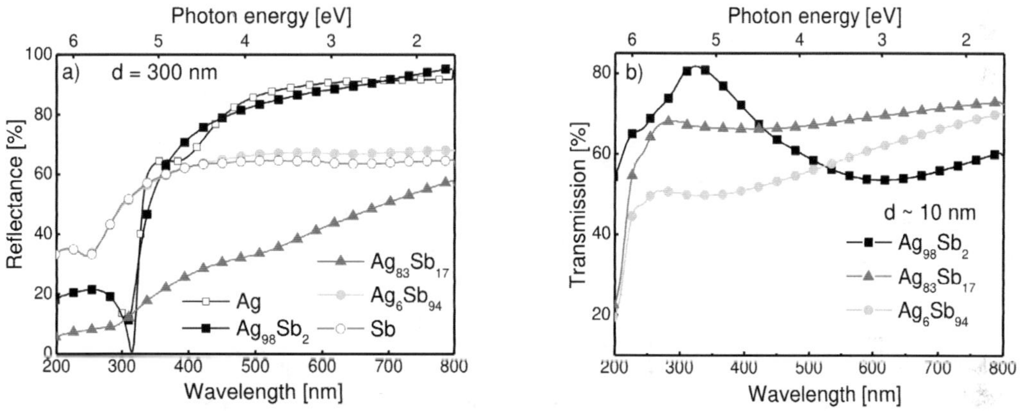

Figure 3. Compositional dependence of the reflectance spectra of thin films with *d* = 300 nm (a) and transmittance spectra of thin films with *d* = 10 nm (b).

Due to the presence of an island structure in the thin layers, proven by AFM, the interaction of the oscillating electrons with the propagating electromagnetic wave causes an LSPR, which is located at higher wavelengths and appears as a maximum in the imaginary part of the complex dielectric function ε'', as determined by the ellipsometric study.

The complex dielectric permittivity $\varepsilon(\omega) = \varepsilon'(\omega) + i\varepsilon''(\omega)$ of the bimetallic films from both systems was derived from spectral ellipsometry measurements by applying the Drude-Lorentz model that is described by the following formula:

$$\varepsilon(\omega) = \varepsilon_\infty - \frac{\omega_p^2}{\omega^2 + i\omega\Gamma_p} + \sum_i \frac{f_i\omega_i^2}{\omega_i^2 - \omega^2 - i\omega\Gamma_i}, \qquad (1)$$

where ε_∞ is the dielectric permittivity at infinite frequency, ω_p is the plasma frequency and Γ_p is the damping coefficient of a free electron. The Lorentzian-like parts are characterized by resonance frequencies ω_i, oscillator strengths f_i and damping factors Γ_i.

Figure 4 (a-d) shows the dispersions of the complex permittivity's real and imaginary parts calculated from ellipsometric measurements of thin Ag-In films with thicknesses of ~ 10 and 100 nm.

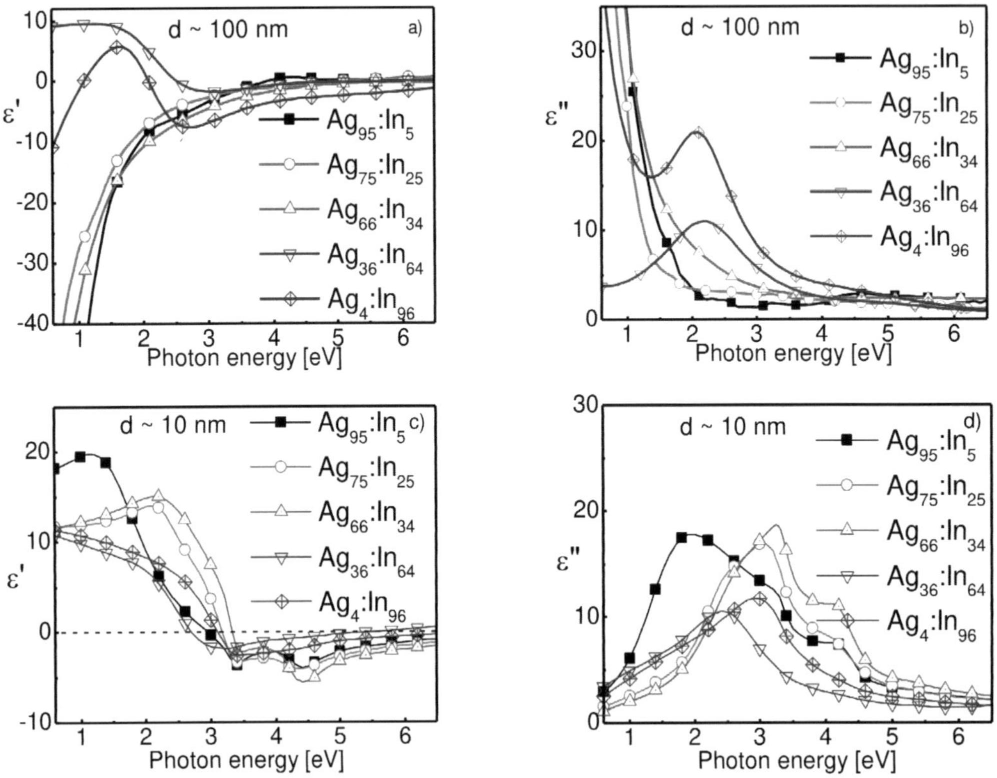

Figure 4. Dispersion of the real ε' and imaginary ε'' parts of the complex permittivity of thin Ag-In films with thicknesses ~ 100 nm (a and b) and ~ 10 nm (c and d).

It is seen that the thick Ag-In films with an indium content below 64 at % are characterized by negative values of ε' in the whole investigated spectral range 0.6 – 6.5 eV and by a decreasing ε'' for photon energies above 2 eV. This suggests an improvement in the excitation efficiency of LSPR in the ultraviolet region, defined, in the case of spherical particles, as the ratio - $Q_{LSPR} = -\varepsilon'/\varepsilon''$. In the case of indium rich layers, a switch of the ε' sign is observed due to the presence of an interband transition of indium at 1.4 – 2.8 eV [9, 10]. These coatings exhibit higher ε'' values in the range of 2 – 5 eV; thus, they have lower Q_{LSPR} values in this spectral region.

As for the thin Ag-In films with thickness of 10 nm, figure 4 (d-c), it is seen that the values of ε' switch their sign in the spectral range of 2.5 – 3.5 eV. The real part of the complex permittivity remains positive because the grains in the thin film never become fully connected. The ε'' possess a maximum in the range of 1.6 – 3.5 eV, which provides information on the interaction of the electromagnetic waves with the transverse oscillations of free electrons in the metallic grains.

The results from the ellipsometric study of the thin Ag-Sb films show that the decrease of the layer thickness causes an increase of the ε' value for all compositions due to the increase of the dielectric component in the nanocomposites. Figure 5 a) presents the thickness dependence of $\varepsilon(\omega)$ of the thin Ag_{83}:Sb_{17} films. The ε'' has a maximum in the range of 1 – 3.3 eV, which shifts to the higher photon energies with the decrease of the thickness and implies a presence of an island structure and a change in the particles' size. As the antimony content increases in the layers with $d = 10$ nm (figure 5 b), ε' increases in the infrared spectral region with the changing antimony content; the maximum of ε'' shifts towards the higher photon energies – for the Ag_6:Sb_{94} layer it is located at 3 eV.

The calculated spectra of the complex permittivity ε were used to determine the dielectric loss function Im{-1/ε}, which gives possibility of visualizing the longitudinal oscillations of free electrons in metals.

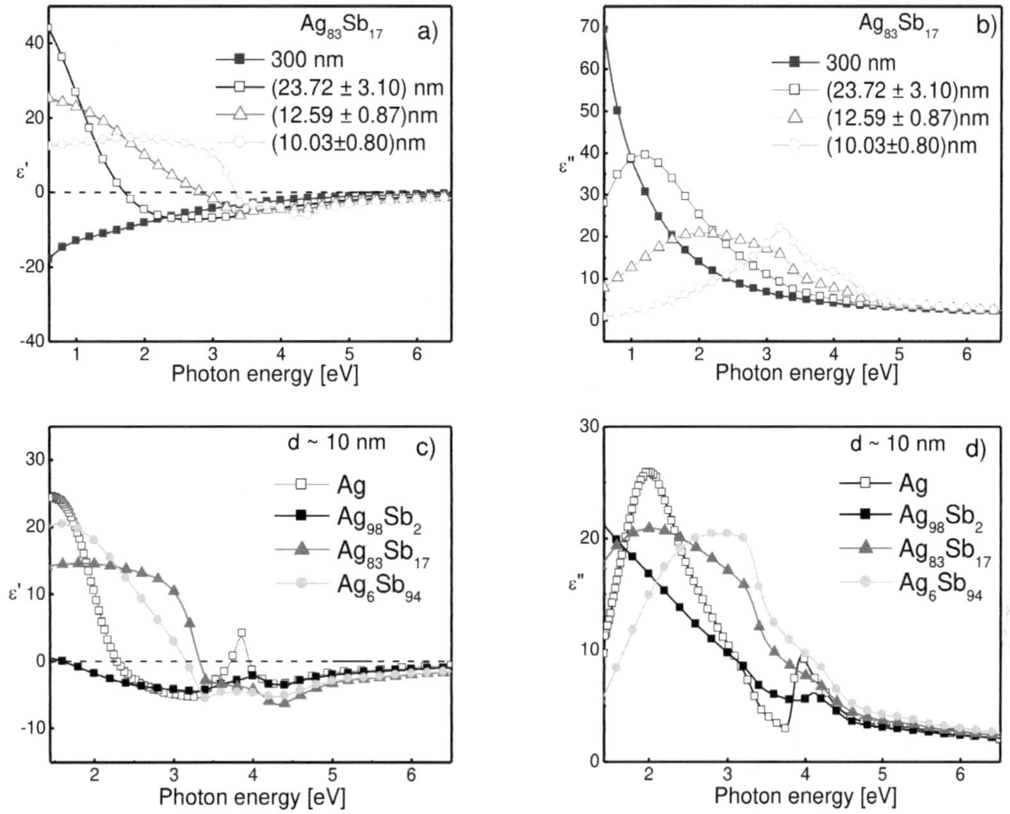

Figure 5. Thickness dependence of the real ε' and imaginary ε'' parts of the complex permittivity of thin Ag_{83}:Sb_{17} films (a and b); compositional dependence of the complex permittivity of Ag-Sb films with thickness ~ 10 nm (c and d).

We used the dispersion parameters obtained to extrapolate the loss function in the spectral region from 6.5 eV to 10 eV. The loss function of thin Ag-In films with indium content ≥ 25 at % In shows a maximum in the spectral range from 7.4 eV to 8.0 eV associated with the longitudinal oscillations of the free electrons. Only the loss function of the silver-rich thin Ag_{95}:In_5 film possesses a typical maximum at 3.8 eV, also observed in the case of pure silver. The results of the loss-function calculation of the samples from the Ag-Sb system showed that a maximum of the $\mathrm{Im}\{-1/\varepsilon\}$ function of the silver-rich thin Ag_{98}:Sb_2 film is observed at 4.10 eV, while the extrapolation yields values of 6.78 eV and 9.7 eV for the thin films with compositions Ag_6:Sb_{94} and Ag_{83}:Sb_{17}, respectively.

Finally, we calculated the spectra of the cross-section of excitation, $\sigma_{exct.}$, of spherical Ag-In nanoparticles with the same composition as the investigated thin films and with a diameter a = 20 nm, embedded in a dielectric medium with a refractive index of 1.5 using a procedure described in [14, 15]. The results show the presence of a maximum of $\sigma_{exct.}$ in the spectral range of 3.5 – 6.0 eV for the compositions Ag_{76}:In_{24} and Ag_{66}:In_{34} from the Ag-In system (figure 6 a). The calculation for the Ag-Sb system (figure 6 b) showed that the thin film with composition Ag_6:Sb_{94} has the highest values of $\sigma_{exct.}$ in the same spectral range.

Fluorescence signal excitation tests were also performed. In the case of Ag-In thin films, tryptophan was used to analyze the efficiency of the localized surface plasmon resonance excited in the UV spectral region. It yields a strong fluorescent band with a maximum at ~ 350 nm when excited by light with a wavelength in the range of 250 – 280 nm. The experiment was performed by dropping 0.5 ml of test solution (tryptophan+water) onto thin coatings with compositions Ag_{66}:In_{34} and Ag_4:In_{96}.

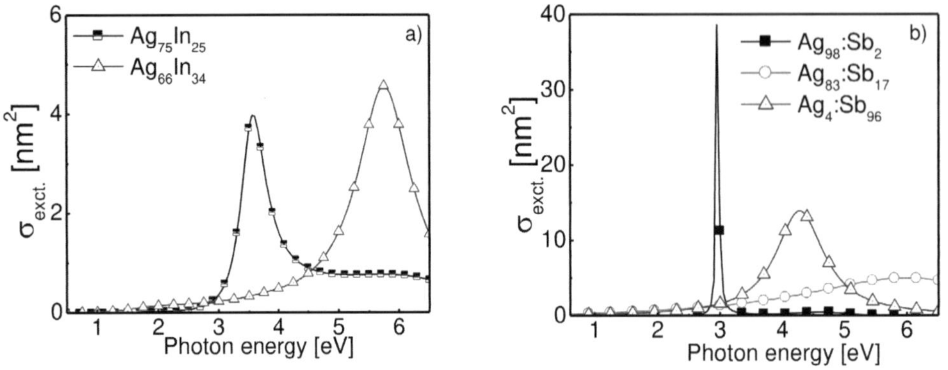

Figure 6. Calculated cross-section of excitation, σ, of spherical nanoparticles with diameter $a = 20$ nm from the Ag-In (a) and the Ag-Sb (b) systems.

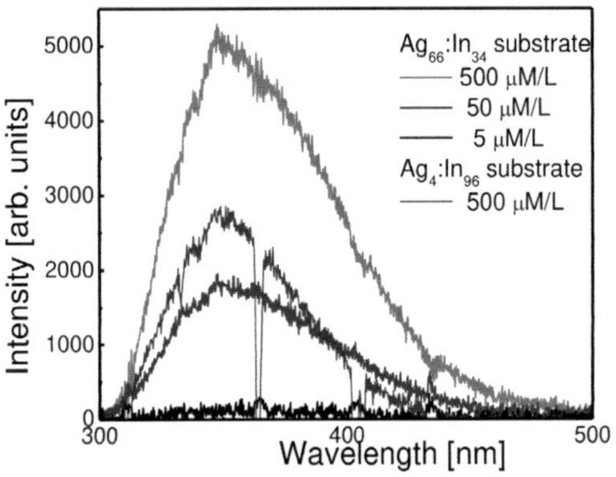

Figure 7. Emission fluorescent spectra of tryptophan excited by 254-nm irradiation on a glass substrate with a pre-deposited Ag_{66}:In_{34} and Ag_{4}:In_{96} layers.

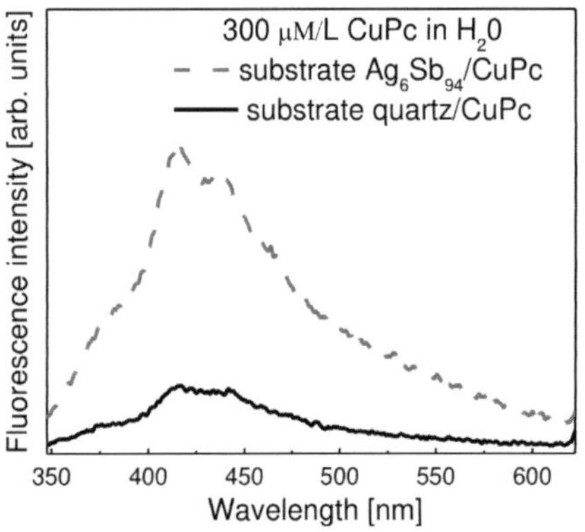

Figure 8. Fluorescence spectra of CuPc on a pure quartz substrate and on a pre-deposited Ag_{6}:Sb_{94} layer.

The emission spectra of tryptophan in aqueous solutions with concentrations of 500 μM/L, 50 μM/L and 5 μM/L are shown in figure 7. Fluorescent bands with a maximum at ~ 350 nm are seen when excited by light with a wavelength of 254 nm (4.88 eV). A significant increase for higher tryptophane concentrations is registered. For comparison, no fluorescent signal is observed even for 500 μM/L tryptophan measured on a clean glass substrate. In the case of thin Ag_{4}:In_{96} coating, whose complex permittivity suggested a lower efficiency for excitation of LSPR, an amplification of the signal was registered only for the highest tryptophan concentration, but to a much lower extent than for the Ag_{66}:In_{34} layer. The latter is a good proof that the amplification is caused mainly by the LSPR present in the films.

Cu (II)-phthalocyanine (CuPc) dye was used to analyze the LSPR efficiency of the Ag_{6}:Sb_{94} layer with a thickness of 10 nm (figure 8). The substance is selected because it emits strong fluorescent bands with a maximum of ~ 430 nm when excited by light with a wavelength in the range of 320 – 330 nm. For this purpose, 0.1 ml of a CuPc + water solution with a concentration of 300 μM/L was dropped on a clean quartz substrate, as well as on such with a pre-deposited bimetallic layer (both pre-protected by a 20-nm thick polymer layer deposited by spin-coating of 0.3-ml solution of PVP polymer in methanol) in order to provide the "zero line spectrum". Figure 8 shows the fluorescence spectra of CuPc obtained at the excitation wavelength of 320 nm. The cleavage of the peak that is observed is caused by the sensitivity of the compound to the solvent. The use of a bimetallic layer is the reason for an amplification of the fluorescent signal of CuPc at a wavelength of 430 nm (2.9 eV), which is in its range of frequency of LSPR. The amplification enhancing coefficient was $k = 4.17$ times.

4. Conclusions

In this work, we investigated the optical properties and the possibility for excitation of a localized surface plasmon resonance (LSPR) in Ag-In and Ag-Sb bimetallic films in terms of varying both the thickness and the composition. The analyses revealed that the layers of both chemical systems are characterized by a granular structure, with the size and the shape of the grains depending on the materials and chemical compounds present in the samples, as well as on the thickness of the films. The decrease of the thickness to ~ 10 nm causes the formation of an island structure, which influences the plasmonic activity of the thin films associated with the interaction of light with the transverse oscillations of electron waves on the surface of the better-pronounced metal nanoparticles. Using the data for the dielectric loss function $\mathrm{Im}\{-1/\varepsilon\}$ and the cross-section of excitation, $\sigma_{exct.}$, derived from the calculated complex permittivity spectra, it was determined that optimal for plasmonic applications in the UV spectral region (3 – 6 eV) are the thinner films with compositions near the Ag-rich compounds in the Ag-In system and the Sb-rich sample from the Ag-Sb layers investigated. To prove the quality of the excited LSPR in these compositions of interest and to evaluate the possibility for their application in surface-enhanced fluorometry, they were used as amplifying substrates for the detection of Cu (II)-phthalocyanine (CuPc) in the case of Ag-Sb and tryptophan for the Ag-In system. The Ag-In sample exhibited a significant amplification of the tryptophan band at ~ 350 nm, which was fully hidden when analyzed on a clean quartz substrate; the Ag_6Sb_{94} substrate increased the signal of the CuPc dye with an amplification enhancing coefficient of 4.17.

Acknowledgements

This work is financially supported by contract KP-06-N38/8 - 05.12.2019 with the Bulgarian National Science Fund. Research equipment of the INFRAMAT Distributed Research Infrastructure, part of Bulgarian National Roadmap for Research Infrastructures supported by the Bulgarian Ministry of Education and Science was used.

References

[1] Cortie M B, Arnold M D and Keast V J 2020 *Adv. Mater.* **32** 1904532
[2] Bauch M, Toma K, Toma M, Zhang Q and Dostalek J 2014 *Plasmonics* **9** 781
[3] Petryayeva E and Krull U J 2011 *Anal. Chim. Acta* **706** 8
[4] Shahbazyan T V and Stockman M I 2013 *Plasmonics: theory and applications* vol 15 (Dordrecht: Springer Netherlands)
[5] Liz-Marzán L M 2006 *Langmuir* **22** 32
[6] Lu X, Rycenga M, Skrabalak S E, Wiley B and Xia Y 2009 *Annu. Rev. Phys. Chem.* **60** 167
[7] McMahon J M, Schatz G C and Gray S K 2013 *Phys. Chem. Chem. Phys.* **15** 5415
[8] Bezerra A G, Machado T N, Woiski T D, Turchetti D A, Lenz J A, Akcelrud L and Schreiner W H 2018 *J. Nanoparticle Res.* **20** 1
[9] De Silva K S B, Keast V J, Gentle A and Cortie M B 2017 *Nanotechnology* **28** 095202
[10] Kim K J, Chen L-Y and Lynch D W 1988 *Phys. Rev.* B **38** 13107
[11] Magruder III R H, Anderson T S, Zuhr R A and Thomas D K 2006 *Nucl. Instrum. Methods Phys. Res.* B **108** 305–312
[12] Chen Y, Wang M, Zheng K, Ren Y, Xu H, Yu Z, Zhou F, Liu C, Qu J and Song J 2021 *Adv. Mater.* **33** 2100039
[13] Okamoto H 2007 *J. Phase Equilibria Diffus.* **28** 403
[14] Todorov R, Hristova-Vasileva T, Atanasova A, Katrova V, Strijkova V, Milushev G and Milanov E 2021 *J. Phys.: Conf. Ser.* **1762** 012023
[15] Novotny L and Hecht B 2012 *Principles of Nano-Optics* (Cambridge University Press) p 369

Effect of laser annealing on the properties of Ag/ZnO nanostructures

M E Koleva[1,2,4], A O Dikovska[1], N N Nedyalkov[1] and D Karashanova[3]

[1]Institute of Electronics, Bulgarian Academy of Sciences, 72 Tsarigradsko Chaussee, Sofia 1784, Bulgaria
[2]Neofit Rilski South-Western University, 66 Ivan Mihailov Str., Blagoevgrad, 2700, Bulgaria
[3]Acad. J. Malinowski Institute of Optical Materials and Technologies, Bulgarian Academy of Sciences, Acad. G. Bonchev Str., bl. 109, Sofia, 1113, Bulgaria

E-mail: mihaela_ek@yahoo.com

Abstract. Ag/ZnO nanostructures were synthesized using pulsed laser deposition (PLD) in open air (at atmospheric pressure). The deposition was carried out by a Nd:YAG laser, with the Ag layer composed of nanoparticles grown on SiO_2 (001) substrates by the laser wavelength of 355 nm; this layer was covered by ZnO deposited by using the laser wavelength of 1064 nm. The samples were laser annealed, which led to a modification of the nanoparticles. The nanocomposites produced were characterized by TEM, UV–Vis and PL spectroscopy. The annealing procedures influence the optical properties of the Ag/ZnO nanocomposites. The laser annealing under certain parameters changes the emission behavior of the Ag/ZnO nanocomposite heterostructures. By employing localized surface plasmon resonance (LSPR), the near band edge (NBE) emission intensity of the ZnO films composed of nanoparticles was varied. An enhancement of the UV emission located at about 383 nm, commonly attributed to an excitonic transition, was registered after laser annealing with one pulse at the wavelength of 355 nm. The results obtained demonstrate that the annealing of Ag/ZnO composite nanostructures plays a key role in tuning the PL performance of a semiconductor material where an LSPR occurs.

1. Introduction

Recently, a particular attention has been focused on composite nanostructures because of their unique properties not observed the separate materials. The optical properties of nanomaterials, such as transmission and light emission, are dynamic and may differ significantly from the properties exhibited by the same bulk materials. New concepts have been developed lately on designing materials with interesting properties [1-3]. Nanocomposite materials based on metal nanoparticles (NPs) in/on oxide matrices have attracted increasing attention thanks to their intriguing chemical and physical properties that can be tailored as a function of the NPs size, shape and mutual interactions [4-7]. The possibility to control their properties paves the way to a broad spectrum of applications, ranging from heterogeneous catalysis, to gas sensing, to non-linear optics.

Enclosing metal NPs in a semiconductor matrix is a method suitable for reducing the aggregation and deterioration of metal NPs [8, 9]. According to previous research, the synthesis of heterostructures

[4] To whom any correspondence should be addressed.

Content from this work may be used under the terms of the Creative Commons Attribution 3.0 licence. Any further distribution of this work must maintain attribution to the author(s) and the title of the work, journal citation and DOI.
Published under licence by IOP Publishing Ltd

of Ag nanocrystals and a semiconductor can often facilitate the charge separation upon generation of electron/hole pairs through photon excitation in the semiconductor, because the Fermi levels of noble metals are typically lower than the conduction band edges of most semiconductor material [10, 11]. The Ag nanocrystals can broaden and enhance the light absorption of semiconductors, because they have extremely large absorption/scattering cross-sections and can strongly focus light close to their surface [12, 13]. Since the collective oscillations of free electrons around metal nanoparticles surfaces attracted much attention, localized surface plasmons (LSPs) have been applied to enhancing the light emission in materials due to selective photon absorptions and enhanced local electromagnetic field around the metal NPs [14, 15]. Many efforts have been directed to applying LSP-enhanced emissions in ultraviolet (UV) optoelectronic devices, such as light emitting diodes [16] and photodetectors [17, 18].

This study is focused on fabricating Ag/ZnO nanostructures by laser deposition in open air and investigating the effect of a laser annealing procedure on the morphology, structure and optical properties of composite nanostructures.

2. Experimental

Ag/ZnO nanostructures was grown by pulsed laser deposition (PLD) in open air by a Nd:YAG laser. The silver layer was deposited on SiO_2 (001) substrates using the wavelength of 355 nm and 1800 pulses. The laser fluence during the growth process was fixed at 3 J/cm^2. By using the same deposition equipment, the PLD grown Ag films were covered by a ZnO layer at atmospheric pressure using the wavelength of 1064 nm of the same laser at a laser fluence of 4.5 J/cm^2. The number of pulses was $N_p = 3000$. All depositions were carried out at room temperature. The samples were laser annealed in the air by the Nd:YAG laser by one pulse ($N_p = 1$) at $\lambda = 355$ nm and a fluence of 0.7 J/cm^2.

The UV–Vis transmission spectra were recorded in the spectral range of 220 – 600 nm by an HR 4000 UV–Vis spectrometer (Ocean Optics).

The morphology, structure and phase composition of the fabricated nanostructures were studied by high-resolution transmission electron microscopy (HRTEM JEOL JEM 2100 operated at 200 kV) and selected area electron diffraction (SAED). The samples were prepared for TEM observation by transferring a small quantity of the material on a Cu-grid preliminary coated with a thin carbon sublayer.

In order to study the influence of the silver nanoparticles on the ZnO photoluminescence before and after the laser annealing, the PL spectra were taken using a Fluorolog spectrofluorometer (HORIBA Jobin Yvon) under excitation at 320 nm.

3. Results

Figure 1 shows transmission electron micrographs and selected area electron diffraction (SAED) patterns of the synthesized Ag/ZnO nanocomposites. The formation is clearly seen of small individual particles with a size distribution in the range of 10 – 30 nm before annealing and 10 – 38 nm after annealing. They predominate over the existing large particles of about 120 nm in size. The size distribution of the small nanoparticles was determined; histograms with the particles diameter are shown in the insets. The mean diameter of 20 nm slightly increases to 24 nm after the annealing. The SAED patterns of the as-deposited samples demonstrate the presence of cubic ZnO and cubic Ag. The HRTEM images confirms the cubic Ag. Cubic ZnO and cubic Ag are also observed in the SAED pattern of the annealed samples. The HRTEM image of the annealed samples shows an interplanar spacing of 2.14 Å, which probably indicates intermixing and is associated with an alloy of hexagonal $AgZn_3$, where $a = 2.8227$, $c = 4.4274$, PDF25-1325 (JCPDS 1998).

According to our previous research [19, 20], laser ablation in open air results in nanoparticle and/or nanoparticle aggregates formation in the plasma plume realized via a fast condensation of the ablated material. The laser-matter interaction leads to material ejection (clusters, atoms, ions, electrons) from the target surface and formation of a plasma plume. The ablated species are subjected to a high environmental pressure (atmospheric pressure) close to the target, which additionally makes the plasma

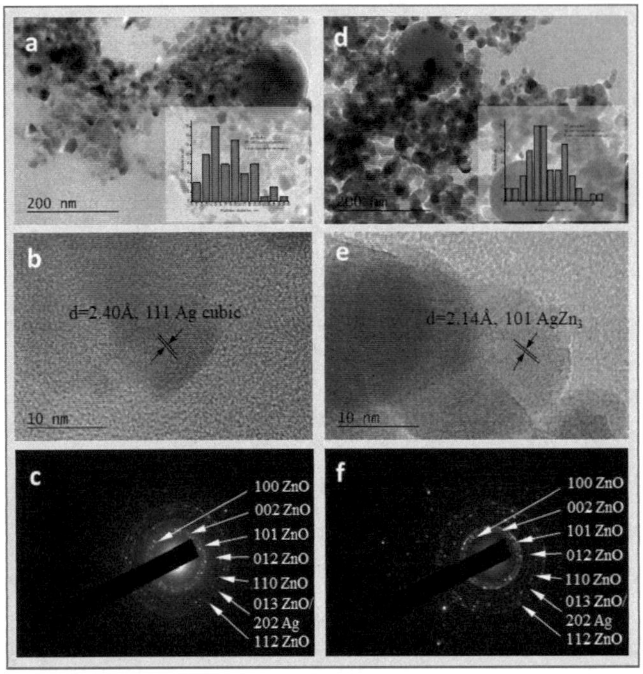

Figure 1. TEM, HRTEM images and SAED patterns of Ag/ZnO nanostructures before (a, b, c) and after laser annealing (d, e, f). The insets show particles size distribution histograms of small nanoparticles.

plume strongly confined. The ablated species have a high density in the plume, which increases the number of inter-particle collisions and the probability for their agglomeration. Thus, the "building blocks" of the samples produced by laser ablation in open air are nanoparticles and/or nanoparticle aggregates, which is confirmed by the microstructure of the samples as well as by our previous research [21]. These nanoparticles and/or nanoaggregates are deposited on a substrate positioned close to the target, since the efficiency of the PLD process in open air is too low. Because the nanoparticles accumulation takes place at room temperature and the arriving particles energy on the substrate is not sufficient, they are not able to occupy an energy favorable position, i.e., they cannot be ordered on the substrate. Therefore, they are "sticking up" incidentally to the already deposited nanoparticles and thereby form a complex 3D structure with a high porosity.

The use of different targets for ablation at atmospheric pressure leads to the creation of uniform structures without changing the particles size. Varying the experimental conditions used for ablation, such as energy density and wavelength, affects the efficiency of the ablation process, but does not lead to a significant change in the structure and morphology. In order to study the possibilities of changing the size of the created nanoparticles, we introduced additional laser annealing of composites already formed in air. In general, the physical properties of metal oxide-noble metal composites created by ablation in air at atmospheric pressure are mainly determined by their morphology and the amount of accumulated material (thickness). When the thickness of the 3D nanostructures is greater (tens of μm), the light is significantly absorbed and the specific plasmon properties of noble metals are not manifested.

The laser annealing of a structure obtained by ablation in air at atmospheric pressure leads to a change in morphology, i. e., the 3D structure is transformed into a 2D structure. The size, shape and composition of the particles formed after annealing are determined by the type of metal oxide, the quantitative ratio of metal oxide-noble metal, as well as the experimental parameters used for annealing (energy per pulse, number of pulses and thickness of the nanoparticles structures to be annealed). The annealing of metal oxide composites, like ZnO, with embedded Ag nanoparticles leads to a shift of the plasmon resonance to the longer wavelengths due to a change in the morphology of the samples [8].

During the laser annealing process, it is possible to reach heating and cooling rates exceeding 10^{10} C/s. In [22], we have discussed the evaluation of the heating and cooling rates. The pulsed laser irradiation affects the surface areas of the processed material by means of very rapid thermal cycles under conditions generally far from thermodynamic equilibrium [23]. As a result, complex mechanisms can be activated as, e.g., local diffusion of species, surface melting, ultra-rapid crystallization, etc. The laser annealing of the films leads to fragmentation and formation of a porous microstructure. The irradiation of these structures by nanosecond laser pulses produces melting and reshaping of the composite Ag/ZnO layer [24].

The UV-Vis spectra of the nanostructures produced are shown in figure 2. A strong resonance absorption of the sample with a deposited Ag layer only cannot be clearly observed (figure 2 a). The addition of ZnO to Ag changes the optical properties (figure 2 b). ZnO exhibits an UV absorption edge at 370 nm, which is a characteristic feature of the ZnO spectrum [25]. An absorption band at 360 nm is observed in the Ag/ZnO spectra only after laser annealing; it is attributed to the plasmonic resonance of silver in Ag/ZnO. However, the Ag/ZnO nanostructure absorption is not a simple superposition of the absorption of the individual constituents; rather, the resonance absorption of the Ag/ZnO nanocomposite is due to a strong electronic coupling between the AgNPs and ZnO matrix [26].

Figure 3 presents the photoluminescence spectra of Ag/ZnO nanostructures before and after laser annealing with one pulse at 355 nm at a laser fluence of 0.7 J/cm^2. Two ZnO peaks are detected, namely, a sharp and narrow peak located at 383 nm and a low and broad peak at approximately 500 nm. It is thus found that the surface plasmon resonance in these composites affects the position of the Vis PL peak and has a strong effect on the intensity of the UV PL peak. The sharp UV emission at 383 nm is attributed to the free excitonic emission near the band edge [27, 28]. The broad emission off 450 – 570 nm in the visible wavelength range is considered as being a deep-level emission caused by oxygen vacancies or Zn vacancy defects [29]. Due to the annealing influence, the peak intensity of the Ag-ZnO heterostructure exhibits an obvious enhancement in the ultraviolet range and a red shift in the visible light range. The intensity of the UV PL emission after annealing (figure 3 b) is significantly stronger compared to that of the as deposited Ag/ZnO nanostructure (figure 3 a).

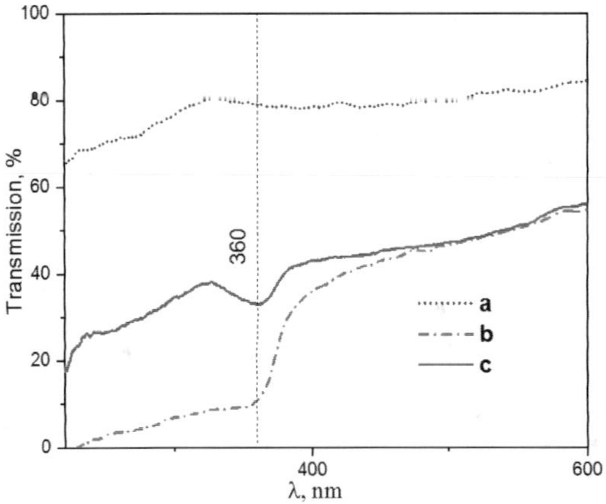

Figure 2. Transmission spectrum of laser grown nanostructures of: a – Ag, b – Ag/ZnO, c – laser annealed Ag/ZnO at 1 pulse.

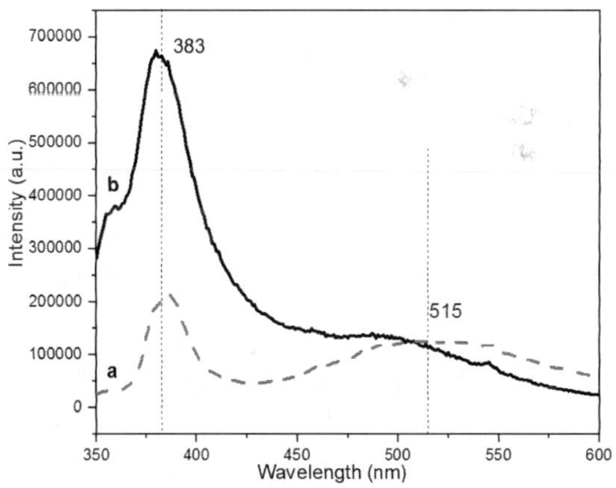

Figure 3. PL spectrum of Ag/ZnO nanostructures a – as grown; b – laser annealed.

4. Conclusions

Ag/ZnO nanocomposites laser-grown in atmospheric pressure by sequential deposition of Ag and ZnO layers were modified by post-deposition laser annealing. The laser processing with one pulse at 355 nm and a fluence of 0.7 J/cm^2 leads to a significant modification of the optical properties of the nanostructures produced. A higher crystallinity is observed after the annealing. The laser modification causes the appearance of a resonance absorption band at about 360 nm, which contributes to an enhancement of the photoluminescence band-edge UV emission at 383 nm. Thus, the control of functional performances of the produced nanocomposites relies on tailoring the material properties by design through a suitable choice of the synthesis and processing routes.

Acknowledgments
This work is financially supported by the Bulgarian National Science Fund under project KP-06-N37/20. Research equipment was used of the INFRAMAT Distributed Research Infrastructure, part of the Bulgarian National Roadmap for Research Infrastructures supported by the Bulgarian Ministry of Education and Science.

References

[1] Drexler K E 1992 *Nanosystems: Molecular Machinery, Manufacturing, and Computation* (New York: J. Wiley & Sons)
[2] Hirasawa M, Shirakawa H, Hamamura H, Egashira Y and Komiyama H 1997 *J. Appl. Phys.* **82** 1404
[3] Schmid G 1998 *J. Chem. Soc., Dalton Trans.* **7** 1077
[4] Wenzel T, Bosbach J, Stietz F and Trager F 1999 *Surf. Sci.* **432** 257
[5] Charton C and Fahland M 2003 *Surf. Coat. Technol.* **174–175** 181
[6] Cai W, Hofmeister H, Rainer T and Chen W 2001 *J. Nanopart. Res.* **3** 443
[7] Jin R, Cao Y-W, Mirkin C A, Kelly K L, Schatz G C and Zheng J G 2001 *Science* **294** 1901
[8] Koleva M E, Nedyalkov N N, Fukata N, Jevasuwan W, Amoruso S, Koutzarova T, Avdeev G V, Georgieva B and Karashanova D 2017 *Superlattices and Microstructures* **109** 886–896
[9] Koleva M E *et al.* 2020 *Applied Surface Science* **508** 145227
[10] Chen Y Z, Zeng D Q, Zhang K, Lu A, Wang L S and Peng D-L 2014 *Nanoscale* **6** 874–881
[11] Dutta S K, Mehetor S K and Pradhan N 2015 *J. Phys. Chem.Lett.* **6** 936–944
[12] Jiang R B, Li B X, Fang C H and Wang J F 2014 *Adv. Mater.* **26** 5274–5309
[13] Sudhagar P *et al.* 2014 *Phys. Chem. Chem. Phys.* **16** 17748–17755
[14] Barnes W L, Dereux A and Ebbesen T W 2003 *Nature* **424** 824–830
[15] Koleva M E, Dikovska A O, Nedyalkov N N, Atanasov P A and Bliznakova I A 2012 *Applied Surface Science* **258** (23) 9181–9185
[16] Lupan O, Pauporte T and Viana B 2010 *Adv Mater* **22** 3298–3302
[17] Yang Q, Wang W, Xu S and Wang Z L 2011 *Nano Lett* **11** 4012–4017
[18] Liu Y, Zhang X, Su J, Li H, Zhang Q and Gao Y 2014 *Opt Express* **22** 30148–30155
[19] Nikov R G, Dikovska A O, Nedyalkov N N, Atanasov P A, Atanasova G, Hirsch D and Rauschenbach B 2017 *Appl. Phys. A: Mater. Sci. & Proc.* **123** 657
[20] Nikov Ru G, Dikovska A Og, Nedyalkov N N, Avdeev G V and Atanasov P A 2017 *Beilstein J. Nanotechnol.* **8** 2438–2445
[21] Atanasova G, Dikovska A Og, Dilova T, Georgieva B, Avdeev G V, Stefanov P and Nedyalkov N N 2019 *Appl. Surf. Sci.* **470** 861–869
[22] Koleva M E, Nedyalkov N N, Karashanova D, Atanasova G B and Stepanov A L 2019 *Appl. Surf. Sci.* **475** 974–981
[23] Wood R F and Giles G E 1981 *Phys. Rev.* B **23** 2923
[24] Koleva M E, Nedyalkov N N, Atanasov P A, Gerlach J W, Hirsch D, Prager A, Rauschenbach B, Fukata N and Jevasuwan W 2016 *Journal of Alloys and Compounds* **665** 282–287
[25] Biswas P, Cho S R, Kim J W, Baek S D and Myoung J M 2017 *Nanotechnology* **28** 225502
[26] Xie W, Li Y, Sun W, Huang J, Xie H and Zhao X 2010 X *Journal of Photochemistry and Photobiology A: Chemistry* **216** 149–155
[27] Guidelli E, Baffa O and Clarke D 2015 *Sci Rep* **5** 14004
[28] Satter S S, Hoque M, Rahman M M, Yousuf M, Mollah A and Susan M 2014 *RSC Adv.* **4** 20612
[29] Lu W W, Gao S Y, and Wang J J 2008 *Journal of Physics and Chemistry* C **112** 16792–16800

Structure and phase composition study of thin TiO_2:C films deposited by r.f. magnetron sputtering

T Milenov[1,8], D Karaivanova[1], O Angelov[2], P Terziyska[3], G Avdeev[4], D Karashanova[5], B Georgieva[5], K Genkov[6], D Dimov[1], K Ivanov[1,7], S Kolev[1] and E Valcheva[6]

[1]Acad. E. Djakov Institute of Electronics, Bulgarian Academy of Sciences, 72 Tsarigradsko Chaussee, 1784 Sofia, Bulgaria
[2]Central Laboratory of Solar Energy and New Sources of Energy, Bulgarian Academy of Sciences, 72 Tsarigradsko Chaussee, 1784 Sofia, Bulgaria
[3]G. Nadjakov Institute of Solid State Physics, Bulgarian Academy of Sciences, 72 Tsarigradsko Chaussee, 1784 Sofia, Bulgaria
[4]R. Kaishev Institute of Physical Chemistry, Bulgarian Academy of Sciences, Acad. G. Bonchev Str., bl. 11, 1113, Sofia, Bulgaria
[5]Acad. J. Malinowski Institute of Optical Materials and Technologies, Bulgarian Academy of Sciences, Acad. G. Bonchev Str., bl. 109, 1113, Sofia, Bulgaria
[6]Faculty of Physics, St. Kl. Ohridski University of Sofia, 5 James Bourchier Blvd., 1164 Sofia, Bulgaria
[7]Institute of Biophysics and Biomedical Engineering, Bulgarian Academy of Sciences, Acad. G. Bonchev Str., bl. 105, 1113 Sofia, Bulgaria

E-mail: tmilenov@ie.bas.bg

Abstract. Thin films of TiO_2 doped with carbon were deposited on $15{\times}25{\times}1\ mm^3$ glass substrates by r.f. magnetron co-sputtering of TiO_2 target and carbon plates on their erosion zone in Ar + air atmosphere (0.5 Pa and 0.6 Pa, respectively). Two different process parameters were varied in different experiments: the total area of the carbon plates (84 mm^2, 560 mm^2, 830 mm^2 or 1480 mm^2) and the radial distance from the center of the circle of the erosion zone with maximum rate of sputtering (2 cm or 4 cm). The as-deposited and annealed (air, 400 °C, 1 h) thin films with thicknesses of 40 – 60 nm were studied by ellipsometry, grazing incidence X-ray diffractometry (GIXRD), Raman spectroscopy, transmission (TEM) and scanning electron microscopies (SEM). The ellipsometric studies revealed the band gap, as well as the tail states and optical constants of the thin films obtained. The GIXRD patterns revealed a mix of amorphous and nanocrystalline anatase and rutile TiO_2 phases in all thin films. The Raman study confirmed this conclusion, but the TEM study showed the presence of some nanocrystalline $C_{20}O_{14}Ti_3$, together with the other TiO_2 phases – brookite and nonstoichiometric TiOx phases (anatase and rutile). The effects of annealing at 400 °C in air for 60 min are also discussed.

[8] To whom any correspondence should be addressed.

Content from this work may be used under the terms of the Creative Commons Attribution 3.0 licence. Any further distribution of this work must maintain attribution to the author(s) and the title of the work, journal citation and DOI.
Published under licence by IOP Publishing Ltd

1. Introduction

Titanium dioxide (TiO_2) is a highly stable and non-toxic compound which is often used in the industry as a pigment, a photocatalyst, as well as thin-film coating in electronics – see for example [1]. The TiO_2 powders and thin films synthesized in various ways exhibit a complex phase composition based on three main titania phases: anatase [2] and rutile [3] with tetragonal symmetry, and brookite [4] with orthorhombic symmetry (taken from Crystallography Open Database [5]), and are built of more or less distorted TiO_6 octahedra centered by Ti^{4+} ions and whose vertices are occupied by six O^{2-} ions. The optical bandgap of these phases in different materials varies between 3.02 eV and 3.6 eV depending on the phase composition and origin of the band transition (direct or indirect). Obviously, the practical application of TiO_2 in different photo-processes requires a slight shift of the absorption edge to the visible region. In the last decade, this has usually been done by doping with carbon due to various reasons [6, 7]. Thin films of TiO_2 are deposited very often by magnetron sputtering (radiofrequency (rf) or direct current (dc)) [7] due to the high reproducibility of the deposited films quality. Two main ways for introducing the carbon dopant are used: from a gas mixture of CO (or CO_2) combined with Ar in different ratios [8, 9], which results in low levels of C doping (0.1 – 1 at % carbon); or by co-sputtering of a target of TiO_2 and carbon plates placed on the target [10], as it is expected that the use of the latter path will result in a larger doping values of carbon, i.e. shifting of the absorption edge to the visible region. The research presented continues and expands our previous research on co-sputtering of TiO_2 and C [10]. It is focused on studying the influence of the carbon plate area and of its position with reference to the center of the TiO_2 target erosion zone. We aimed at synthesizing heavily C-doped TiO_2 films (in the range of up to 3 at % carbon) in order to identify and study the distribution of the carbon phases in a nano-crystalline oxide medium. Some initial results on the influence of low-temperature (400 °C) annealing in air on the structure and electro-physical properties of the deposited films, and on the carbon phases separation are also presented.

2. Experimental

Thin films of TiO_2:C were deposited on heat-resistant glass substrates (15×25×1 mm^3) by r.f. magnetron co-sputtering of a TiO_2 target and carbon plates with different areas (S_G): 84 mm^2, 830 mm^2 or 1480 mm^2. The radial distance from the erosion zone center (the zone with a maximum rate of sputtering) R_D was also varied in different experiments, namely, 2 cm or 4 cm. A gas atmosphere of Ar (0.5 Pa) + air (0.6 Pa) was used. All other parameters of the r.f. sputtering process are described in more detail in [10]. The TiO_2:C films thus obtained were characterized by grazing incidence X-ray diffractometry using a PANalytical Empyrean apparatus in (ω-2θ) scans. The surface morphology of the specimens was examined by scanning electron microscopy (SEM) in both secondary electron and backscattered electron imaging (SEI and BEI, respectively) on a LYRA TESCAN microscope, while the chemical composition was determined by energy-dispersive X-ray analysis carried out on a Brucker Quantax 200 Spectrometer unit. The microstructure and phase composition of the specimens were studied by transmission electron microscopy (TEM) by an HR STEM JEOL JEM 2100 microscope. The Raman spectra were measured by a HORIBA Jobin Yvon Labram HR 800 Visible micro-Raman spectrometer with a He-Ne (λ = 633 nm) laser. The optical constants n and k (refraction and extinction coefficients), as well as the films' thickness, were determined by ellipsometry measurements using a Woollam M2000D rotating compensator spectroscopic ellipsometer with a wavelength range from 193 nm to 1000 nm in a reflection mode. The data acquisition and analysis software used was CompleteEASE 5.10 J.A.Woollam Co., Inc. [11]. The spectroscopic ellipsometry data on the angles Ψ and Δ were taken at room temperature at angles of incidence of 50°, 60°and 70°. The spectral dependence of the absorption coefficient (α), the energy of interband transition(s) (E_g) (indirect, where $N = 2$ (eq.2) or direct, where $N = 0.5$ (eq.2)), as well as the excitation energy of the defect Urbach-tail states (E_u) can be calculated from these datasets by using the relations (see [12, 13, 14]):

$$\alpha = 4\pi k/\lambda \tag{1}$$

$$\alpha E = A(E-E_g)^N \tag{2}$$

$$\alpha(E)=\alpha_0 exp((E-E_g)/E_u) \tag{3}$$

3. Results and discussion

The different samples were denoted as follows: I, II, III, IV, V, VI and VII and IA, IIA, IIIA, IVA, VA, VIA and VIIA for as deposited and annealed TiO_2:C films, respectively. The film's thickness was determined by measurements using a TALYSTEP profilometer, by measurements from the cross-section SEI and BEI images, as well as by ellipsometric measurements – see the inset in figure 1 a). The results are very similar and the deviations do not exceed ± 5 nm in the different samples. The results of the different experiments (including R_D and S_G) are summarized in table 1. The films obtained are smooth and no differences in the contrast in the BSE images are observed – figures 1 a) and b). The composition of the TiO_2:C films cannot be determined precisely by EDX as their thickness is too small, but the carbon content in the most heavily doped specimens (from experiment IV and VII) does not exceed 3 at %, as obtained by the EDX measurements.

The GIXRD patterns obtained by all experiments are presented in figure 2: the most prominent reflections of anatase planes ($d^A_{(101)}$) appear at about 25.15° – 25.30° [2], while those of the rutile phase ($d^R_{(110)}$), at about 27.2°– 27.6° [3]. It should be noted that reflections from the (210) planes of the brookite phase [4] can be also observed in the patterns taken from I and IA specimens. The most intensive XRD reflection of brookite ($d^B_{(021)}$) should appear at about 25.2°, i.e., it should be overlapped with those of the anatase phase – see figure 2.

Figure 1. a) BSE image taken from specimen II. **b)** BSE image taken from specimen IIA. Inset: SEI cross-section image from sample I.

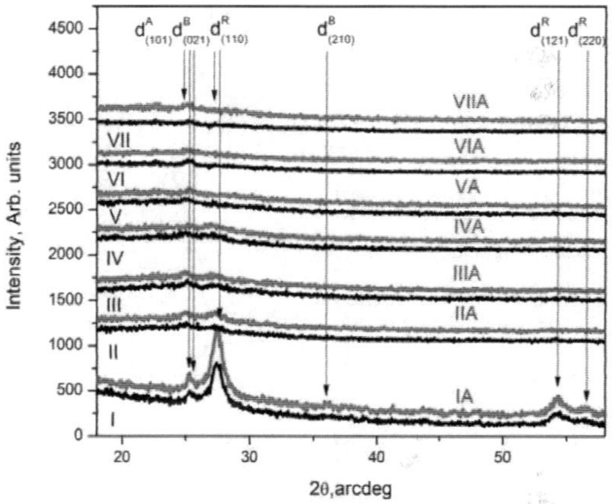

Figure 2. GIXRD patterns taken from all experiments: I, IA, II, IIA, III, IIIA, IV, IVA, V, VA, VI, VIA, VII and VIIA in (ω-2θ) scans, ω = 0.5°.

The phase composition and the size of crystallites for specimens I, IA and IIA could be calculated by the Debye-Scherrer equation, as the crystal size was high enough in these films only; the results are shown in table 2.

The annealing slightly increases the relative intensity of the (101) peaks of anatase phase in the IA and IIA samples, and additionally increases the mean crystallite size in the IA and IIA experiments (table 2). The slight downshift observed of the 2θ values for the interplanar distance in both the anatase and rutile planes point to the existence of tensile stress in the films after annealing.

The spectral dependence of the basic optical parameters (n and k) of I, IA, II and IIA films determined as indicated in the Experimental section are plotted in figure 3. The values found for the energy of the interband transition(s) (E_g) and the excitation energy of the defect-driven Urbach-tail states are summarized in table 1. Indirect interband transitions with energies in the range 3.18 – 3.55 eV are revealed in all thin films (table 1); the examples for samples I and IA, II and IIA and for sample IV (the results for IV, IVA, VII and VIIA practically coincided) are presented in figure 4 a). The energy of the defect-driven Urbach-tail states (E_U) varies in the different specimens between 1.55 eV and 3.37 eV (table 1); the examples for samples I, IA, II, IIA, III and IIIA are given in figure 4 b).

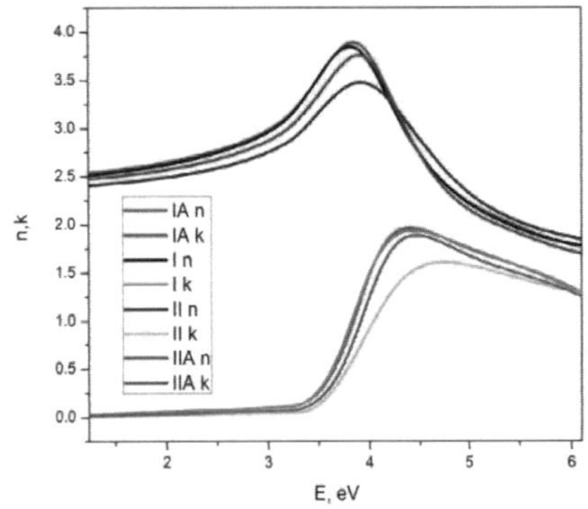

Figure 3. Spectral dependence of n and k measured in samples I, IA, II and IIA.

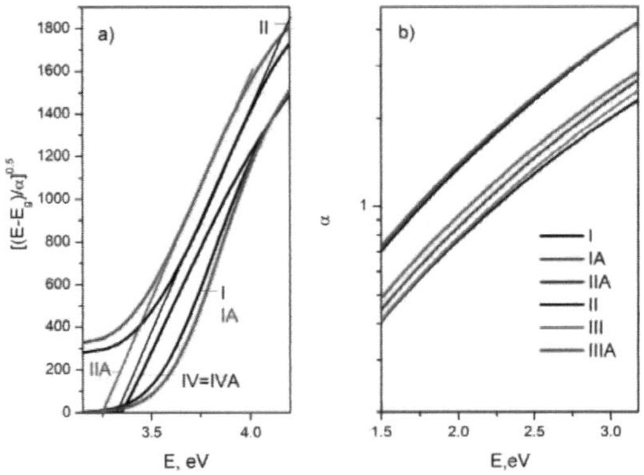

Figure 4. Calculated energy of: **a)** indirect interband transitions (equation (2)) for samples I, IA, II, IIA and IV; **b)** Urbach-tail states (equation (3)) for samples I, IA, II, IIA, III and III A.

Table 1. Summarized data for the experimental details, the thickness and for the calculated interband transitions and defect-driven Urbach-tail states (E_g and E_U, respectively).

Experiment	R_D, cm	S_G, mm^2	Thickness, nm	E_g, eV	E_U, eV
I	-	-	120	3.55	1.76
IA	-	-	120	3.44	1.73
II	2	84	65	3.33	2.20
IIA	2	84	65	3.28	2.05
III	2	850	70	3.26	2.91
IIIA	2	850	70	3.25	1.59
IV	2	1480	55	3.37	3.37
IVA	2	1480	55	3.30	1.55
V	4	84	47	3.30	2.51
VA	4	84	47	3.30	1.66
VI	4	850	48	3.22	2.94
VIA	4	850	48	3.20	1.57
VII	4	1480	40	3.37	3.37
VIIA	4	1480	40	3.18	1.56

The TEM study of the samples from experiments I (figure 5 a), b) and c), IA (figure 5 d), e) and f), II (figure 6 a), b) and c) and IIA (figure 6 d, e and f) shows that the areas studied of the as-deposited films and of the annealed sample IA consist of anatase and brookite [2, 4] predominantly (according to the SAED patterns), but some nano-sized crystals (size of 5 – 10 nm) of the $C_{20}O_{14}Ti_3$ phase (see [15]) can be observed in the HRTEM images – figures 5 c), d) and 6 c). The complex $C_{20}O_{14}Ti_3$ phase can be observed more frequently in the annealed samples IIA – their patterns are clearly distinguished in the SAED patterns (see figure 6 e). Additionally, a rutile phase is also observed in sample IIA.

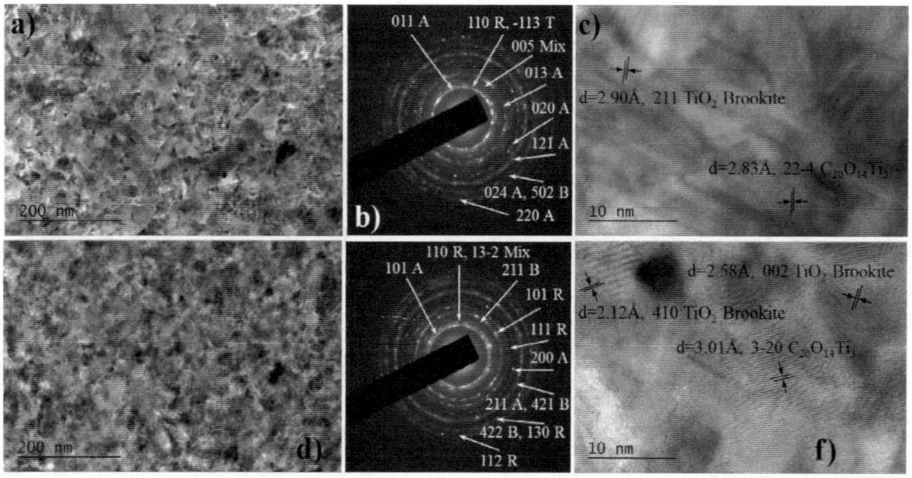

Figure 5. Low-magnification TEM image taken from sample I **a)** and the corresponding SAED image **b)**; HRTEM image from sample I **c)**; low-magnification TEM image taken from sample IA **d)** and the corresponding SAED image **e)**; HRTEM image from sample IA **f)**. The indices "A" denote anatase patterns, "R", rutile and "Mix", $C_{20}O_{14}Ti_3$ in the SAED patterns shown in panel **d)**.

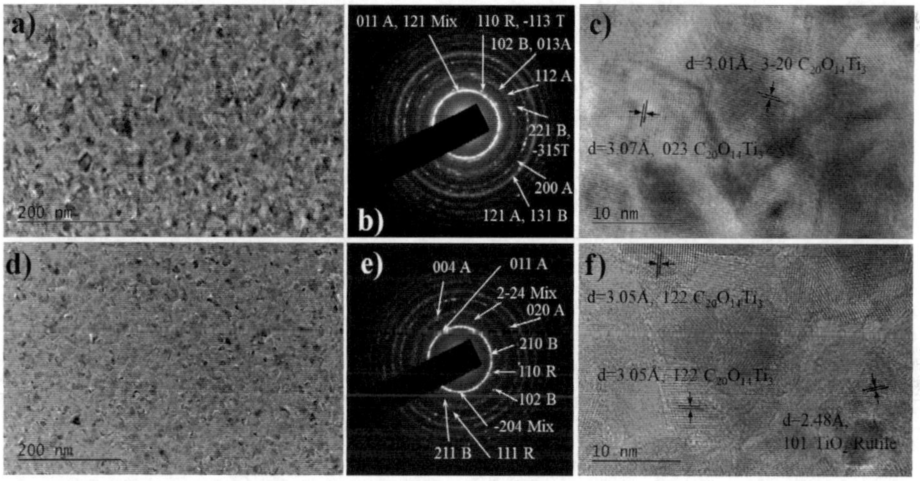

Figure 6. Low magnification TEM image taken from sample I **a)** and the corresponding SAED image **b)**; HRTEM image from sample II **c)**; low magnification TEM image taken from sample IIA **d)** and the corresponding SAED image **e)**; HRTEM image from sample IA **f)**. The indices "A" denote anatase patterns, "R", rutile and "Mix", $C_{20}O_{14}Ti_3$ in the SAED patterns shown in panels **b)** and **d)**.

Obtaining the Raman spectra of the titania phases (brookite, anatase and rutile) can facilitate the identification of their presence. For example, the most prominent peak of the brookite phase (A^B_{1g} mode of brookite phase at 153 cm^{-1}) is a very strong Raman feature; no other Raman feature of anatase and rutile phases appear at this frequency [16]. This mode (denoted as A^B_{1g} in figure 7) dominates the Raman spectra measured in all thin films of TiO_2:C (figure 7), which undoubtedly proves the presence of brookite. Few weak features, namely those denoted as E^R_g and A^R_{1g}, as well as their overtones, can also be distinguished in figure 7. It should be emphasized that the mode denoted as A^B_{1g} is slightly shifted to the lower frequencies by 3 – 8 cm^{-1} after annealing, which points to a probable transformation of the brookite to anatase/rutile phases, as the E^A_g mode of anatase and the B^R_{1g} mode of rutile appear at 143 $cm^{-1}$and 144 cm^{-1}, respectively [16].

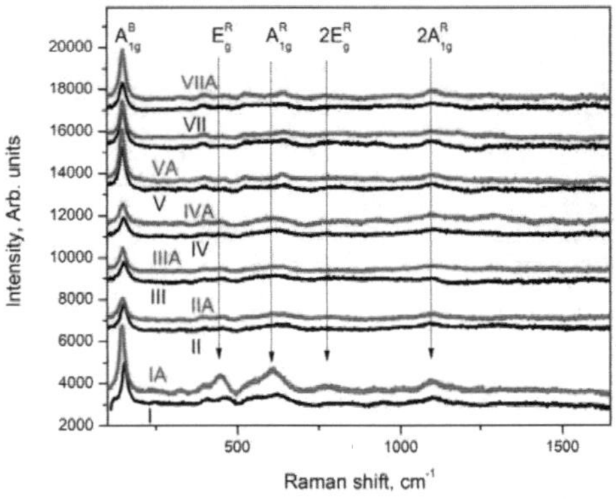

Figure 7. Raman spectra of samples from all TiO2:C films.

Table 2. Summarized results for the phase composition and the mean crystal size determined by GIXRD measurements for samples I, IA and IIA.

	Phase composition, mol %		Mean crystal size, nm	
	Anatase (A)	Rutile (R)	A	R
I	24.3	75.7	12	7
IA	8.4	91.6	20.7 (101)	8.2
IIA	81.7	18.3	20.7 (101)	8.2

Bearing in mind the above observations, it appears that the as-deposited thin films contain predominantly anatase, rutile and brookite phases of TiO_2. Some small quantities of the complex $C_{20}O_{14}Ti_3$ phase are observed as nano-crystals in practically all thin films deposited; therefore, it can be assumed that it is synthesized from the carbon oxide/s from the gas atmosphere (as already noted above, the deposition processes were carried out in an atmosphere consisting of air (0.6 Pa) and Ar (0.6 Pa) and the sputtered TiO_2 ensembles. It also appears that the formation of this complex Ti-C-O compound is enhanced by the thermal annealing at 400 °C, as it is observed even in the SAED pattern of the annealed TiO_2:C films, which should be related also to the carbon oxide/s from the ambient air. The results of the calculated spectral dependencies of $\underline{E}_g$ and E_U (table 1) show that the absorption edge is shifted to the visible part of the electromagnetic spectrum. The increased quantity of dopant shifted the tail states energy to the UV region in the as-deposited samples, but the annealing significantly downshifted E_U. A similar behavior of the C-doped TiO_2 samples has been observed very often [6, 7, 15].

4. Conclusions

Carbon-doped thin films of TiO_2 (thickness of 40 – 120 nm) were deposited by r.f. magnetron sputtering in Ar (0.5 Pa) + air (0.6 Pa)) atmosphere. The residual quantities of carbon oxides in the working atmosphere werr sufficient for some doping with carbon to occur of the TiO_2 films. Doping with higher dopant quantities by co-sputtering of a carbon plate and a TiO_2 target did not affect significantly the interband transitions energy. The annealing at 400 °C shifted significantly the energy of the defect-related Urbach-tail states; it is expected that this should stimulate the use of TiO_2:C films in photo-catalytic processes.

Acknowledgements

The authors gratefully acknowledge the financial support by the National Science Fund of Bulgaria under grant DN18/9-11.12.2017. The work of D. K. was partially supported by the Bulgarian Ministry of Education and Science under the National Research Program "Young scientists and postdoctoral students" approved by DCM #577/17.08.2018. Research equipment of the INFRAMAT Distributed Research Infrastructure, part of the Bulgarian National Roadmap for Research Infrastructures and supported by the Bulgarian Ministry of Education and Science was used in this investigation. E.V. acknowledges the support of the European Regional Development Fund within the Operational Program "Science and Education for Smart Growth 2014 – 2020" under Project CoE "National center of mechatronics and clean technologies" BG05M2OP001–1.001–0008-C01.

References

[1] Janus M ed 2017 *Application of titanium dioxide* (London: Intechopen) DOI: 10.5772/intechopen.70121

[2] # 96-720-6076, Crystallography Open Database
[3] # 96-900-4138, Crystallography Open Database
[4] # 96-900-4144, Crystallography Open Database
[5] Vaitkus A, Merkys A and Gražulis S 2021 *Journal of Applied Crystallography* **54** (2) 661
[6] Hua L, Yin Z and Cao S 2020 *Catalysts* **10** (12) 1431
[7] Wang Y-H, Rahman K-H, Wu C-C and Chen K-C 2020 *Catalysts* **10** (6) 598
[8] Irie H, Washizuka S and Hashimoto K 2006 *Thin Solid Films* **510** (1-2) 21
[9] Wei B, Li H G, Cao H T and Wen F 2019 *Key Engineering Materials* **807** 41
[10] Angelov O 2020 *J. Phys.: Conf. Ser.* **1492** 012055
[11] CompleteEASE® Software Manual (Lincoln: J. A. Woollam Co.) USA 2021
[12] Dresselhaus M 2001 *Solid State Physics, p. II, Optical Properties of Solids* 36
[13] Tauc J, Grigorovici R and Vancu A 1966 *Phys. Status Solidi* **15** (2) 627
[14] Urbach F 1953 *Phys. Rev.* **92** 1324
[15] Di Valentin C, Pacchioni G and Selloni A 2005 *Chem. Mater.* **17** (26) 6656

Formation and characterization of CuO coatings deposited by reactive magnetron sputtering

M Ormanova[1,3], G Kotlarski[1], S Valkov[1,2], D Dechev[1], N Ivanov[1] and P Petrov[1]

[1]Acad. E. Djakov Institute of Electronics, Bulgarian Academy of Sciences, 72 Tsarigradsko Chaussee, 1784 Sofia, Bulgaria
[2]Technical University of Gabrovo, 4 H. Dimitar Str., 5300 Gabrovo, Bulgaria

E-mail: m.ormanova@ie.bas.bg

Abstract. In this work, CuO coatings were deposited on 304 L stainless steel substrates by reactive magnetron sputtering. During the process, the substrate temperature was varied in the range from 100 °C to 400 °C. The crystallographic structure of the samples thus prepared was characterized by X-ray diffraction, and the results are discussed with respect to the technological conditions applied. It was found that the phase composition of the deposited CuO coatings is in the form of a monoclinic crystal structure. Also, the growth of the coatings was accompanied by a change in the orientation of the crystallites from (022) to (110) and a decrease in the concentration of imperfections.

1. Introduction

The progress in the development of new biomaterials places new requirements on their properties, such as enhanced antibacteriality, good wear and corrosion resistance, and excellent biocompatibility. During the last decades, multiple studies have been carried out based on topics related to manufacturing new biomaterials with improved functional properties [1, 2].

Metals and their alloys, including stainless steel, titanium, and cobalt alloys, are widely used as implant materials [3]. However, after the implant is placed in the human body, inflammatory and/or infection processes could occur, which result in disadvantageous reactions and implant failure [4]. One of the approaches to overcoming these drawbacks is applying various methods for surface modification. Among them, the physical vapor deposition (PVD) methods, such as magnetron sputtering, are widely used for deposition of both nitride (TiN, CrN, VN, etc.) and oxide (TiO_2, CuO, HfO, etc.) coatings, which are capable of improving the surface properties of materials [5, 6].

In particular, metal oxides are used to increase the resistance to wear and corrosion and to improve the biocompatibility of materials [7]. For example, TiO_2 coatings can be used as protective non-toxic coatings that improve the wear and corrosion resistance of materials, with additional benefits such as self-cleaning and antibacterial effects [8-10]. Copper oxide (CuO) in the form of a coating or nanoparticles is used in medical applications [11-13]. The authors of [14] have successfully synthesized a new nanocomposite CuO-TiO_2 using an oxalate method for use as nanofillers for epoxy coatings to protect the steel surface against corrosion and bacterial growth. The results obtained showed an improved corrosion resistance behavior of the epoxy coatings with nanocomposites inclusions.

[3] To whom any correspondence should be addressed.

Content from this work may be used under the terms of the Creative Commons Attribution 3.0 licence. Any further distribution of this work must maintain attribution to the author(s) and the title of the work, journal citation and DOI.
Published under licence by IOP Publishing Ltd

Moreover, the antibacterial properties of epoxy coatings were improved after the addition of CuO/TiO_2 nanocomposite.

It is clear that copper oxide (CuO) can be characterized as a promising material for application in the modern biomedicine. However, the detailed study of CuO coatings obtained by reactive magnetron sputtering is still less well investigated. The main purpose of the present study is to study the possibility of successful deposition of CuO coatings on stainless steel substrates by reactive magnetron sputtering and to follow the effects of the technological deposition process conditions on the important structural parameters, such as phase composition, crystallographic texture, crystal lattice perfection, and grains size.

2. Experimental procedures

The experiments were performed on 304 L stainless steel specimens with a diameter of 20 mm and a thickness of 4 mm and the following chemical composition in wt % – 0.03 C; 17.5– 19.5 Cr; 2.00 Mn; 1.00 Si; 8 – 10.5 Ni; balance Fe. The substrates were polished mechanically with grinding papers of sizes 320, 400, 600, 800, 1000, and 1500, and then subjected to a polishing by a paste with a grain size of 400 nm. The surface roughness obtained by this treatment is in the nanometric range.The phase composition of the substrate is in the form of a double-phase structure of γ-Fe and martensitic α-Fe.

Copper oxide (CuO) coatings were deposited using reactive magnetron sputtering (RMS). The diameter of the sputtered target was 100 mm, and the Cu purity was 99.8%. Before the deposition of CuO films, the substrates were etched by Ar ions for 10 minutes to remove surface contaminations. The process took place in an Ar-O_2 atmosphere at a working pressure of 5×10^{-2} Pa and a deposition time of 60 minutes. The coatings were deposited at four diffrrent substrate temperatures, namely, 100 °C, 200 °C, 300 °C and 400 °C. The thicknesses of the deposited coatings were estimated to be about 400 nm for all investigated samples, corresponding to a deposition rate of about 1 Å/sec.

The samples were analyzed by X-ray diffraction (XRD, Empyream system, Malvern Panalytical), with Cu Kα radiation (1.54 Å). The measurements were carried out within the range of 30 – 80° in a symmetrical Bragg-Brentano mode (θ-2θ). The chemical composition of the obtained samples was investigated using energy-dispersive X-ray spectroscopy (EDX) analysis.

3. Results and discussion

The X-ray diffraction measurements of the phase composition were carried out in Brëgg-Brentano (B-B) geometry. Figure 1 presents the XRD patterns of the CuO coatings deposited at different temperatures (100 – 400 °C). All diffraction maxima are indexed and all four patterns exhibit a relatively low background. The phase identification was performed according to the International Center for Diffraction Data (ICDD) Database file Powder Diffraction File (PDF) #05-0661 for the CuO crystal phase. Figure 1 shows the presence of a monoclinic structure with space group C2/c (15). The diffraction maxima of the 304 L stainless steel substrate were also identified. Thus, the phase composition of the substrate is in the form of a double-phase structure of γ-Fe and martensitic α-Fe. The relative intensities the registered diffraction maxima of the gamma and alpha phases correlate with those available in the ICDD database, meaning that the existence of a strong crystallographic texture at the substrate is not observed. The chemical composition of the surface of the samples with the deposited CuO coating was investigated by EDX. The results are presented in figure 2 and summarized in table 1. The presence of copper (Cu) and oxygen (O) was detected in all samples, which is a confirmation of the results obtained by X-ray diffraction for the phase composition of the deposited coatings.

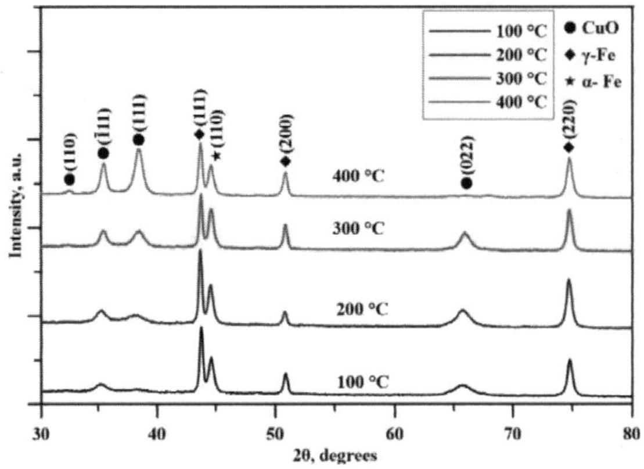

Figure 1. XRD patterns of CuO coatings deposited at different temperatures on 304 L substrates.

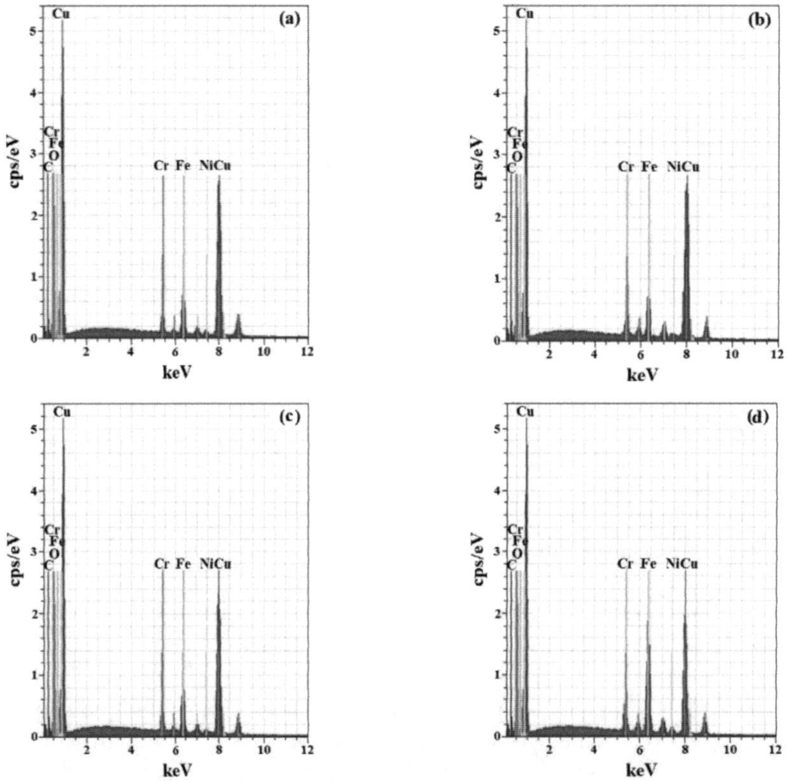

Figure 2. EDX spectra for samples with deposited CuO coating at different substrate temperatures: (a) 100 °C; (b) 200 °C; (c) 300 °C; (d) 400 °C.

Table 1. Chemical composition of the surface of the samples with CuO coatings.

Specimen	*Chemical composition, wt.%*					
	Cu	*O*	*Fe*	*Cr*	*Ni*	*C*
100 °C	66.27±1.8	14.06±2.4	11.13±0.3	3.47±0.1	1.29±0.1	3.76±0.6
200 °C	66.73±1.9	13.55±2.8	10.96±0.3	3.51±0.1	1.28±0.1	3.97±0.8
300 °C	65.36±1.8	13.45±2.3	12.63±0.4	3.93±0.1	1.43±0.1	3.2±0.6
400 °C	52.16±1.5	12.21±2.0	23.28±0.7	6.94±0.2	2.59±0.1	2.81±0.5

The formation of a preferred crystallographic orientation was studied by the texture coefficient, which measures the relative degree of preferred orientation of a specific (*hkl*) plane. This coefficient is calculated by (1):

$$T_c(hkl) = \frac{I(hkl)}{I_0(hkl)}\left(\frac{1}{n}\sum_{i=1}^{n}\frac{I(h_ik_il_i)}{I_0(h_ik_il_i)}\right)^{-1}, \tag{1}$$

where I_0 is the standard intensity of the considered peak taken from a PDF card, *n* is the number of diffraction maxima, and *I* is the measured intensity of the $(h_ik_il_i)$ plane. The crystallite size (*D*) is calculated using Scherer's equation (2):

$$D = \frac{0.9\lambda}{\beta\cos\theta}, \tag{2}$$

where λ is the X-rays wavelength, θ is the Braggs diffraction angle, and β is the full width at half-maximum intensity (FWHM) of the diffraction peaks. The results obtained are summarized in Table 2.

Table 2. Shares of micro-volumes belonging to different texture components, FWHM of $(\bar{1}11)$ diffraction peak and *D* (crystallite size).

Specimens	$T_c(hkl)$, %				*FWHM, degrees*	*D, nm*
	$T_{c\,110}$	$T_{c\,\bar{1}11}$	$T_{c\,111}$	$T_{c\,022}$		
100 °C	-	11.3	-	88.7	1	8.6
200 °C	-	9.8	8.3	81.9	0.75	11.2
300 °C	-	10.0	10.2	79.8	0.55	14.6
400 °C	20.6	32.6	46.8	-	0.5	16.1

It is seen that at a substrate temperature of 100° C the preferred orientation is (022), where its contribution is about 89%. The increase in the temperature from 100 °C to 200 °C leads to the appearance of (111), but (022) is still dominant (82%). A further increase in the substrate temperature to 300 °C is accompanied by a slight decrease in the contribution of the (022) plane to about 80%. Raising the temperature to 400 °C leads to a change in the orientation of the crystallites – (022) disappears and (110) appears. In this case, the largest contribution is (111) – about 47%. The change in the texture coefficient of the resulting coatings can be explained by the temperature increase during the deposition process (100 °C – 400 °C), which is accompanied by lower cooling rates and a greater atoms mobility. The crystallographic texture and the formation of a preferred crystallographic orientation play an important role in the resultant coatings' functional properties. For example, it has a major influence on the corrosion resistance. Materials that exhibit crystallographic orientation towards the densest planes are characterized by significantly better corrosion properties in comparison with those oriented towards less-dense planes [15]. Therefore, the preferred crystallographic orientation of the considered materials has a significant influence on the resultant corrosion properties, that are of major importance for the implant. Also, coatings textured with highly-dense planes exhibit higher hardness and wear resistance.

The lattice imperfections, including dislocations, stacking faults, etc., can be characterized by the broadening of the diffraction maximum, as it is a function of the crystallographic imperfections [16]. In this work, the crystallographic perfection was evaluated via the FWHM of $(\bar{1}11)$ peak; the results are given in fig. 3 and Table 2. It is seen that the increase in the temperature during the deposition process is accompanied by a decrease in the full width at half-maximum. Therefore, the concentration of the crystallographic imperfections decreases with an increase in tempertature of the substrate.

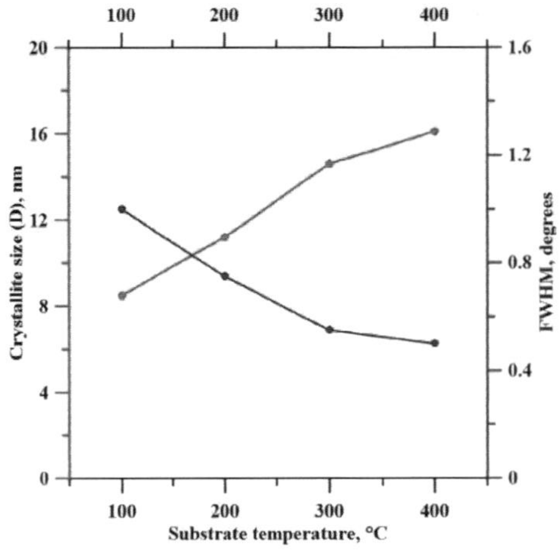

Figure 3. Full width at half-maximum (FWHM) and crystallite size (*D*) for CuO coatings deposited at different temperatures.

The crystallite size (*D*) as evaluated from the ($\bar{1}11$) diffraction peak was ~8.6 nm and 16.1 nm at substrate temperatures of 100 °C to 400 °C, respectively. Therefore, the crystallite size increased with the increase in the substrate temperature during deposition. The increase in the temperature of the substrates, respectively the mobility of the atoms, leads to agglomeration of the crystallites and an increase in their size [17,18]. It is well known that the grain size has a significant influence on the mechanical properties. Lower dimension of the crystalline size corresponds to higher hardness and wear resistance [19]. On the other hand, finer microstructure corresponds to deterioration in the corrosion properties. According to the authors of [20], the grain refinement resulted in an increase in the corrosion current density, corresponding to an increased corrosion rate.

4. Conclusions

This study demonstrated the possibility of successful deposition of copper oxide (CuO) coatings on 304 L stainless steel substrates by reactive magnetron sputtering. The influence was investigated of the technological conditions during the deposition process on the phase composition, crystallographic texture, crystal lattice perfection, and grains size of the produced coatings. It was found out that the temperature of the substrates has a significant effect on the studied structural parameters of the copper oxide coatings. The results demonstrate that all CuO coatings are in the form of a monoclinic structure. The increase of the substrate temperature during the deposition of the coatings was accompanied by a change in the orientation of the crystallites from (022) to (110), and by a decrease in the concentration of imperfections. Also, the higher substrate temperatures raised the crystallite size. The possibilities of controlling important structural parameters, such as preferred crystallographic orientation, grain size, etc., are of major importance for the resultant functional properties of the deposited coatings, for example, corrosion and wear resistance, which, in turn, are of major importance for the development of modern implant materials.

Acknowledgements

This work was supported by the Bulgarian Ministry of Education and Science under the National Research Program "Young scientists and postdoctoral students" approved by DCM #577/17.08.2018.

References

[1] Omran A-N, Ali M M and Moatasem M K 2020 *Appl. Phy.* A **126** 942
[2] Grabarczyk J, Gaj J, Pazik B, Kaczorowski W and Januszewicz B 2021 *Trib. Int.* **153** 106560
[3] Zhang E, Zhao X, Hu J, Wang R, Fu S and Qin G 2021 *Bioactive Mat.* **6** 2569
[4] Zhang D, Chen Q, Shi C, Chen M, Ma K, Wan J and Liu R 2020 *Adv. Funct. Mater.* **31** 2007226

[5] Du J W, Chen L, Chen J and Du Y 2020 *Vacuum* **179** 109468
[6] Nikolova M P, Nikolova V, Ivanova V L, Valkov S, Petrov P and Apostolova M D 2020 *Mat. Tod.: Proc.* **33** (4) 1781
[7] Grigore M E, Biscu E R, Holban A M, Gestal M C and Grumezescu A M 2016 *Pharmaceut.* **9** (4) 75
[8] Valkov S, Parshorov S, Andreeva A, Bezdushnyi R, Nikolova M, Dechev D, Ivanov N and Petrov P 2019 *Coatings* **9** (8) 513
[9] Nikolova M P, Ormanova M, Nikolova V and Apostolova M D 2021 *Int. J. Mol. Sci.* **22** 6369
[10] Peron M, Cogo S, Bjelland M, Afif A B, Dadlani A, Greggio E, Berto F and Torgersen J 2021 *J. Magn. & Alloys* In Press
[11] Akhavan O and Ghaderi E 2010 *Surf. & Coat. Tech.* **205** 219
[12] Barua S, Chattopadhyay P, Phukan M M, Konwar B K and Karak N 2014 *Mat. Res. Expr.* **1** 045402
[13] He X, Zhang G, Wang X, Hang R, Huang X, Qin L, Tang B and Zhang X 2017 *Cer. Int.* **43** (18) 16185
[14] Kumar A M, Khan A, Suleiman R, Qamar M, Saravanan S and Dafalla H 2018 *Progr. Org. Coat.* **114** 9
[15] Dwivedi D, Lepkova K and Becker T 2017 *RSC Advances* **8** 4580
[16] Iordanova I, Kelly P J, Burova M, Andreeva A and Stefanova B 2012 *Thin Solid Films* **520** 5333
[17] Sawicka-Chudy P, Wisz G, Sibiński M, Starowicz Z, Głowa Ł, Szczerba M and Cholewa M 2020 *Optik – Int. J. Light Electr. Opt.* **206** 164297
[18] Subramanian B, Ananthakumar R, Vidhya V S and Jayachandran M 2011 *Mat. Sci. & Eng.* B **176** 1
[19] Petrov P 1997 *Vacuum* **48** 49–50
[20] Soleimani M, Mirzadeh H and Dehghanian C 2020 *Mater. Res. Express* **7** 016522

Study of the structure, roughness and optical properties of HfO_2 coatings deposited on microscopic glass substrates

S Rabadzhiyska[1,4], M Ormanova[1], S Valkov[1,2], D Dechev[1], P Terziyska[3] and P Petrov[1]

[1]Acad. E. Djakov Institute of Electronics, Bulgarian Academy of Sciences, 72 Tsarigradsko Chaussee, 1784 Sofia, Bulgaria
[2]Technical University of Gabrovo, 4 H. Dimitar Str., 5300 Gabrovo, Bulgaria
[3]Acad. G. Nadjakov Institute of Solid State Physics, Bulgarian Academy of Sciences, 72 Tsarigradsko Chaussee, 1784 Sofia, Bulgaria

E-mail: s_rabadjiiska@ie.bas.bg

Abstract. In this work, we deposited HfO_2 films on microscopic glasses by direct current (DC) magnetron sputtering for different deposition times, namely 120 min and 180 min. The phase composition of the coatings was assessed by X-ray diffraction (XRD). The surface roughness, the optical constants, the refractive index *n*, the extinction coefficient *k* and the thickness of the prepared HfO_2 coatings were estimated by ellipsometric measurements. The XRD results pointed to the presence of a polycrystalline monoclinic phase. A slight difference was found in the surface roughness of the two HfO_2 films, which was 5.1 nm and 5.7 nm, respectively, for the deposition times of 120 minutes and 180 minutes. The optical constants *n* and *k* were determined at the wavelength of 630 nm by using the Cauchy model. The results showed almost the same value of the refractive index for both films (1.88 and 1.89), but a higher extinction coefficient (0.01) for the film grown for 180 minutes, compared to a value of 0.007 for the one deposited for 120 minutes. The films thickness was 410 nm and 780 nm for deposition times of 120 minutes and 180 minutes, respectively. The deposited HfO_2 films were transparent, with the longer deposition time corresponding to an improved reflection (57%), compared to the 28% reflection for the film deposited for 120 minutes.

1. Introduction

Hafnium dioxide (HfO_2) is an important optical coating material with a high dielectric constant, a low absorption coefficient, a large transparent spectral range from the ultraviolet to the infrared ranges, and good thermal, chemical and mechanical stability [1-6]. Moreover, the unique combination of structural, optical and electrical properties of the HfO_2 films finds various electronics and optoelectronics applications in memory devices, chirped mirrors, band pass filters, UV mirrors with a high damage threshold, and heat mirrors for energy-efficient windows [7-9]. In optoelectronics, hafnium oxide thin films are very useful as anti-reflection coatings in optical waveguide devices in view of reducing the Fresnel losses [10]. On the other hand, optical applications require uniform, smooth, dense and stoichiometric films, since the extrinsic causes of optical losses include surface roughness, impurities, film composition and structural non-uniformity. The overall performance of an optical coating depends

[4] To whom any correspondence should be addressed.

Content from this work may be used under the terms of the Creative Commons Attribution 3.0 licence. Any further distribution of this work must maintain attribution to the author(s) and the title of the work, journal citation and DOI.
Published under licence by IOP Publishing Ltd

on choosing a suitable deposition technique and technological conditions resulting in high-quality HfO_2 films. Good optical properties can be achieved by maximizing the achievable reflection coefficient of the material with as low optical losses as possible.

Various physical and chemical methods have been employed to grow HfO_2 thin films, such as electron-beam evaporation [11], reactive ion plating [12], metal-organic molecular beam epitaxy [13], atomic layer deposition [14, 15], etc. Each method has its own merits and demerits. The direct current (DC) reactive magnetron sputtering method provides many advantages, like uniform coating, good adhesion, reproducibility, high deposition rate, versatility and proven reproducibility in growing thin films [16, 17]. Zhibing et al. reported the effect of the O_2/Ar ratio on the structure and properties of HfO_2 films [17]. Other authors [18] investigated the optical properties from the visible to the far-infrared regions of HfO_2 thin films with thicknesses from about 180 nm to 500 nm. The optical properties are discussed with correlation to both the film thickness and the surface roughness. The authors of [19] deposited HfO_2 films on Si substrates and studied the influence of low temperatures on the structure and the optical properties of the produced films. In our previous work [20], the influence was studied of the bias voltage on the films' structure and optical properties. The results indicated a good correlation between the surface roughness and films' structure and the improved optical properties, but the reflection coefficient was not sufficiently high for uses in modern optical applications. However, the effect of the deposition time on the phase composition and the optical properties are currently less well investigated.

Therefore, this work aims at studying the deposition time effect on the phase composition and the optical properties of HfO_2 coatings deposited by DC magnetron sputtering.

2. Experimental part

HfO_2 films were deposited for 120 minutes and 180 minutes by DC magnetron sputtering. The sputtered target was of high-purity Hf (about 99.8%) with a diameter of 100 mm. Deltalab microscopic polished glass substrates with dimensions 40×26×1 mm were used as substrates. The distance between the target and the substrate was 100 mm. Before the deposition process, the vacuum chamber was evacuated to a low pressure of 1.5×10^{-3} Pa. Before deposition, the substrates were sputter-cleaned by Ar^+ plasma for 10 min to eliminate contaminations from their surface. The process took place in Ar-O_2 atmosphere. The technological conditions for the HfO_2 coatings deposition are summarized in Table 1. For better uniformity and density of the HfO_2 films, the substrates were placed at an angle of 15°. The technological conditions were optimized according to our previous investigations published in Ref. [20].

Table 1. Experimental conditions for HfO_2 films deposition

Deposition parameters	Value
Working pressure	8×10^{-2} Pa
Substrate bias voltage	–115 V
Substrate temperature	300 °C
Deposition time	120, 180 minutes

The phase composition of the coatings was assessed by X-ray diffraction (XRD). The measurements were realized in a grazing mode (GIXRD, ω- 2θ), with the angle of incidence of 2°. The experiments were performed with Cu Kα radiation (1.54 Å) at 2θ scale from 20° to 60° with a step size of 0.05° and a scan step time of 1.5 s.

The ellipsometric measurements were performed using a Woollam M2000D rotating compensator spectroscopic ellipsometer with a wavelength range from 193 nm to 1000 nm. The involved data acquisition and analysis software was Complete EASE 5.10 J. A. Woollam Co., Inc. To determine the thicknesses of the layers, the spectroscopic ellipsometry data Ψ and Δ (taken at angles of incidence of 55°, 65° and 75°) were analyzed using a model consisting of a Si substrate and a Cauchy layer (HfO_2 layer) as a second layer in the transparent wavelength range from 500 nm – 1000 nm.

The optical properties of the as-deposited HfO_2 coatings were determined based on transmittance measurements in the ultraviolet, visible and near infrared region by an SF-46 spectrophotometer operating in the range from 190 nm to 1100 nm. The reflection coefficients were measured at an incidence angle of 5° by an optical system developed by us using a diode laser emitting the wavelength of 630 nm, and an ARM Cortex STM 32 microprocessor allowing measurements of high accuracy. The system developed performs simultaneous measurements of the incident, transmitted and reflected power and calculates the losses in the films deposited. A current-voltage converter (230 V) with a linear characteristic was utilized, with pin diode sensors working at 630 nm.

3. Results and discussion

The thicknesses measured were 410 nm and 780 nm for the coatings deposited for 120 min and 180 min, respectively. The films thickness grows as the substrate deposition time is increased. X-ray diffraction measurements were performed in order to determine the structural properties of the samples. Figure 1 shows the grazing angle XRD patterns of HfO_2 films deposited for 120 minutes and 180 minutes. The intensity peaks are indexed using the powder diffraction file for monoclinic HfO_2 from the International Centre for Diffraction Data Database (ICDD PDF#43-1017) [21]. The results show the presence of polycrystalline monoclinic crystal structure with space group P21/c for both HfO_2 films deposited. As can be seen in the XRD patterns, there is a significant difference in the crystallinity of the two films. The main peaks appeared at about 27° and 33°, and are attributed to the orientations (−111) and (020) of the monoclinic HfO_2 phase for both samples. Also, the XRD patterns exhibit a lack of amorphous-like halos.

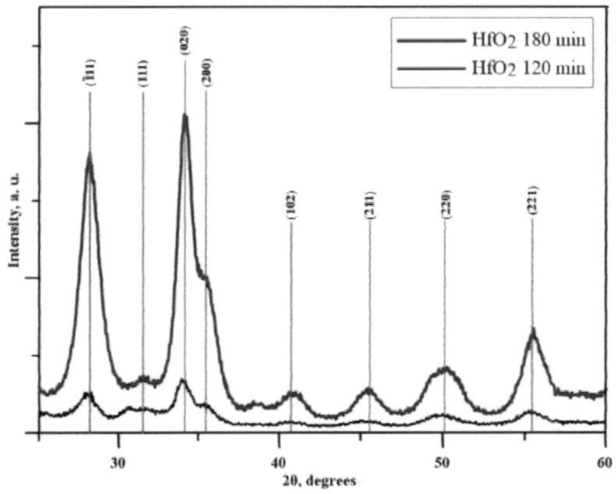

Figure 1. XRD patterns of the HfO_2 films deposited at 120 and 180 minutes.

The surface roughness of both HfO_2 films was evaluated by ellipsometric measurements. The results show a relatively small difference in the average surface roughness, namely, 5.1 nm and 5.7 nm for the samples deposited at 120 minutes and 180 minutes, respectively. According to [22], the surface roughness is strongly related to the grain size, with larger crystallites corresponding to a rougher surface, and vice-versa. The increase in the deposition time to 180 minutes leads to an increase in the grains size due to the enhanced ion bombardment and the formation of a larger number of defects blocking the grain boundaries migration.

The recorded transmittance spectra of the HfO_2 films deposited for different times on microscopic glass substrates are presented in figure 2. The results show that the films are well transparent down to a wavelength of 300 nm. The highest optical transmittance is 89% and 87% for the HfO_2 films grown for 120 minutes and 180 minutes, respectively. The films are homogeneous since the average transmittance is almost equal to the transmittance of the uncoated substrate (92%). The results show that the transmittance decreases slightly as the film thickness increases. However, the change in the roughness of the deposited films is very small (~0.6 nm), so that roughness cannot be the only factor for the slight change (1%) in the transparency. We assumed that another factor, such as the extinction coefficient (figure 4), is responsible for this change. The oscillations in the spectrum are due to interference effects [23].

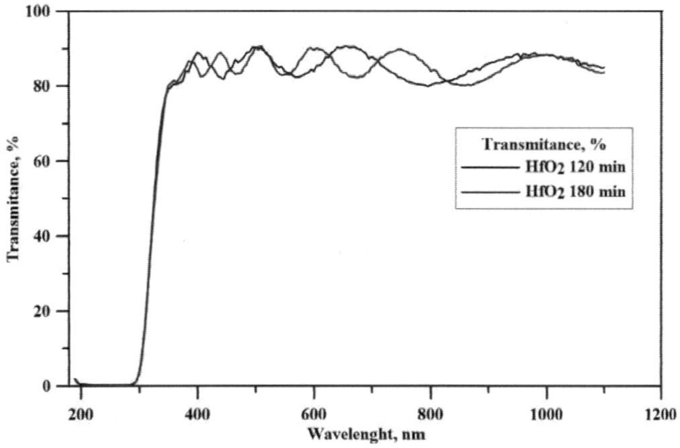

Figure 2. Transmittance spectra of HfO_2 films deposited for different deposition times.

The reflection coefficients were measured in the visible spectral region at the wavelength of 630 nm for both HfO_2 samples. The results show a reflection coefficient of 28% for the film grown for 120 minutes and a higher value of 57% for the one grown for 180 minutes, which points to a more homogeneous composition of the film and improved stoichiometry. The reflection coefficients obtained are close to those quoted in [24].

Figure 3 and figure 4 show the variation of the refractive index *n* and the extinction coefficient *k* for the films deposited for 120 minutes and 180 minutes. The optical constants were determined for the wavelength of 630 nm by the Cauchy model.

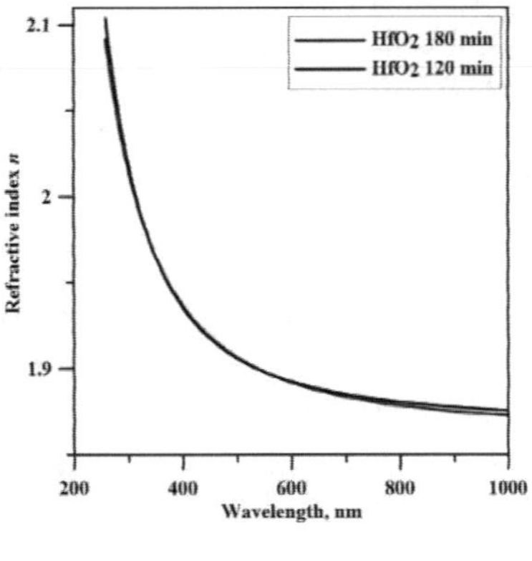

Figure 3. Variation of the refractive index *n* with the wavelength.

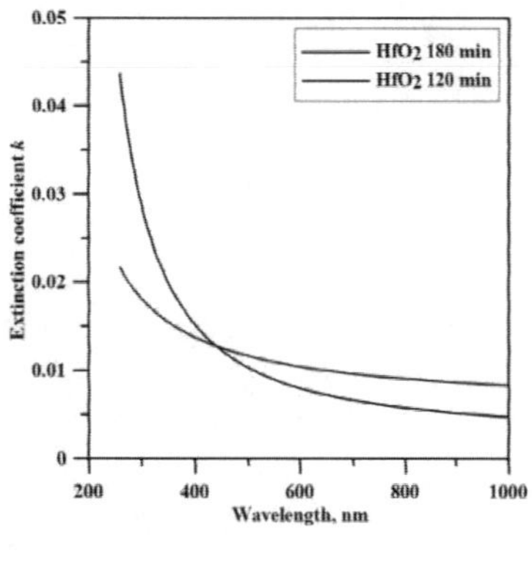

Figure 4. Variation of the extinction coefficient *k* with the wavelength.

The results show close values of the refractive index (figure 3) for both deposition time, namely, 1.88 and 1.89, as well as a higher extinction coefficient for film grown for 180 minutes (0.01) compared to the value of 0.007 for the one deposited for120 minutes. As is known, the substrate temperature affects the refractive index of the grown HfO_2 films. In general, a higher refractive index means that a more compact film structure is obtained [23]. The refractive index n decreases at higher substrate temperatures, which may be due to decreasing density, grain size and crystallinity. The extinction coefficient k of the HfO_2 films was calculated using the following relation [25]:

$$k = \alpha\lambda/4\pi \tag{1}$$

where α is an optical absorption coefficient.

The extinction coefficient is largely responsible for the increase or decrease in the films' transparency. The light scattering off grains is the main process determining the extinction coefficient of the films [22]. Besides, as seen in figure 4, the films's extinction coefficient is very low and approaches zero in both cases, which testifies to very small optical losses. The optical properties of hafnium oxide films indicate that these films are dense with a high refractive index. They meet the requirement of high transparency over the entire solar spectrum. The HfO_2 film deposited for 180 minutes has a significantly higher reflection coefficient and thus could be used in modern optical applications.

4. Conclusions

HfO_2 films were deposited on microscopic polished glass substrates by direct current magnetron sputtering at deposition times of 120 minutes and 180 minutes with regard to studying the deposition time effect on the structure, surface roughness and optical properties of the films. The XRD results are indicative of polycrystalline HfO_2 films with a monoclinic phase for both deposition times. A slight difference was found in the surface roughness of the HfO_2 films, namely, 5.1 nm and 5.7 nm, for the deposition times of 120 minutes and 180 minutes, respectively. The optical constants n and k were determined at the wavelength of 630 nm by the Cauchy model. The results showed almost equal refractive index values for the two films (1.88 and 1.89) and a higher extinction coefficient value (0.01) for the film grown for 180 minutes, compared to 0.007 for the one deposited for 120 minutes. The thickness of the films was 410 nm and 780 nm for the deposition times of 120 minutes and 180 minutes, respectively. The HfO_2 films deposited by DC magnetron sputtering were transparent; also, the films were smooth, uniform and dense, which makes them suitable for the development of optical elements and devices based on wedge-shaped structures for laser and optoelectronics equipment.

Acknowledgements

This work was supported in part by the Bulgarian National Science Fund at the Ministry of Education and Science under the competition for financing fundamental scientific research – 2017, project DN 17/07.

References

[1] Chow R, Falabella S, Loomis G, Rainer F, Stolz C and Kozlowski M 1993 *Appl. Opt.* **32** 5567
[2] Feng L, Liu Z and Shen Y 2009 *Vacuum* **83** 902
[3] Li F, Bayer B, Hofmann S, Speakman S, Ducati C, Milne W and Flewitt A 2013 *Phys. Status Solidi* B **250** 957
[4] Pervak V, Krausz F and Aplonski A 2007 *Thin Solid Films* **515** 7984
[5] Vargas M, Murphy N and Ramana C 2014 *Opt. Mater.* **37** 621
[6] Quinteros C, Zazpe R, Marlasca F, Golmar F, Casanova F, Stoliar P, Hueso L and Levy P 2014 *J. Appl. Phys.* **115** 024501
[7] Torchio P, Gatto A, Alvisi M, Albrand G, Kaiser R and Amra C 2002 *Appl. Opt.* **41** 3256
[8] Liu Z, Tibuleas S, Shin D, Young P and Magnusson R 1998 *Opt. Lett.* **23** 1556
[9] Alvisi M, Tomasi F, Ferrone M, Protopapa M, Rizzo A, Sarto F and Scaglione S 2001 *Thin Solid Films* **396** 44
[10] Wang Y, Lin Z, Cheng X, Xiao H, Zhang F and Zou S 2004 *Appl. Surf. Sci.* **228** 93
[11] Du Q, Wang W, Li S, Zhang D and Zheng W 2016 *Thin Solid Films* **608** 21
[12] Waldorf A, Dobrowolski J, Sullivan B and Plante L 1993 *Appl. Opt.* **32** 5583

[13] Moon T, Ham M and Myoung J 2005 *Appl. Phys. Lett.* **86** 102903
[14] Ritala M, Leskelä M, Niinistö L, Prohaska T, Friedbacher G and Grasserbauer M 1994 *Thin Solid Films* **250** 72
[15] Gope J, Batra N, Panigrahi J, Singh R, Maurya K, Srivastava R and Singh P 2015 *Appl. Surf. Sci.* **357** 635
[16] Ravish J, Yogendra G, Vikramaditya D, Amit C and Ramesh C 2013 *Appl. Surf. Sci.* 2**83** 332
[17] Ohring M 1992 *The Materials Science of Thin Films* (San Diego, CA: Academic Press)
[18] Zhibing H, Weidong W, Hua X, Zhang J and Tang Y 2006 *Vacuum* **81** 211
[19] Khosman J, Khan A and Kordesch M 2008 *Surf. Coat. Technol.* **202** 250
[20] Rabadzhiyska S, Ormanova M, Valkov S, Ivanov N, Terziyska P, Ivanov K and Petrov P 2021 *Journal of Physics: Conference Series* **1859** 012066
[21] International Centre for Diffraction Data 1999 *JCDS Powder Diffraction* File №34-104
[22] Zhang T, Gan B, Park S, Wang Q and Kim K 2014 *Surf. Coat. Technol.* **253** 115
[23] Basu S, Lee B and Zhang Z 2010 *J. Heat Transfer* **132** 023301
[24] Al-Kuhaili M F 2004 *Optical Materials* **27** 383–387
[25] Wu K, Wang J, Liu W, Chen Z and Wu J 2006 *Appl. Surf. Sci.* **252** 5829

Comparative analysis of cavitation erosion of Cr-C carbide coatings depending on the carbon content

V Safonov[1,5], A Zykova[1], J Steller[2], T Seremak[3] and N Donkov[4]

[1]National Science Centre "Kharkov Institute of Physics and Technology" 1 Akademicheskaya Str., 61108 Kharkiv, Ukraine
[2]The Szewalski Institute of Fluid-Flow Machinery of the Polish Academy of Sciences 14, J. Fiszera St., 80-231 Gdansk, Poland
[3]Gdansk University of Technology, 11/12 G. Narutowicza Str., 80-233 Gdansk, Poland
[4]Emil Djakov Institute of Electronics, Bulgarian Academy of Sciences, 72 Tsarigradsko Chaussee, 1784 Sofia, Bulgaria

E-mail: safonov600@gmail.com

Abstract. The aim of the study was to conduct a comparative analysis of the cavitation erosion resistance of thick (up to 20 μm) Cr-C coatings by means of the rotating-disk testing method. An experimental study was carried out of the process of cavitation destruction of $Cr_{18}Ni_{10}Ti$ stainless steel samples with Cr-C thick coatings depending on the carbon concentration in the coating. Nanocrystalline thick coatings of Cr-C were deposited by using the cathodic arc evaporation (Arc PVD) method. The results showed that the erosion rate of the coated samples was more than four times less than that of the uncoated sample. The Cr-C samples coated with 28.8 at % carbon content demonstrated less surface damage compared to the other coated samples and the reference samples.

1. Introduction

Application of protective coatings is an effective method of reducing the impact of destructive factors, such as corrosion and erosion, i.e., for increasing the lifetime of turbine components [1-4]. The thick Cr-C coatings are promising candidates for a variety of anti-erosion applications due to their ability to form amorphous and nanocrystalline structures [5].

The cavitation erosion behavior of the coated materials is significantly influenced by the main properties of the substrate and the coatings [6]. The surface roughness, scratches and pits on the coated substrate surface and defects in the coating structure, such as inclusions and droplets, have been found to be preferred initiation sites for cavitation [7]. A coating degradation mechanism was previously observed for films deposited by using cathodic arc [7, 8] and magnetron sputtering [3]. The aim of this study was a comparative analysis of the protective efficiency of thick (up to 20 μm) Cr-C coatings depending on the carbon content by means of the rotating-disk testing method.

[5] To whom any correspondence should be addressed.

Content from this work may be used under the terms of the Creative Commons Attribution 3.0 licence. Any further distribution of this work must maintain attribution to the author(s) and the title of the work, journal citation and DOI.
Published under licence by IOP Publishing Ltd

2. Materials and Methods

An experimental study was carried out of the process of cavitation destruction of stainless steel $Cr_{18}Ni_{10}Ti$ samples with Cr-C thick coatings depending on the carbon concentration in the coatings. Nanocrystalline coatings of Cr-C were deposited by cathodic arc evaporation (Arc PVD). The arc current was 130 A at a substrate bias voltage of –80 V. For deposition, a cathode of a low-alloyed chromium alloy (La 0.3 – 0.4%, Ta 0.3 – 0.5 %, V 0.3 – 0.5 %) was used. The Cr-C coatings were deposited in a toluene (C_7H_8) atmosphere with 50 % Ar. The carbon content was in the range from 20 –50 at %. A negative bias voltage was applied to the samples during the coating formation process. The temperature of the substrate was controlled by an IR pyrometer and was kept within 350 – 400 °C. The distance between the samples and the cathode was 270 mm. The erosion resistance was evaluated by tests conducted on the rotating-disk facility (figure 1) in the IMP PAN (Szewalski Institute of Fluid-Flow Machinery) laboratory on stainless steel $Cr_{18}Ni_{10}Ti$ samples of 30-mm diameter with Cr-C thick coatings of 20-μm thickness.

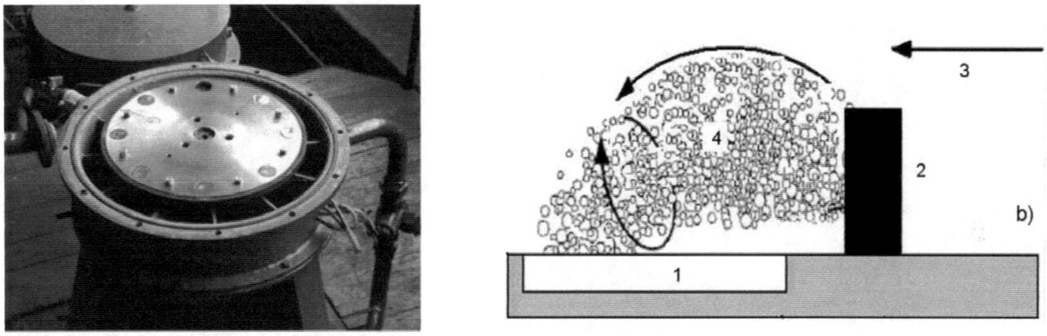

a) b)

Figure 1. **a)** Cavitation disk at the IMP PAN rotation facility, **b)** Cavitation process scheme of the rotation disk method: 1 - sample, 2 - cylindrical cavitation agent, 3 - water flow direction, 4 - cavitation zone.

The deposited coatings characteristics were studied using various techniques. The thickness of the Cr-C coatings was measured by a Calotest device. The adhesion properties were measured by standard methods with the use of a Revetest scratch tester (CSM Instruments). The phase composition of the coatings was studied using a DRON-4-07 X-ray diffractometer with Cu-Kα radiation using a selectively absorbing Ni filter. The structure of the samples was studied by scanning electron microscopy (SEM). The cavitation erosion damage was analyzed using optical imaging and 3D X-ray micro tomography (Phoenix v/tome/x m, GE Inspection Technologies GmbH, Germany). The kinetic dependences were obtained of the erosion destruction of samples with different carbon content in the Cr-C coatings. The main details of the deposition conditions and the coatings properties are presented in table 1.

Table 1. Deposition parameters, thickness, hardness, and adhesion strength of Cr-C coatings.

Deposition and coatings parameters	Type of coating				
	Cr-C 0 [9]	Cr-C 1	Cr-C 2	Cr-C 3	Cr-C 4
Pressure of residual gases, Pa	3×10^{-3}	3.5×10^{-3}	4×10^{-3}	4×10^{-3}	4×10^{-3}
Working pressure of Ar, %	-	50	50	50	50
Working pressure of C_7H_8 +Ar, Pa	0.6	0.15	0.095	0.26	0.43
Concentration C, at %	49.8	28.8	22.2	38.7	42.3
Coatings thickness, μm	19	20	18.3	19	18.5
Adhesion strength, N	74	56	98	54	68
Hardness, GPa	23.1	21.3	15.3	16.0	18.9

3. Results and Discussion

Magnusson et al. [10] showed that the $Cr_{1-x}C_x$ films are amorphous in a wide range of compositions; however, for $x \approx 0.47$, nanocrystalline grains with a diameter of < 3 nm could be detected using high-resolution transmission electron microscopy.

An X-ray amorphous phase is observed in all samples. The diffractogram of sample Cr-C 2 shows an X-ray amorphous Cr phase with a coherent scattering region (CSR) size of 9.3 nm. Only one low-intensity diffraction line was observed in the Cr-C 0 coated sample, which corresponds to chromium with a lattice parameter $a = 2.863$ Å (less than the values cited in the literature, which may be due to the presence of macrostresses) (figure 2 a). The size of the CSR of chromium is $D = 14.5$ nm estimated according to the Scherrer formula. The diffraction pattern revealed two diffraction halos in the angles regions $2\theta = 39 - 46°$ and $2\theta = 72 - 89°$. This fact indicates the presence of an X-ray amorphous phase in the sample in addition to the crystalline phase (that is, either the amorphous or the CSR size is less than 3 nm). Sample Cr-C 1 contains two phases, Cr and Cr_7C_3 (figure 2 b), which correspond to the Cr-C phase diagram [11]. The CSR size of Cr was 52.2 nm; and the CSR size of the Cr_7C_3 phase was 27.3 nm.

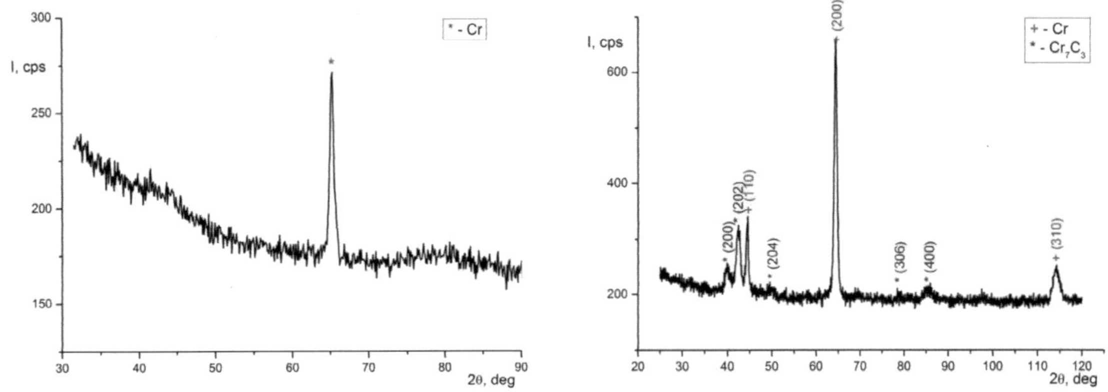

a) **b)**

Figure 2. Diffraction patterns of coatings **a)** Cr-C 0, **b)** Cr-C 1.

The Cr-C coatings demonstrated a fine multilayer structure. As shown in the SEM cross section of the Cr-C 0 coating (figure 3 a), b), each layer's thickness was about 100 nm [5].

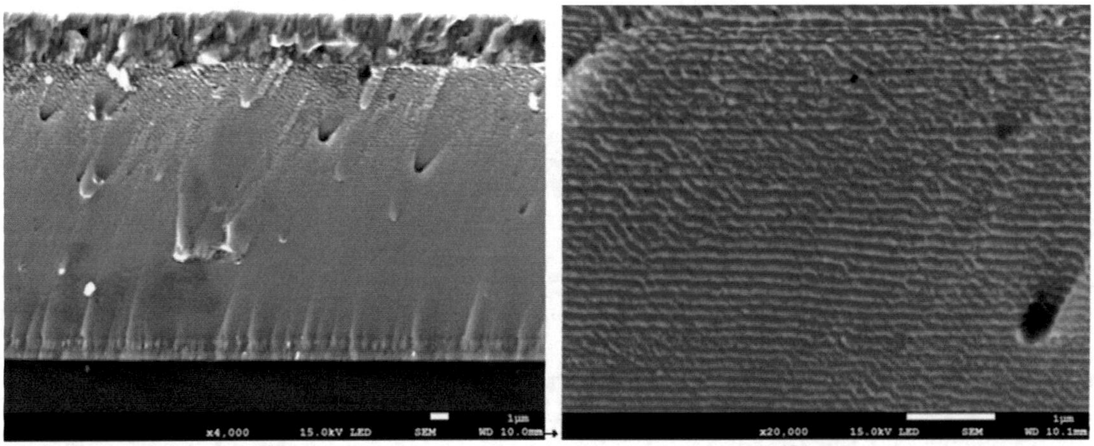

a) **b)**

Figure 3. SEM cross section of the Cr-C 0 coating. Magnification: **a)** ×4 000; **b)** formation of a nanolayer structure, ×20 000.

The cumulative mass loss curves of Cr-C coatings and of the $Cr_{18}Ni_{10}Ti$ reference sample were determined by up to 6-hour testing on the facility. The cavitation tests results for the Cr-C coated samples and the uncoated stainless steel $Cr_{18}Ni_{10}Ti$ reference sample are presented in figure 4.

The character of the samples' weight losses depending on the cavitation exposure duration (figure 4) indicates that all coatings studied are characterized by a rather low cavitation wear. The most efficient coating was the one with carbon content of about 28.8 at %. All coatings exhibited low wear, 4 – 5 times less than that of the uncoated steel. This fact was clearly detected by means of 3D X-ray micro tomography (figure 5 a).

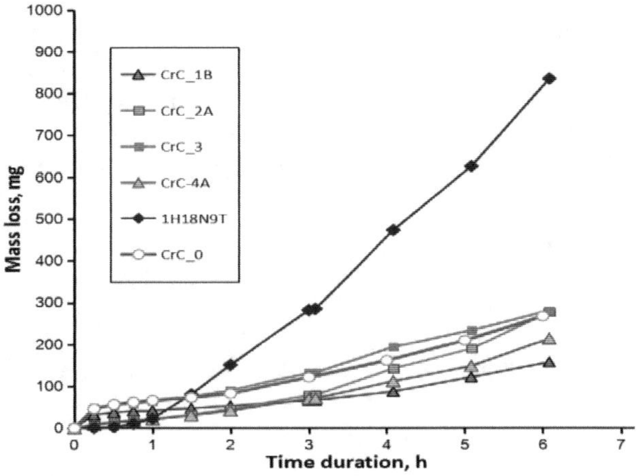

Figure 4. Cavitation wear kinetic curves of Cr-C coated samples and of the uncoated $Cr_{18}Ni_{10}Ti$ stainless steel reference ones.

The results of cavitation destruction of the samples after six hours of testing are shown in figure 5 b). Under the influence of microjets created by an asymmetric collapse of the cavitation cavities, the accumulated material around the craters is removed. This process leads to the formation of deep caverns and further craters elongation and merging with neighboring ones [1]. The higher cavitation resistance of the Cr-C coating with a carbon concentration of 28.8 % may be due to the hard Cr_7C_3 chromium carbide phase in the two-phase structure of the coating.

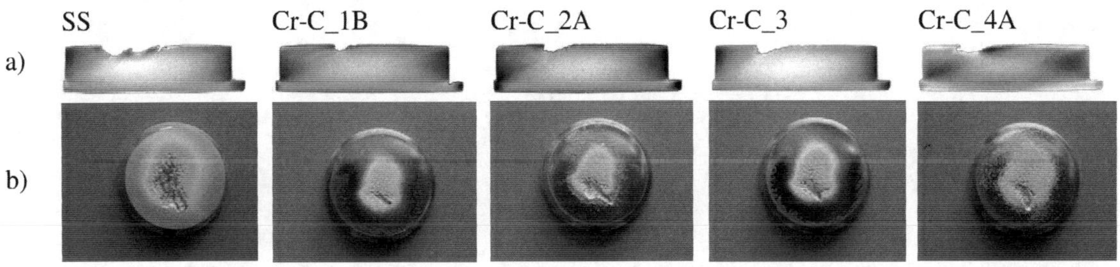

Figure 5. a) 3D X-ray micro tomography cross-sectional view of the samples, **b)** optical photographs of samples after six-hour test duration. Uncoated $Cr_{18}Ni_{10}Ti$ stainless steel (SS) and Cr-C coatings.

The results demonstrate the possibility to decrease the cavitation erosion rate by using nanostructured coatings. After six hours of testing, the mass loss of a Cr-C coated sample with a carbon content of about 28.8 at % was lower by a factor of 4 – 5 than that of the $Cr_{18}Ni_{10}T$ reference material.

4. Conclusions

The kinetic dependences were obtained of the erosion destruction of the Cr-C coated samples with different carbon content. The results showed that the erosion rate of the coated samples was more than four times less than that of the uncoated sample. The Cr-C coated sample with 28.8 at % carbon content demonstrated less surface damage compared to the other coated and the reference samples. The higher cavitation resistance of this coating may be due to its two-phase structure that includes the hard Cr_7C_3 chromium carbide phase.

Acknowledgments
The present study was conducted under the collaboration project "Enhancement of turbomachinery blade erosion resistance by means of carbide and nitride protective coatings" between the Polish Academy of Sciences and the National Academy of Sciences of Ukraine (Project 46); the research was also supported by the program for international scientific cooperation between the Bulgarian Academy of Sciences and the National Academy of Sciences of Ukraine.

References

[1] Azar G T, Yelkarasi C and Urgen M 2017 *Surf. Coat. Technol.* **322** 211
[2] Krella A 2009 *Surf. Coat. Technol.* **356** 263
[3] Cai F, Huang X and Yang Q 2015 *Wear* **324–325** 27
[4] Szala M, Walczak M, Pasierbiewicz K and Kamiński M 2019 *Coatings* **9** 340
[5] Safonov V, Zykova A, Krella A, Steller J, Gajowiec G, Szkodo M and Ghaemi H 2016 *Abstracts of conference contributions, Polish Hydropower Conference RENEXPO*, Warsaw 52
[6] Krella A 2013 *Wear* **297** 992
[7] Sebastiani M, Piccoli M and Bemporad E 2013 *Surf. Coat. Technol.* **215** 407
[8] Tkadletz M, Mitterer C, Sartory B, Letofsky-Papst I, Czettl C and Michotte C 2014 *Surf. Coat. Technol.* **257** 95
[9] Safonov V, Zykova A, Steller J and Seremak T 2020 *Funct. Mater.* **27** (4) 744
[10] Magnuson M, Andersson M, Lu J, Hultman L and Jansson U 2012 *J. Phys.: Condens. Matter* **24** 225004
[11] Venkatrsman M and Neumann J P 1990 *Bulletin of Alloy Phase Diagrams* **11** (2) 152

Surface treatment of magnetron-sputtered tantalum pentoxide coatings and its effect on the phenotype and function of immune cells *in vitro*

N Donkov[1,5], A Zykova[2,3], V Safonov[2,3], S Dudin[3], S Yakovin[3], A Goltsev[4] and T Dubrava[4]

[1]E. Djakov Institute of Electronics, Bulgarian Academy of Sciences, 72 Tsarigradsko Chaussee, 1784 Sofia, Bulgaria
[2]Kharkov Institute of Physics and Technology National Science Center 1 Akademicheskaya Str., 61108 Kharkiv, Ukraine
[3]V.N. Karazin Kharkiv National University, 4 Svobody Sq., 61022 Kharkiv, Ukraine
[4]Institute for Problems of Cryobiology and Cryomedicine NASU, 23 Pereyaslavskaya Str., 61015 Kharkiv, Ukraine

E-mail: nikolaidd@abv.bg

Abstract. The effect was analyzed of surface treatment by argon ions and electron beam on the properties of tantalum pentoxide coatings deposited by reactive magnetron sputtering and the treatment's correlation with immune cells response. The characteristics of the coatings were determined by means of energy dispersive X-ray (EDX) scanning electron microscopy (SEM), X-ray photoemission spectroscopy (XPS), and atomic force microscopy (AFM). The results demonstrate that the electron-irradiation post-treatment process stimulates the adhesive potential and phagocytic activity of antigen-presenting cells.

1. Introduction

Antigen presenting cells, such as macrophages may affect the immune response after implantation. Implanted biomaterials are in a close contact with macrophages so that characteristics such as surface chemistry and topography may play a critical role in initiating an anti-inflammatory response [1-3]. The immune response to artificial implants leads to postoperative complications, inflammatory processes and a risk of repeated surgical interventions. Inflammatory and anti-inflammatory reactions can be regulated via modifying the biomaterial's properties and surface [4, 5]. Surface modifications are a direct way of changing the cells' phenotype and function with the aim of reducing immune reactions and promoting a beneficial healing response.

This study was devoted to the effect of surface treatment by argon ions and electron beam on the structure and surface properties of Ta_2O_5 coatings deposited by reactive magnetron sputtering and its correlation with immune cells' response in *in vitro* tests.

---

[5] To whom any correspondence should be addressed.

Content from this work may be used under the terms of the Creative Commons Attribution 3.0 licence. Any further distribution of this work must maintain attribution to the author(s) and the title of the work, journal citation and DOI.
Published under licence by IOP Publishing Ltd

2. Materials and Methods

Ta_2O_5 coatings were deposited by ion-source-assisted magnetron sputtering on Petri-dish glass substrates in a high-vacuum pumping system with a base pressure of about 10^{-2} Pa. Oxygen for the reactive deposition was delivered through an ICP plasma source with $q = 60$ sqcm; the magnetron voltage was $U_m = 700$ V and the magnetron current was about $I_m = 5.7$ A. For part of the substrates, the deposition process was carried out through simultaneous bombardment of the growing film by argon ions using an ion source. The ion-source parameters were as follows: magnetic field current 1.5 A, ion acceleration voltage 2.5 keV, ion source current 30 mA. A primary electron beam was created by an UL-119 electron gun type. The reflected electrons energy was 20 keV, the current density on the sample surface, 14 mA/cm^2, the irradiation time, 1500 s. The coatings thickness was measured by a Calotest device. The surface morphology and topography were observed by a JSM 5500 LV electron scanning microscope and an atomic force microscope. The chemical composition of the coatings was analyzed by energy dispersive X-ray (EDX) spectroscopy (Oxford Link ISIS 300). The coated surfaces' advancing contact angles and wettability were evaluated by tensiometric measurements (Kruss 12). The total surface free energy and its polar and dispersion parts were estimated by using Wu and Owens-Wendt-Rabel-Kaeble methods.

The immune cells' adhesive and proliferative activity was evaluated by standard protocols. The study was conducted on the phagocytes of the peritoneal cavity of 15-week old CBA/H mice. The phagocytes cells were cultured with a density of $1{\times}10^7$ cells/cm^2 on control and oxide-coated glass Petri dishes in a CO_2 incubator (5 % CO_2) at 37 °C and 95 % relative humidity. The cells' adhesive potential was evaluated after 30 minutes of cultivation on coated/uncoated substrates. After the indicated time, unattached cells were removed by washing the tested substrates twice with Hanks solution (PanEco, Russia). The adherent cells were fixed with methanol for 5 min and counted. The statistical analysis of the test results was performed using the nonparametric Mann-Whitney method. To study the phagocytic activity, a suspension of cells ($1{\times}10^7$) in 2 ml of medium 199 with 3% ETS was placed in a Petri dish with coated and control samples; 0.9 ml of inactivated staphylococcus (*Staphylococcus aureus*) with $1{\times}10^9$ cell/ml dilution were added and incubated in a thermostat at 37 °C for 45 minutes. The dishes were then washed twice with Hanks solution, fixed with methanol, colored, and the percentage of phagocyte cells and the number of staphylococcal particles absorbed by each cell were calculated. Visual control of the cell cultures was carried out using a Primo Star light phase-contrast microscope (Zeiss, Germany) and an Axiovert 40C inverted microscope (Zeiss, Germany).

3. Results and Discussion

The volt-ampere characteristics of the magnetron discharge with tantalum target in argon for different values of the reactive gas flow were presented previously [6]. The deposition process carried out with simultaneous bombardment by argon ions resulted in structural changes of the growing films. The coating thickness and cross-sectional structure were obtained by SEM cross-section measurements. The chemical composition of the magnetron-sputtered coatings grown with simultaneous bombardment by argon ions were analyzed by energy dispersive X-ray (EDX) spectroscopy (Oxford Link ISIS 300 (figure 1)

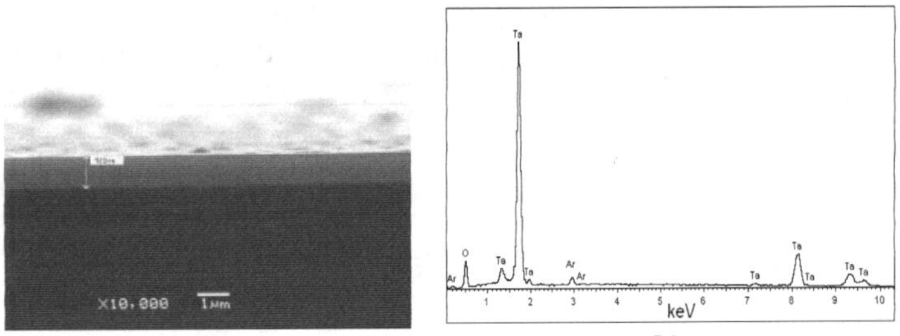

a) b)

Figure 1. SEM cross-section image **a)** and EDX spectrum of Ta_2O_5 coatings deposited by simultaneous bombardment with argon ions **b)**.

Bombarding the surface of the growing coatings with high-energy argon ions during the deposition process led to considerable changes in the relief due to sputtering which increases the surface roughness. For comparison, the other coated samples were treated by an electron beam after deposition. The process parameters (electron energy and dose) did not affect significantly the morphology of the deposited coatings, i.e., they were identical with the original ones. A possible effect of this process would be a change in the surface free energy and the surface charge state. These factors have a significant impact on the biological interaction and the cell/material response.

The changes in the ceramic oxide coatings' surface topography and roughness parameters were evaluated by atomic force microscopy (figure 2 a), b).

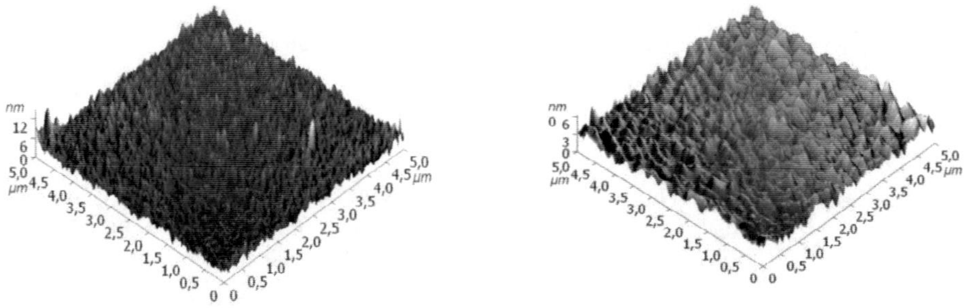

a) **b)**

Figure 2. AFM images **a)** surface topography of Ta_2O_5 coatings, **b)** surface topography of Ta_2O_5 coatings after surface treatment by argon ions.

The ion bombardment resulted in a change of the surface roughness parameters from $S_a = 0.9$ μm, $S_q = 1.2$ μm in the case of Ta_2O_5 coatings to $S_a = 0.6$ μm, $S_q = 0.7$ μm in the case of Ta_2O_5 coatings deposited with simultaneous bombardment by argon ions. In contrast, the electron-beam post-treatment process did not result in any significant changes in the oxide ceramic coatings' surface and structural parameters. The structure of the electron-beam irradiated coatings was similar to that of the as-deposited coatings – roughness parameters $S_a = 0.8$ μm, $S_q = 1.1$ μm

The changes in the structural parameters affected the surface characteristics and functional properties of the deposited coatings. As previously demonstrated and analyzed in [6], the X-ray diffraction profiles of as-deposited Ta_2O_5 coatings prove the amorphous nature for as-deposited magnetron sputtered Ta_2O_5 coatings – no peaks are observed. The Ta 4f lines of the deposited films are in a good agreement with the Ta 4f doublet representative of the Ta-O bond in Ta_2O_5. The Ta/O ratio estimated from the spectra is about 0.4 for all coatings investigated.

The surface free energy (SFE) plays an important role in the mechanism of antigen presenting cell/biomaterial response [7, 8]. The SFE and its polar part were estimated by Owens-Wendt-Rabel-Kaeble method for the following liquid system: α-bromonaphthalene-formamide-ethylene glycol-diidomethane-glycerol-water (figure 3 a) at a temperature of 20 °C. The data demonstrated that the Ta_2O_5 coatings' SFE calculated by means of the quoted method was in the range of 40 – 50 mN/m, with the SFE polar part in the range 11 – 16 mN/m. The SFE minimal values were exhibited by the tantalum pentoxide ceramic coatings deposited with simultaneous bombardment by argon ions. The oxide ceramic coatings' properties shifted to the more hydrophilic region; also, following the electron beam-irradiation post-treatment of the oxide-coated substrates, their SFE increased. The topographic and physico-chemical characteristics of the surface act as factors that modulate the cells' adhesive potential [9, 10]. Figure 3 b) shows the results of studying the adhesive potential of the phagocytes of the peritoneal cavity of CBA/H mice on glass (control) and Ta_2O_5 coatings before and after surface treatment by electrons and argon ions.

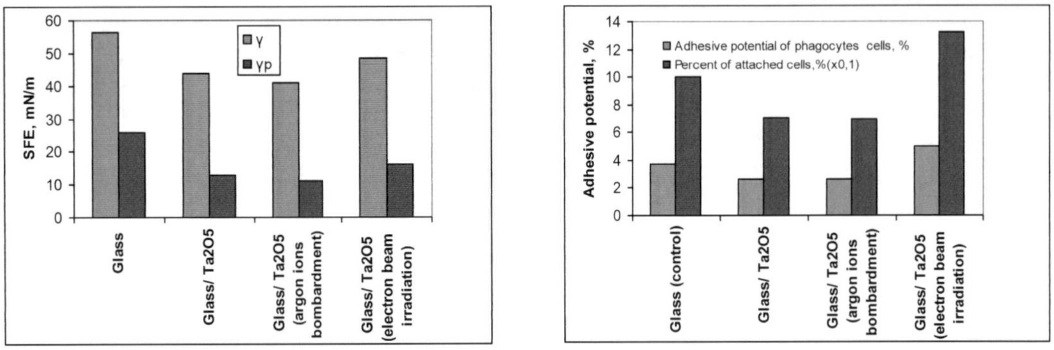

a) b)

Figure 3. **a)** Surface free energy (SFE) (γ and its polar part γ_p; **b)** adhesive potential of the phagocytes cells on glass and on Ta_2O_5-coated substrates.

The Ta_2O_5 coating reduced the adhesive potential of phagocytic cells by 29 % in comparison with the control (glass). The deposition process with simultaneous bombardment by argon ions did not increase the cells' adhesive ability in comparison with the Ta_2O_5 coatings. In contrast, the electron-beam treatment of the Ta_2O_5 coating contributed to a 32.7-% increase in the adherent cells content, in a good agreement with previous results [11].

Further, the phagocytic activity of phagocyte cells was studied. Figure 4 shows phagocytes stained by Romanovsky after a 45-minute exposure to inactivated staphylococcus. In figure 4, the macrophages cells with sizes in the range 15 – 20 µm show a foamy cytoplasm with jagged edges; the lymphocytes with sizes in the range 5 – 10 µm have a compact core; and the staphylococcus particles with sizes in the range 0.6 – 1.0 µm are absorbed by the macrophages cells. The main quantitative results, namely, percentage of phagocytic cells (PI) and number of staphylococcal particles absorbed by each cell (PN) are presented in table 1.

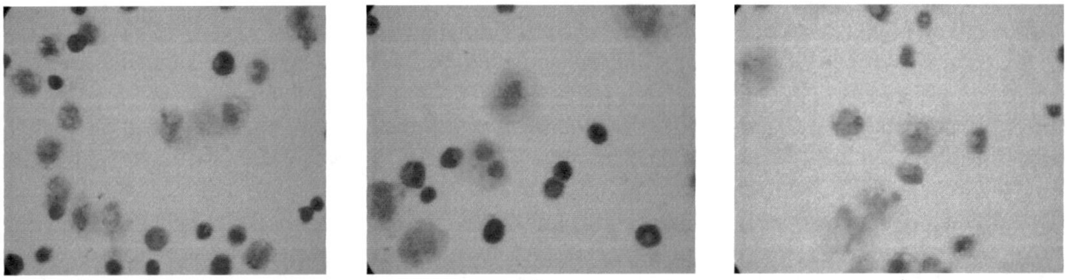

a) b) c)

Figure 4. **a)** Phagocyte activity after a 45-minute exposure with inactivated staphylococcus glass (control), **b)** Ta_2O_5 after ion treatment, **c)** Ta_2O_5 after electron irradiation, magnification ×100.

Table 1. Phagocytic activity on glass and Ta_2O_5 coated substrates.

Substrate/coatings	Phagocytic index, %	Phagocytic number, %
Glass (control)	92.0±1.43	8.8±0.52
Glass/ Ta_2O_5	89.4±1.60	8.4±2,41
Glass/ Ta_2O_5 (argon ion bombardment)	77.0±1.35*	4.7±0.38*
Glass/ Ta_2O_5 (electron-beam irradiation)	97.0±2.48*	10.8±0.56*

*Difference statistically significant to control, P <0.05.

The analysis of the data in table 1 shows that the PI and PN of cells cultured on the substrate with a Ta_2O_5 coating did not have statistically significant differences from the control indicators (glass). The Ta_2O_5-coated surface deposited with simultaneous argon ions bombardment reduced both the percentage of phagocytic cells and the number of staphylococcal particles absorbed by each cell in comparison with the control. In contrast, the Ta_2O_5-coated substrate demonstrated a statistically significant increase of the main indicators of phagocytosis after electron-beam irradiation. The effect of material surface modification on several inflammatory events *in vivo* was previously reported in [8, 12]. In the present study, the surface roughness changes after argon ions bombardment did not lead to a significant increase of the adhesive potential and on the activity of phagocytes in *in vitro* tests. Conversely, the changes in the surface free energy and the shift to the more hydrophilic region induced by electron-beam processing had the effect of increasing the cells' adhesive potential and phagocytic activity.

4. Conclusions

The effect was analyzed of surface treatment of Ta_2O_5 nanostructured coatings by argon ions and electron beam on the functional potential of antigen-presenting cells of the monocyte-phagocytic system. It was shown that the electron irradiation post-treatment process leads to a stimulation of the adhesive potential and the phagocytic activity of antigen-presenting cells on the Ta_2O_5 coated surfaces. On the contrary, the surface treatment by argon ions significantly reduces the functional activity of the cells, which can be important in vascular surgery. The changes in the parameters of the surface free energy and its polar part and the shift to the more hydrophilic region after electron-beam irradiation results in an increase of the functional activity of cells of the monocytic-phagocytic system. The modification of biomaterial surfaces with respect to their immune response can pave the way to a new generation of materials with immunomodulatory properties for a wide range of clinical applications.

Acknowledgments

The research was supported by the international scientific cooperation project # 08-03-20 of the National Academy of Sciences of Ukraine and by the program or international scientific cooperation between the Bulgarian Academy of Sciences and the National Academy of Sciences of Ukraine.

References

[1] Rostam H M, Fisher L E, Hook A L, Winkler D A, Alexander M R and Ghaemmaghami A M 2020 *Matter* **2** 1564
[2] Higgins D M, Basaraba R J, Hohnbaum A C, Lee E J, Grainger D W and Gonzalez-Juarrero M 2009 *Am. J. Pathol.* **175** 161
[3] Vishwakarma A, Bhise N S, Evangelista M, Rouwkema J, Dokmeci M R, Ghaemmaghami A M, Vrana N E and Khademhosseini A 2016 *Trends Biotechnol.* **34** 470
[4] MacLauchlan S, Skokos E A, Meznarich N, Zhu D H, Raoof S, Shipley J M, Senior R M, Bornstein P and Kyriakides T R 2009 *J. Leukoc. Biol.* **85** 617
[5] Jenney C R and Anderson J M 2000 *J Biomed Mater.Res.* **49** 435
[6] Zykova A, Safonov V, Goltsev A, Dubrava T, Rossokha I, Donkov N, Yakovin S, Kolesnikov D and Goncharov I 2016 *J. Phys. Conf. Series* **700** 012027
[7] Hallab N J, Bundy K J, O'Connor K, Moses R L and Jacobs J J 2001 *Tissue Eng.* **7** 55
[8] Rostam H M, Singh S, Vrana N E, Alexander M R and Ghaemmaghami A M 2015 *Biomater. Sci.* **3** 424
[9] Anselme K, Linez P, Bigerelle M, Le Maguer D, Le Maguer A, Hardouin P, Hildebrand H F, Alain I and Leroy J M 2000 *Biomaterials* **21** 1567
[10] Diener A, Nebe B, Luthen F, Becker P, Beck U, Neumann H G and Rychly J 2005 *Biomaterials* **26** 383
[11] Yakovin S, Dudin S, Zykova A, Safonov V, Goltcev A, Dubrava T and Rassokha I 2020 *Springer Proceedings in Physic* **244** 151
[12] Werthen M, Sellborn A, Kalltorp M, Elwing H and Thomsen P 2001 *Biomaterials* **8** 827

Low temperature quaternary hard coatings with Cr- to Ti-based transition composition

V Chitanov[1,3], St Kolchev[1], T Cholakova[1], L Kolaklieva[1], R Kakanakov[1], Ch Pashinski[1,2] and N Hristeva[1]

[1]Central Laboratory of Applied Physics, Bulgarian Academy of Sciences, 61 Sankt Peterburg Blvd., 4000 Plovdiv, Bulgaria
[2]Department of Mechanics, Technical University – Sofia, Plovdiv Branch, 25 Tsanko Diustabanov Str., 4000 Plovdiv, Bulgaria

E-mail: vchitanov@gmail.com

Abstract. A new quaternary hard coating based on Cr/CrN adhesion layers and a TiCrAlN main coating layer was developed for low-temperature applications. It was deposited in a closed-field unbalanced magnetron sputtering equipment. The nanohardness, adhesion and thickness of the coating were characterized by Anton Paar's nano-indenter tester, scratch tester and calotester. The chemical composition and surface morphology were investigated by a Hitachi SU 500 scanning electron microscope equipped with an energy dispersion analyzer. The coating showed a low coefficient of friction and a good adhesion strength between the substrate and the Cr/CrN transition layer, as well as between the transition layer and the TiCrAlN layer. The nanohardness achieved was 33 GPa. The research results showed that the quaternary Cr/CrN/TiCrAlN hard coating deposited at 200 °C has mechanical properties comparable to coatings with a similar composition deposited at standard deposition temperatures.

1. Introduction

Hard coatings of CrN are widely used for corrosion resistance and protection applications because of their good wear properties. The PVD magnetron sputtering is a quite common method for deposition of CrN thin layers [1]. The Cr contact layers and the CrN coatings are characterized by improved scratch resistance and microhardness [2]. The adhesion, hardness and toughness of quaternary coatings, like CrAlSiN, are improved by adding Cr/CrN transition layers. The nanohardness of the Cr/CrN/CrAlSiN coating has been measured to be 21.54% higher in comparison with the CrAlSiN coating system. The critical load during scratch tests increased from 38.1 N up to 72.5 N when a Cr/CrN interlayer was included, leading also to one order of magnitude lower wear rate during tribology tests [3]. Because of these properties, the Cr/CrN stack is widely applied for implementation as adhesive layers in hard coatings with multi-element composition structures. The performance and reliability of the binary hard coatings, like TiN and CrN, are limited by their mechanical properties [4]. The addition of Al improves the thermal stability and oxidation resistance of binary coatings; at present, in multicomponent and multilayer design it is an important way of increasing the hardness of Cr and Ti-based nitride coatings for cutting tools applications [5]. The compound materials are capable of combining the advantages of individual hard materials. Hard coatings, like TiCrAlN and CrTiAlN, deposited by closed-field

[3] To whom any correspondence should be addressed.

Content from this work may be used under the terms of the Creative Commons Attribution 3.0 licence. Any further distribution of this work must maintain attribution to the author(s) and the title of the work, journal citation and DOI.
Published under licence by IOP Publishing Ltd

unbalanced magnetron sputtering showed high hardness (29 – 34 GPa), good adhesion strength up to 60 N, high wear resistance and excellent cutting performance [6]. A major advantage of this technology is the possibility of depositing hard coatings at a temperature of 200 °C [7]. New hard coatings for temperature-sensitive substrates, like mild steels, martensitic stainless steels, some medical tool steels and other materials, are essential and need to be developed [8]. The standard temperature for PVD deposition of CrN and TiN coatings is over 200 °C [9]. Special PVD deposition techniques, like high-power impulse magnetron sputtering (HIPIMS) or closed-field unbalanced magnetron sputtering (CFUBMS), where no additional heating of the substrate is applied, are being used for deposition at lower temperatures [10]. The TiN layers have a higher hardness (20 – 25 GPa) in comparison with the CrN layers (14 – 18 GPa) [11]. The CrN coatings have better wear and adhesion properties than TiN. The Cr/CrN structure demonstrates an improved performance than the CrN single coating in terms of abrasion and sliding wear [12]. Bearing in mind the properties of these well-developed coating layers, in this work a new quaternary hard coating with a transition composition from Cr/CrN sublayers to quaternary TiCrAlN coating layer was developed for special temperature-sensitive substrates that do not withstand being exposed to temperature exceeding 200 °C, such as beryllium bronze and medical stainless steel. It was expected that the coating structure investigated should have the adhesive properties of CrN-based layers, but improved mechanical properties typical of the quaternary TiCrAlN coating structures deposited at temperatures over 200 °C. The mechanical properties and composition of the proposed Cr/CrN/TiCrAlN structure deposited under 200 °C are still not well studied. Obtaining such new results could support the implementation of this structure for real low-temperature industrial applications. A CFUBMS deposition technique was applied for fulfilling this research.

2. Experiment

The hard coatings were deposited using the UDP850/4 CFUBMS industry-size equipment. Four rectangular unbalanced magnetrons were placed with two Ti (99.99 %), one Al (99.99%) and one Cr (99.99%) targets used in a closed-field configuration with direct water-cooling. High-speed steel (HSS) substrates were chosen for the experiments. The substrates were first cleaned from contaminants and then rinsed in de-ionized water and immersed in an ultrasonic bath with ethanol. The vacuum chamber was pumped down to 1.4×10^{-3} Pa. Six samples were prepared; the technological parameters of their deposition are given in table 1. Samples 01 and 02 were preliminarily heated in the vacuum chamber up to a substrate temperature of 200 °C. Before sputtering, argon gas was introduced to the chamber by a mass flow controller (MFC) at a flow rate of 25 sccm to ignite and sustain the plasma during the deposition process. Additional cleaning was performed by applying a –500 V bias potential to the substrates for 30 minutes. Cr/CrN layers were first deposited on the substrates to provide good adhesion and smooth transition to the main coating layers. The nitrogen flow was also controlled by the MFC. For the deposition of the TiCrAlN layers, the Cr and Ti targets were controlled in a DC mode, while the Al target was controlled in a pulsed mode at 150 kHz with a pulse duration of 5.17 μs. The working pressure was varied from 0.19 Pa to 0.33 Pa by varying the nitrogen flowrate. The rotation speed of the sample holder was fixed at 5 rpm. The deposition time of the transition layers was 40 minutes; for the TiCrAlN layers it was varied between 120 and 260 minutes.

Table 1. Technological parameters of the main TiCrAlN layer deposition.

Sample	Target current			Bias	Ar	N_2	T	P	Deposition time
No	**Cr [A]**	**Ti [A]**	**Al [A]**	**[V]**	**[sccm]**	**[sccm]**	**[°C]**	**[Pa]**	**[min]**
#01	6	8	4	-70	25	19	200	0.19	120
#02	6	8	4	-70	25	21	200	0.20	120
#03	4	8	6	-70	25	25	190	0.19	120
#04	4	8	6	-70	25	28	193	0.33	150
#05	4	8	6	-70	25	28	190	0.33	230
#06	4	8	6	-70	25	26	180	0.20	260

The coating thickness was measured by the ball-cratering method using a Compact Calotester CAT[2] (Anton Paar, Austria). Based on the formed craters diameters, the sublayer thickness and the total coating thickness were calculated by the specialized software of the equipment.

The mechanical properties of coated samples were characterized using a Compact Platform CPX MHT/NHT device (Anton Paar, Austria). The data for the nanohardness and Young's modulus were determined by the Oliver and Pharr method applying a 5-s pause. The Poisson ratio of the samples was set at 0.25. The adhesion of the coating and interfaces was evaluated by a micro-scratch tester (MHT) (Anton Paar) with a load progressively increasing from 0.5 N to 30 N. The scratch length was 1 mm at a velocity load of 1 N/min.

The surface observation and composition analysis were performed on a Hitachi SU 5000 Schottky field emission scanning electron microscope (Hitachi, Japan) equipped with an energy dispersive X-ray microanalysis system (Thermo Scientific, USA). Surface morphology images were obtained in secondary electrons (morphology contrast) and back-scattered electrons (density contrast).

3. Results and discussion

Figure 1a) shows the result from the calotest conducted on sample 03. The sublayer stack of Cr/CrN and the TiCrAlN layers were identified after the test. Calotest measurements were carried out on all six samples, with the thickness measured of the coatings varying between 1.4 μm and 2.5 μm. The maximal nanohardness and the modulus of elasticity measured as depending on the indentation depth are illustrated in figure 1 b) at loading steps of 10 mN, 20 mN, 30 mN, 40 mN, 50 mN and 100 mN. As the trends of the graphs shows, increasing the indentation load reduces the nanohardness and the modulus of elasticity because of the influence of the substrate. Indentation depths over 250 nm exceed the 10 – 20% region where the influence of the substrate could be suppressed. The very shallow depths under 10% also have to be analysed bearing in mind the roughness of the coatings. This is why the values obtained at indentation depths between 145 nm and 245 nm by a 20-mN load were chosen for comparing the different coatings; the results are summarized in table 2. The standard deviation of the measured nanohardness was also calculated.

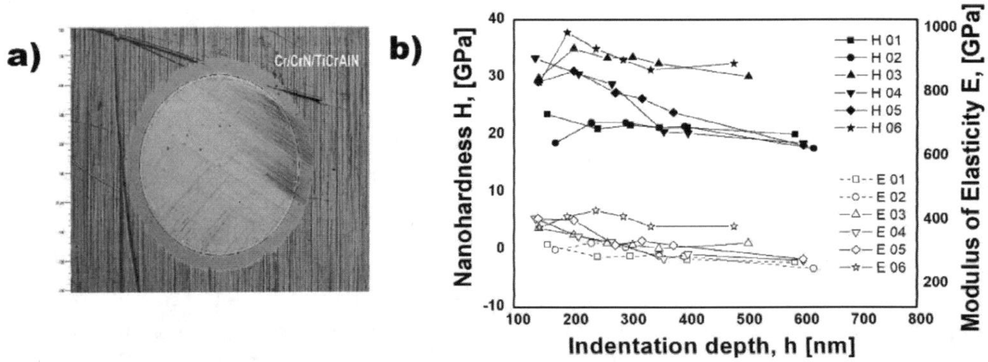

Figure 1. a) Calotest of sample 03 determining the coating thickness, b) nanohardness and modulus of elasticity of the TiCrAlN hard coatings.

Samples 06 and 03 have the highest nanohardness of 37.8 GPa and 35 GPa. The lowest measured nanohardness is 21 – 22 GPa of samples 01 and 02. The calculated average nanohardness in the load interval is also added to the table. The modulus of elasticity is following the trend of the nanohardness. The measured values are between 258 GPa and 384 GPa. The calculated strain-to-failure capability (H/E^*) and the resistance to plastic deformation (H^3/E^{*2}) are also presented. They could be used for preliminary evaluation of the coating's friction and wear behavior. The higher values of the ratios H/E^* and H^3/E^{*2} for samples 03 and 06 indicate their higher wear resistance [13]. The samples deposited with a lower power on the Cr targets and with an increased nitrogen flow have higher values of these ratios, which points to their better mechanical properties.

Table 2. Measured nanohardness of the samples.

Sample	Nanohardness at 20 mN load	Standard deviation	Average nano-hardness 10 – 100 mN	Modulus of elasticity	H/E^*	H^3/E^{*2}
No	[GPa]	[GPa]	[GPa]	[GPa]		[GPa]
#01	21.1	1.2	21.5	258	0.08	0.12
#02	22.1	1.4	20.5	302	0.07	0.10
#03	35.0	3.1	32.4	327	0.10	0.35
#04	30.6	1.3	25.4	322	0.09	0.24
#05	31.1	1.8	26	373	0.09	0.19
#06	37.8	4.5	33.1	384	0.09	0.32

Figure 2 shows the traces of the six scratches made by a progressive load from 0.5 N to 30 N taken at objective magnification ×5. For samples 01 and 02, some slight initial cracks at 12 N and 22 N can be seen; these turn to thin banding cracks at 26 N. The test for sample 03 showed one initial single tensile crack at 20.8 N and serial conformal and tensile edge cracks after 27 N. The tests of samples 04 and 05 produced no cracks and deformations in the range 18 – 30 N. The track of the scratch test of sample 06 is indiscernible on the coating surface and one single crack at 10 N and a few semi-spherical cracks at 23.4 N are visible at higher magnification. The coating surfaces and interfaces were without delamination from the substrate and did not show any buckling or spallation deformations. All the samples have very good adhesion to the substrate. The dependence of the friction coefficient on the nitrogen flow applied during deposition is visualized by the bar graph in figure 2. The samples deposited at a higher nitrogen flow have a lower coefficient of friction. The lowest coefficient of friction with a value of 0.11 was achieved for sample 05.

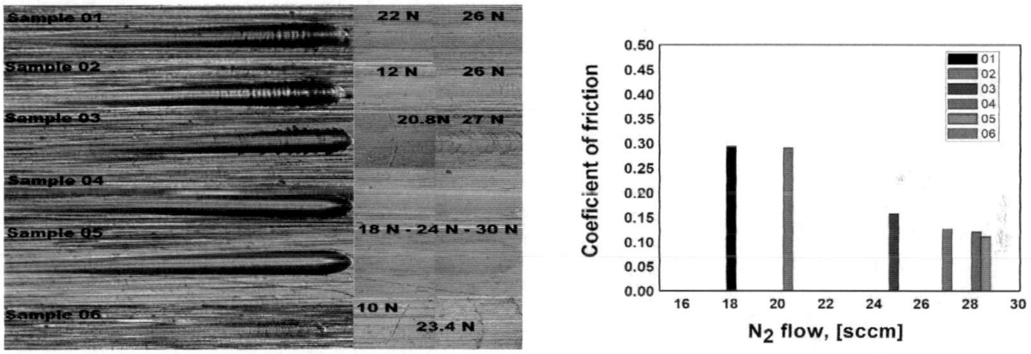

a) b)

Figure 2. a) Photographs of the scratches and **b)** friction coefficient as a function of the nitrogen flow.

The surface morphology of the coatings was studied by SEM (figure 3). The result shows a well-defined grain structure, which is typical of such a type of coatings. The elemental composition of the first four samples is presented in table 3. The almost identical technological regimes of sample 05 and sample 04, and of sample 06 and sample 03 is a premise for their equivalent elemental composition, so that the samples can be divided in three groups. The samples of the first group (samples 01 and 02) have a similar content of Ti, Cr and N and a

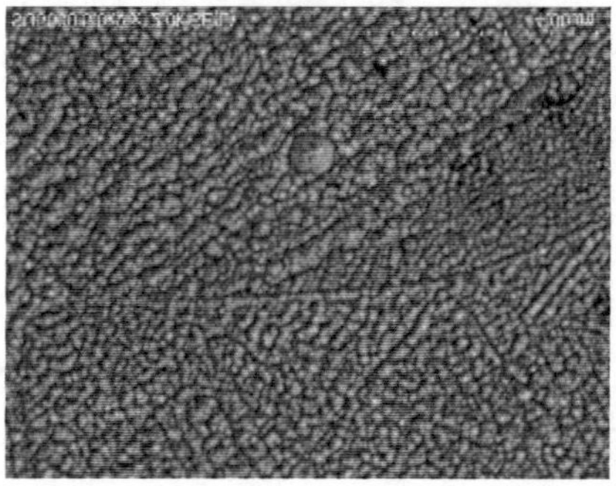

Figure 3. Surface morphology obtained by SEM.

small amount of Al. These samples have the lowest content of nitrogen indicating obvious sub-stoichiometric composition. This could be the reason for the nanohardness of these samples being near the nanohardness of the binary CrN, TiN and AlN layers and for these samples having poor mechanical properties in comparison with the coatings from the other groups.

Table 3. Elemental composition from SEM-EDS analysis.

Sample	Ti [at. %]	Cr [at. %]	Al [at. %]	N [at. %]	SUM [at. %]
01	28.3	41.7	8.2	21.8	100
02	25.5	39.2	8.1	27.2	100
03	23.7	21.6	17.2	37.5	100
04	14.55	34.15	6.55	44.75	100

Compared to the previous one, the samples of the second group (samples 03 and 06) have a slightly larger content of Ti, half as much Cr and an increased nitrogen amount. The higher nanohardness of Ti-based layers and the nitrogen content nearer stoichiometry leads to the increased nanohardness of these samples. The samples of the third group (samples 04 and 05) show a considerable prevalence of Cr over Ti and Al and the highest nitrogen content. The CrN-based layers have improved adhesion and wear properties and, consequently, this group showed the excellent scratch tests results and the lowest coefficient of friction. As the above comparison demonstrates, a Cr- to Ti-based transition was achieved in sample 03.

4. Conclusions

Six samples with quaternary coating layers were deposited and their elemental composition was optimized as regards the technological deposition parameters. A smooth transition from Cr/CrN adhesion layers to TICrAlN layers was achieved. Average nanohardness of 33 GPa was measured. Good adhesion and cohesion strength of the coatings to the substrates was proved. These mechanical properties were reached at the low deposition temperatures of 180 – 200 °C and are comparable with those achieved at deposition temperatures exceeding 200 °C.

Acknowledgments

This work was supported by the European Regional Development Fund within the OP "Science and Education for Smart Growth 2014 - 2020", Project CoC "Smart Mechatronic, Eco- and Energy Saving Systems and Technologies ", No BG05M2OP001-1.002-0023.

References

[1] Pradhan S, Nouveau C, Vasin A and Djouadi M 2005 *Surf. Coat. Technol.* **200** 141–145
[2] Li G, Deshpande P, Li J H and Lin R Y 2005 *Tsinghua Science and Technology* **10** (6) 690–698
[3] Ye Y, Liu Z, liu W, Zhang D, Wang Y, Zhao H and Li X 2018 *RSC Advances* **8** 5596
[4] Sikder A, Irfan I, Kumar A, Durvin R, McDonough M and Smith M 2002 *Mat. Res. Soc. Sym. Proc.* **697** 133–139
[5] Chang Y and Chuang C 2020 *Coatings MDPI* **10** (7) 605
[6] Yang S, Wiemann E and Teer D 2004 *Surf. Coat. Technol.* **88–189** 662–668
[7] Bao M, Xu X, Sun H, Teer D and He J 2007 *Materials Science* **353–358** 1720
[8] Hoche H, Pusch C and Oechsner M 2020 *Surf. Coat. Technol.* **391** 125659
Khadem M, Penkov O, Yang H and Kim D 2017 *Friction* **5** (3) 248–262
Ghufran M, Uddin G, Khan A, Hussein H, Khrshid K and Arafat M 2019 *Advances in manufacturing II* **4** 451–460
[9] Gåhlin R, Bromark M, Hedenqvist P, Hogmark S and Håkansson G 1995 *Surf. Coat. Technol.* **76–77** 174–180
[10] Luo Q, Yang S, Cooke K and Magowan S 2012 *20th congress of international federation for heat*

treatment and surface engineering Beijing, China, 23-25 October 2012 Physics Procedia **50** 508–504

[11] Yasar P and Barnett S 1998 *J. Vac. Sci. Technol* A **16** 2913

[12] Martinez E, Romero J, Lousa A and Esteve J 2003 *Surf. Coat. Technol.* **163–164** 571–577

[13] Chauhan A and Vaish R 2013 *Materials and design* **44** 240–245

Spectral readout of SPR excited in diffraction grating

E Belina[1,3], V Mankov[1], H Kisov[1], T L Dimitrova[2] and G Dyankov[1]

[1]Acad. Jordan Malinowski Institute of Optical Materials and Technologies, Bulgarian Academy of Sciences, bld. 109, Akad. G. Bontchev Str., 1113 Sofia, Bulgaria
[2]Paisii Hilendarski University of Plovdiv, 24 Tzar Asen Str., 4000 Plovdiv, Bulgaria

E-mail: evdokiyabelina@yahoo.de

Abstract. The surface plasmon resonance (SPR) sensing has emerged as a powerful tool for measuring the binding of analytes to functionalized surfaces and immobilized (bio)molecules. We describe the development of a grating for spectral readout of SPR excited in a diffraction grating. The SPR chips based on a metallized diffraction grating have two main problems: undesired reflection from the non-metallic substrate and the half-width of the resonance curve, which determines the accuracy of the SPR sensors [1]. In this report, we demonstrate that a few metal layers with a fixed thickness deposited on a polycarbonate substrate in a certain sequence represent a structure as required for high-quality SPR chips. The structure can provide correct spectral readout, which was confirmed by our numerous experiments. Also, the layers deposition can be performed in one technological step by sputtering or by thermal evaporation in vacuum. We discuss the dependence of the SPR parameters on the multilayer structure parameters.

1. Introduction

SPR chips based on a diffraction grating can have two types of structures: metal or metallized [2]. The former is expensive due to the cost of the precious metals used, such as gold and silver, and are also difficult to manufacture. The second consists of a non-metallic base on top of which a thin metal layer is deposited for SPR excitation. Commonly, polycarbonate substrates are used of commercial CD-R or DVD disks covered with a metal layer with a thickness of 80 – 100 nm, [3]. The SPR grating chips based on polycarbonate are very suitable since they are produced by the inexpensive injection molding technology that is the state of the art for production of compact disks. However, this structure requires a variable angle readout, which implies some technical and experimental difficulties related to the accuracy and the measurement time. Spectral readout is not possible since a spectrally-modulated reflection from the polycarbonate-gold interface poses concerns about the correctness of the measurement [4]. This problem motivated us to search for possible solutions for improving the characteristics of such SPR chips. As was confirmed by our numerous experiments, used phthalocyanine-based dyes in CD and DVD disks having a proper absorption band can provide a correct spectral readout [5].

[3] To whom any correspondence should be addressed.

Content from this work may be used under the terms of the Creative Commons Attribution 3.0 licence. Any further distribution of this work must maintain attribution to the author(s) and the title of the work, journal citation and DOI.
Published under licence by IOP Publishing Ltd

2. Results and discussion

2.1. SPR chips based on metallized polycarbonate diffraction grating

The detection of SPR in diffraction gratings is based on measuring the angle at which the resonance occurs and its change in the process of detecting the analyte; the excitation is performed by monochromatic light.

In our experiments, we performed spectral detection, which has not been done so far. In this case, the SPR is excited by polychromatic light and the spectral resonance shift is measured when the binding reactions of the ligand and the analyte are detected. The main advantage is the speed of measurement and the simple experimental realization – there is no need for a precise mechanical device to set up the angle of incidence and the angle of reflection.

The experiment included applying gold coatings on the surface of polycarbonate pads for CD-R discs, which are used as diffraction gratings. The grating period was about 1.5 µm and the height of the diffraction slits, 50 nm. In several experiments, we deposited by vacuum evaporation gold layers on the polycarbonate base. The layers obtained had thicknesses in the range 20 – 200 nm and different adhesion, having been deposited under different conditions – evaporation time and rate and substrate temperature. In the working process, we established an evaporation regime that provided adhesion allowing the diffraction gratings to be reused. The main problem was that no plasmon resonance was observed in the diffraction gratings thus prepared. Figure 1 shows the measured normalized reflection under which conditions SPR should be observed in such gratings. Instead of reflection minima, peaks are observed at wavelengths close to those at which an SPR is expected to occur.

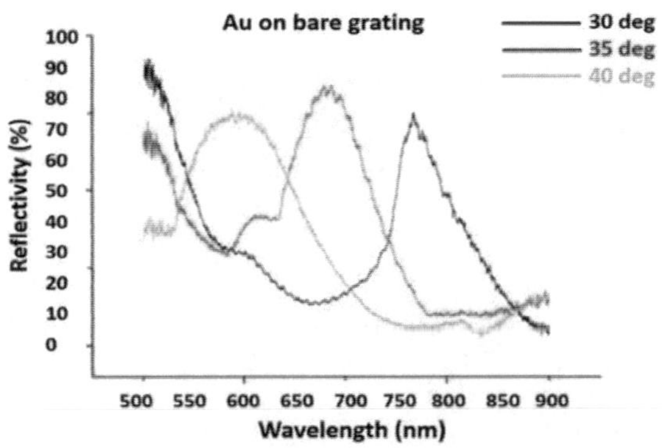

Figure 1. Normalized reflection measured on the surface of a gilded polycarbonate diffraction grating. The applied gold layer is 110-nm thick. Measurements are performed at angles of incidence of 30°, 35° and 40°.

It has been suggested that the underlying reason for this is the SPR excitation by polychromatic light [6]. Probably, a plasmon wave is excited in the gold layer and reflected by the polycarbonate substrate. Since the plasmon wave does not change due to external influences on the surface, it has been assumed that it is excited at the metal-polycarbonate interface. In any case, the presumed cause is the reflection from the polycarbonate substrate.

2.2. Elimination of undesired reflection from the substrate

To eliminate the unwanted reflection, experiments were performed by depositing layers of absorbent on the polycarbonate [7]. A number of dyes were studied – figure 2 shows the absorption spectrum in the visible-light range of the most suitable one (xylene cyanol).

Figure 3 shows the reflection from the same polycarbonate CD-R pad with a deposited dye layer and a vaporized gold layer with a thickness of 110 nm. Obviously, there is an effective SPR excitation when the reflection from the substrate is eliminated. It turned out that commercial CD-Rs / DVDs with deposited dyes have similar characteristics [6]. We currently use commercial disks coated with a dye layer, on which we deposit the gold coating.

However, this does not solve the problem about the measurement accuracy, since the excited SPR has an FWHM of about 45 nm, while theoretical analyses have shown that it should be possible to excite a plasmon resonance with a half-width of the resonance curve of about 9 nm. The accurate SPR sensing by spectral readout requires gratings optimized for a given spectral region.

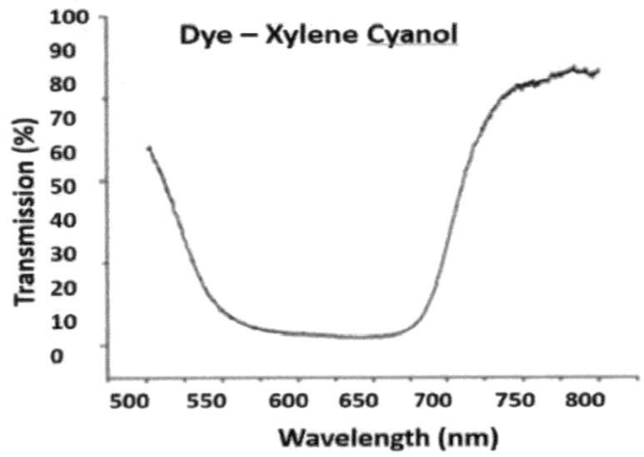

Figure 2. Transmission spectrum of the xylene cyanol dye deposited by spin coating on the surface of polycarbonate diffraction grating; the layer thickness is approximately 5 nm.

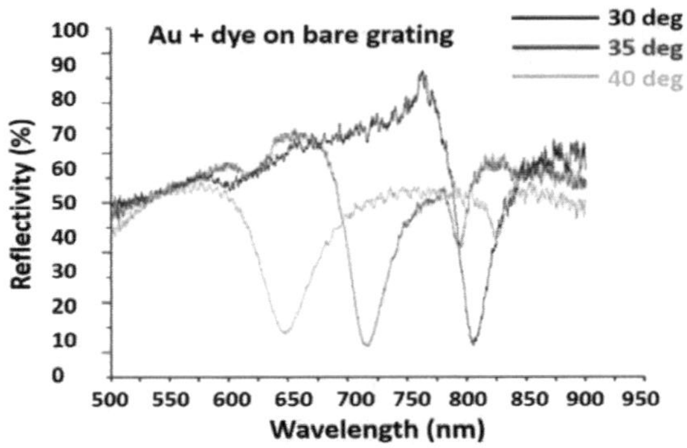

Figure 3. Effective SPR excitation on the surface of a polycarbonate diffraction grating coated with absorbing layer topped by a 110-nm thick gold layer; the measurements are performed at angles of incidence of 30°, 35° and 40°.

2.3. Spectral readout of SPR using metal instead of dye layer

The biosensors' accuracy is determined by the resonant half-width. In addition, depositing a dye layer requires spin-coating, and that of metal layers, such as gold, vacuum evaporation. This makes the production of such sensors a multistage time-consuming process, which prompted us to search for a suitable dye substitute.

We found that by depositing a thin layer of chromium instead of a dye, we could eliminate the unwanted reflection from the polycarbonate substrate. In addition, among the precious metals, silver is also very suitable for SPR excitation. Thus, in this study, several polycarbonate diffraction gratings were coated with a thin layer of chromium topped by a thin silver layer. The deposition of these layers was performed by thermal evaporation in vacuum. After tempering at room temperature for about 20 minutes, the first measurement was performed. The results for the excited plasmon resonance are shown in figure 4 – black curve. The resonance depth was measured to be 85 % and the half-width, 8 nm, which represents an excellent result. The measurement was repeated one week later under the same conditions: the result is also shown in figure 4. The red resonance curve differs from the one measured immediately after evaporation. The difference is expressed in both the intensity and the half-width of the curve. As can be seen, resonance parameters have worsened.

After several repetitions of Cr and Ag evaporation and measurements, we obtained the same results. Therefore, we concluded that the SPR excitation immediately after evaporation of the two metal layers is highly efficient, which is very suitable for detection. However, this efficiency is not maintained over time, probably due to the characteristic oxidation of silver.

This led us to perform another experiment – using thermal evaporation in vacuum we deposited sequentially the following layers: chromium with a 3-nm thickness, silver with a 110-nm thickness and, on top, a gold layer with a thickness ranging from 20 nm to 60 nm. The excitation and measurement of SPR were performed following the same procedure – immediately after evaporation and tempering (figure 5, black curve) and one week later (figure 5, red curve). As can be seen, a high-efficiency plasmon resonance (85 %) is excited and the half-width of the resonance curve is about 19 nm. These results are significantly better than those shown above when using a dye (45 nm); although not as good as the case with silver layers (9 nm), they are quite acceptable for detection. In addition, the resonance parameters do not change over time, which is extremely important and is due to the fact that the problem of silver oxidation has been eliminated.

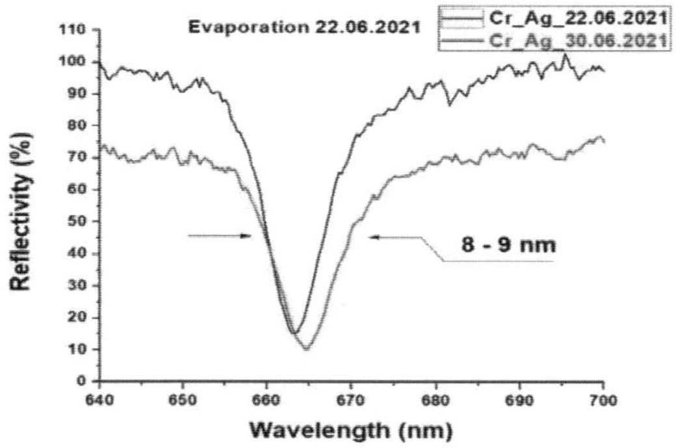

Figure 4. SPR excited on the surface of a polycarbonate diffraction grating with deposited chromium (100 nm) and silver (30 nm) layers; the measurements are performed after evaporation (black curve) and one week later (red curve).

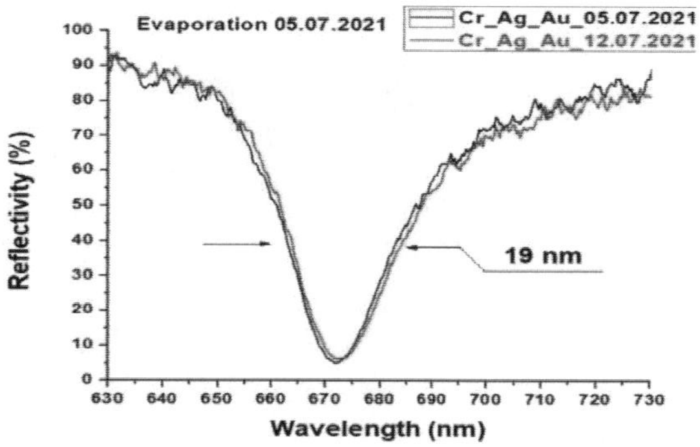

Figure 5. SPR excited on the surface of a polycarbonate diffraction grating with deposited layers of chromium (3 nm), silver (110 nm) and gold (30 nm) layers; measurements are performed after evaporation (black curve) and one week later (red curve).

3. Methods and equipment

3.1. Deposition methods
The metal layers were deposited by using a high-vacuum installation for thermal evaporation (Hochvacuum Dresden B.55.3-3) with a computer control developed in IOMT-BAS. The evaporation parameters were the following: total evaporation time 27 – 32 minutes, including reaching the required vacuum level (2×10^{-5} – 3×10^{-5} mbar). Before evaporation, the substrates were heated to 80 °C; their temperature was monitored and maintained throughout the evaporation process.

The deposition of dye layers was performed by spin coating using a WS-650 Laurell Technologies device with a maximum of 12 000 rpm and vacuum fixation of the samples. In the present work, a four-step procedure was followed: 1. Placing 80 microliters of dye on the sample via drop casting; 2. Rotating the sample at 350 rpm for 2 seconds; 3. Accelerating up to 1700 rpm for 70 seconds; 4. Drying of the layer for 3 minutes.

3.2. Thickness measurement
The thickness of the obtained layers was measured as follows: Smooth polycarbonate plates (2 pieces) of the same dimensions were covered simultaneously with each series of diffraction gratings (20 pieces). After drying or tempering, depending on the deposition method, the thickness of the layers deposited on these plates was measured using a Zeta 203D Optical Profiler and a Talystep Filmetrics F20 (Taylor Hobson,112/1037 M-S).

3.3. SPR excitation and measurement
A G1.5 type goniometer θ-2θ was used for exciting the SPR and measuring the reflection (figure 6).

The deposition of layers of chromium, silver and gold was repeated several times; the results obtained were processed and summarized in figure 5. This gives us a reason to conclude that the SPR excitation is not only highly efficient, but its optical characteristics remain stable over time.

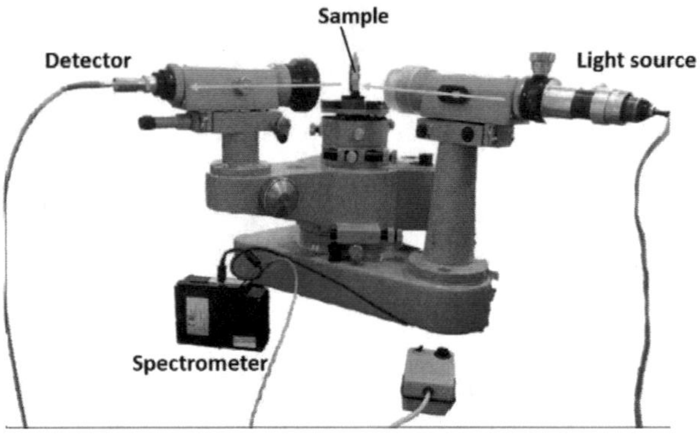

Figure 6. G1.5 goniometer used to measure the reflectivity from the SPR chip; angles of incidence of 30°, 35° and 40°.

4. Conclusions

In summary, the unwanted reflection from the polycarbonate substrate was successfully eliminated through Cr deposition; the parameters of the SPR chips were improved and stable over time. This fulfilled the purpose of the present study, namely, spectral detection of SPR. In addition, the deposition of metal layers can be performed in one cycle, which significantly reduces the time and costs for production of such SPR chips.

Acknowledgments

This work was supported by the Bulgarian Ministry of Education and Science under the National Scientific Program for young scientists and postdoctoral fellows, DCM 577/2018 and 271/2019.

References

[1] Homola J 2006 *Surface Plasmon Resonance Based Sensors* (Berlin: Springer)
[2] Estrela P, Bhalla N, Jolly P and Formisano N 2016 *Essays Biochem* **60 (1)** 1–8
[3] Dou X, Phillips B M, Chung Pei-Yu and Jiang P 2012 *Opt. Lett.* **37** 3681–3683
[4] Sun Y, Sun S, Wu M, Gao S and Cao J 2018 *RSC Advances* **8** 48
[5] Long S, Cao J, Wang Y, Gao S, Xu N, Gao J and Wan W 2020 *Sensors and Actuators Reports* **2** 1 100016
[6] Su W, Luo Y, Ding Y and Wu J 2021 *Sensors and Actuators A: Physical* **330** 112858
[7] Tedesco A C, Primo F L and Beltrame M 2016 *Reference Module in Materials Science and Materials Engineering* (Elsevier ISBN 9780128035818)

Improvement in the gas-sensing properties of a $ZnO:Zn_2TiO_4$ composite sensor by light irradiation

G Atanasova[1,3], T Dilova[1], A Dikovska[2], Ru Nikov[2] and N Nedyalkov[2]

[1]Institute of General and Inorganic Chemistry, Bulgarian Academy of Sciences, Acad. G. Bonchev Str., bld. 11, 1113 Sofia, Bulgaria
[2]Institute of Electronics, Bulgarian Academy of Sciences, 72 Tsarigradsko Chaussee, Sofia 1784, Bulgaria

E-mail: genoveva@svr.igic.bas.bg

Abstract. In the present study, we demonstrate the application of highly porous $ZnO:Zn_2TiO_4$ nanocomposites as gas sensors. The composite sensor elements were prepared by pulsed laser deposition performed in air at atmospheric pressure. Gas detection was achieved through irradiating the sensor surface with UV light. The gas-sensing properties of the composite sensor were studied under exposure to ammonia, carbon monoxide, ethanol, and acetone. The highest response rate of the sensor element was observed in the case of exposure to carbon monoxide, with detection below 1 ppm being demonstrated. The influence was investigated of the additional light irradiation on the response and the times of response and recovery of the sensor elements. It was found that illuminating the sensor surface with blue light increased the sensor response to all tested gases, while a red-light irradiation enhanced the sensor element selectivity to carbon monoxide.

1. Introduction

In recent years, semiconductor metal-oxide gas sensors have attracted a great deal of attention due to the large number of advantages, such as simple manufacturing techniques, low cost, fast response and recovery [1]. However, the commercial semiconductor metal-oxide gas sensors usually operate at higher temperatures (150 – 500 °C), which limits their wide application. New approaches have been explored to keep the gas sensor running at lower or room temperature [2, 3]. Light irradiation is a widely used such an approach improving the gas-sensing properties and reducing the operating temperature [4-6]. Although ZnO has been studied extensively compared to other metal-oxide materials for gas sensors [7], a large number of studies are still being conducted dealing with the zinc oxide gas sensors performance under light irradiation in order to reduce the operating temperature. Fan et al. found that UV light improves the sensor sensitivity and rates of response and recovery [8]. By studying the photo-responsive behavior of ZnO, they observed that electrons generated by UV light promote oxygen adsorption and form photo-induced oxygen ions [O_2^- (*hv*)]. These oxygen ions are responsible for gas sensing at room temperature and promise improved sensor performance through further optimization [8]. In addition to UV light and visible light, such as blue, red, green and white, are also being used to enhance the sensor performance. Our previous research has shown that irradiating the Ag-ZnO sensor surface by UV and red light improves the response to CO exposure

[3] To whom any correspondence should be addressed.

Content from this work may be used under the terms of the Creative Commons Attribution 3.0 licence. Any further distribution of this work must maintain attribution to the author(s) and the title of the work, journal citation and DOI.
Published under licence by IOP Publishing Ltd

while suppressing the response to NH_3, ethanol and acetone [9, 10]. Several studies have reported on the properties of the ZnO-based gas sensor being remarkably improved by introducing a TiO_2 dopant into the sensor films [11]. The formation of composite oxide nanostructures with specific structure and morphology can favor a lower temperature gas-sensing performance [12]. Wongchoosuk et al. used three nanowire sensors to detect NO_2, CO, O_2 and O_3 at room temperature [13]. Ultraviolet illumination of oxide surfaces can generate oxygen species (O_2^-) that are adsorbed on the surface of nanowires and interact with gas molecules. The results show that pure ZnO nanowires demonstrate a cross-sensitivity at room temperature, while $ZnAl_2O_4$ nanowires show high sensitivity and selectivity to NO_2. Zn_2TiO_4 nanowires show improved gas response to NO_2 and CO at low concentrations (100 – 300 ppb) due to the large surface area [13]. In a previous work, we reported on the fabrication of a composite ZnO:Zn_2TiO_4 nanostructure and demonstrated its photocatalytic properties [14].

In the present work, we focused our attention on the gas-sensing properties of ZnO:Zn_2TiO_4 composites under light irradiation. The sensors were activated by irradiating their surface with UV light. The gas-sensing properties of the composite sensors were studied under exposure to NH_3, CO, ethanol, and acetone. The effect was investigated of an additional light irradiation on the response and the response and recovery times of the sensor elements. It was found that additional irradiation of the sensor surface with blue light increased the sensor response to all tested gases, while a red-light irradiation enhanced the sensor element selectivity to CO exposure.

2. Experimental

The sensor elements were prepared by PLD in open air. Details for their fabrication have already been presented in [9, 10, 15]. A target used for ablation was synthesized from a mixture of ZnO and TiO_2 powder with a ratio of 10 wt % TiO_2 in ZnO. The experimental setup used for the gas-sensing measurements was reported in [9, 10]. Briefly, the gas-sensing was carried out by measuring the resistance changes at room temperature and relative humidity of 46 % under exposure to ammonia (NH_3), ethanol (C_2H_6O), acetone ($(CH_3)_2CO$), and carbon monoxide (CO) with a concentration of 30 ppm. The test gas was injected by a microsurgical syringe in the reference gas flow (N_2). The samples were irradiated by a UV LED ($\lambda = 396$ nm) with intensity in the range of 34 – 76 mW/cm^2 as the distance between the LED and the sample surface was kept at 1 cm. A set of experiments were performed by additional irradiation of the sensor surface by a red ($\lambda = 633$ nm) or a blue ($\lambda = 470$ nm) LED simultaneously with the UV radiation. The intensity used for additional irradiation was in the range of 18 – 46 mW/cm^2 for the red and 25 – 46 mW/cm^2 for the blue LED. The gas response of the sensor element to the test gas was defined as $S = (R_g–R_0)/R_0) \times 100$, where R_g and R_0 are the sensor resistance in the test gas and in the reference gas. The response time was defined as the time needed to reach 90 % of the maximum response when the gas was turned on, with the recovery time being the time necessary to reach 10 % of the maximum response when the test gas was turned off.

3. Results and discussion

The composites were prepared by PLD in open air which ensured the formation of highly porous 3D structures composed of nanoparticles and/or nanoaggregates (figure 1) [15-17]. These nanoparticles and/or nanoaggregates are crystalline and their phase composition was identified as ZnO and Zn_2TiO_4 [14]. The ratio between the ZnO and Zn_2TiO_4 phases is 85 % to 15 % [14]. Typical for metal oxides (produced by PLD in open air) with such a surface composition is an increase in the amount of components assigned to the oxygen deficiency states and to the adsorbed species [15]. In summary, the sensor element had a composite structure consisting of ZnO and Zn_2TiO_4 nanoparticles with ZnO being the predominant phase. The nanoparticles (with the highest surface to volume ratio) were arranged into highly porous structures with an increased amount of oxygen deficiency states and adsorbed species on their surface, meaning that the as-prepared samples were suitable for gas sensor application. The gas sensor properties of the samples were studied under light irradiation. The gas detection regime was achieved after UV light irradiation. As previously reported and discussed, the light creates electron-hole pairs in the semiconductor [18]. The photo-generated electrons enhance the conductivity by transiting to the conducting band, i.e., the sample's resistance decreases, while the photo-generated holes react with the negatively charged oxygen ions adsorbed on the surface; the oxygen species are thus photo-desorbed from the surface [19].

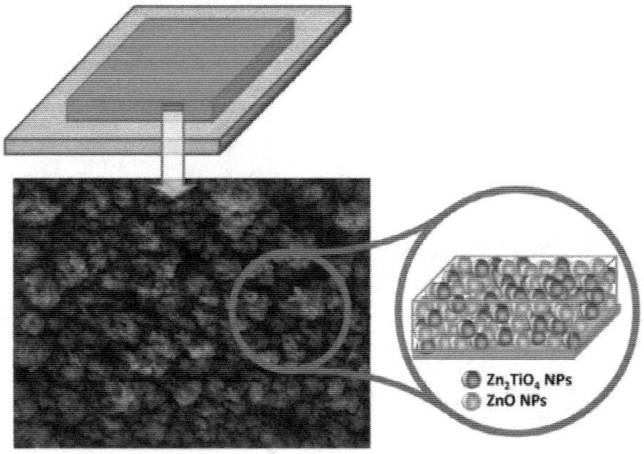

Figure 1. SEM image and schematic view of the sensor element.

The dependence of the $ZnO:Zn_2TiO_4$ sensor response to different gas exposures on the intensity of the UV light irradiation is presented in figure 2. As seen, the sensor reacts to all tested gases, but its response to CO exposure is significant. The highest sensor response to CO and the lowest response to NH_3, acetone and ethanol occurred at the UV intensity of 60 mW/cm^2. Therefore, the highest sensor element selectivity with respect to CO was achieved at 60 mW/cm^2 UV irradiation intensity. This value was chosen as an optimal for UV irradiation and all further gas-sensing experiments were performed with it.

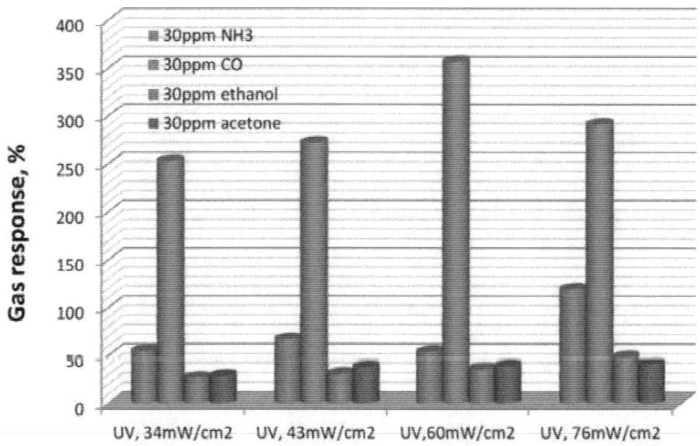

Figure 2. Dependence of the $ZnO:Zn_2TiO_4$ sensor gas response on the UV light irradiation intensity.

Figure 3 presents the sensor response to different gases under simultaneous irradiation with UV and blue light. The dependence of the gas sensor response on the blue light intensity is also shown. As seen, increasing the blue light intensity increases the sensor response to NH_3, acetone and ethanol compared to the sensor response under UV irradiation only. In the case of CO exposure, the sensor response reached its highest value at 36 mW/cm^2. The increase of the sensor response under additional blue light irradiation is associated with the release of surface active sites by desorption of oxygen [6, 20]. Further, figure 3 illustrates the sensor response to different gases under simultaneous irradiation with UV and red light. The influence of the red-light intensity on the sensor response is also shown. In general, the red light suppresses the sensor response to all tested gases. It is well known that the metal oxides exhibit different reaction rates to different gases, so that by varying the sensor temperature one could find conditions suitable for enhanced responses to different gases [21]. However, the sensor selectivity to CO is higher (i.e., the sensor response to CO with respect to the response to the other gases) compared to the sensor selectivity obtained by UV only or UV and blue light irradiation. The value of 20 mW/cm^2 was found to be the optimal one regarding the sensor selectivity to CO. It should be mentioned that the sensor response to all tested gases was improved substantially by irradiating the sensor surface with red light with intensity of 46 mW/cm^2. This improvement is associated with the local heating effect, which probably enhances the rates of adsorption/desorption of gas molecules on/from the sensor surface.

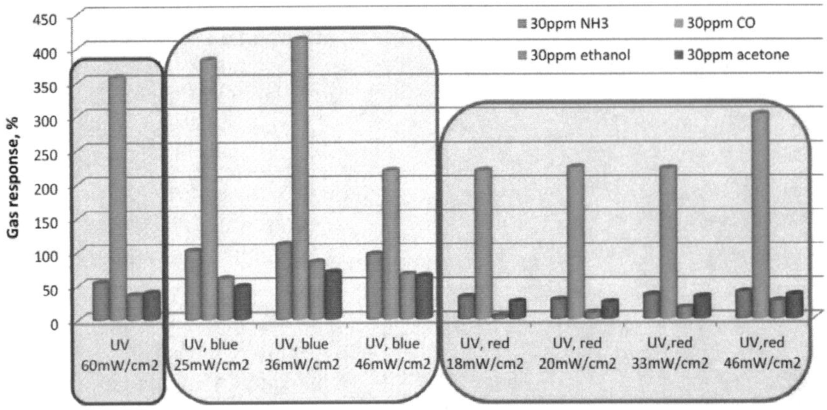

Figure 3. Gas response of $ZnO:Zn_2TiO_4$ sensor element under light irradiation.

The response of the sensor element to different concentrations of CO under optimized UV (60 mW/cm^2) and red (20 mW/cm^2) light irradiation is demonstrated in figure 4. As seen, this allows a concentration as low as 250 ppb to be detected by the $ZnO:Zn_2TiO_4$ sensor element at room temperature. As the CO concentration is raised from 250 ppb to 30 ppm, the sensor response increases from 4 % to 284 %.

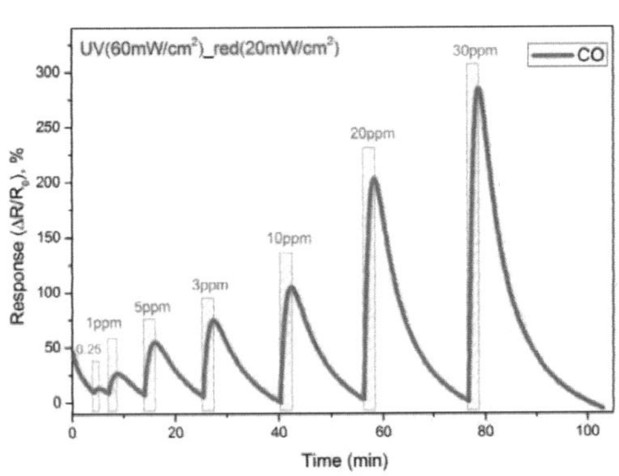

Figure 4. $ZnO:Zn_2TiO_4$ sensor response to different CO concentrations under simultaneous UV and red light irradiation.

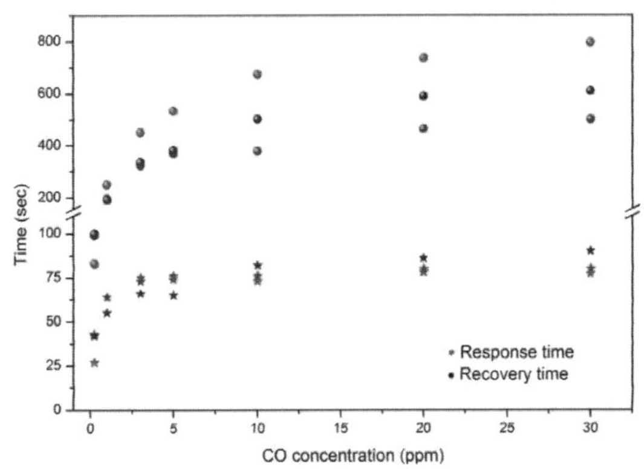

Figure 5. $ZnO:Zn_2TiO_4$ sensor response and recovery times for different CO concentrations under UV (violet), simultaneous UV and blue (blue), and simultaneous UV and red (red) light irradiation.

Furthermore, we also studied the light irradiation effect on the response and recovery times of the sensor element. As seen in figure 5, the response and recovery times increase with the increase of CO concentration, since the higher number of the gas molecules need more time to react with the adsorbed ionized oxygen on the sensor surface. It seems that the time for response of the sensor element under CO exposure is not affected by the light irradiations applied. In the case of sensor recovery time after CO exposure, a gradual rise was seen of the recovery time under UV irradiation compared with the recovery time registered under UV and additional light irradiation. Irradiating simultaneously the sensor surface with UV and red light increased the sensor recovery time after CO exposure.

4. Conclusions

Composite structures based on ZnO:Zn_2TiO_4 nanoparticles were fabricated by PLD in open air and the possibility for their application as gas sensor elements was demonstrated. The sensors were activated by irradiation of their surface with UV light. The effect was studied of additional light irradiation of the sensor surface on the gas-sensing properties. It was found that additional blue light irradiation increases the response to all tested gases, which is likely due to the release of surface active sites by desorption of oxygen. When the sensor surface was additionally irradiated by red light, the selectivity of the sensor element to CO was substantially improved. A CO concentration as low as 250 ppb could be detected by the ZnO:Zn_2TiO_4 sensor element at room temperature under optimized UV and red light irradiation.

Acknowledgments

The authors acknowledge the financial support of project KP-06-N37/20 "Formation and physical properties of composite nanostructures of metal oxides and noble metals" under the Competition for Financial Support of Basic Research Projects 2019 Program of the Bulgarian National Science Fund.

References

[1] Dey A 2018 *Materials Science & Engineering* B **229** 206–217
[2] Sowmya B, John A and Panda P K 2021 *Sensors International* **2** 100085
[3] Li Z, Li H, Wu Z, Wang M, Luo J, Torun H, Hu PA, Yang C, Grundmann M, Liu X and Fu YQ 2019 *Mater. Horiz.* **6** 470
[4] Geng Q, He Z, Chen X, Dai W and Wang X 2013 *Sensors and Actuators* B **188** 293–297
[5] Zhu L and Zeng W 2017 *Sensors and Actuators:* A **267** 242–261
[6] Kumar R, Liu X, Zhang J and Kumar M 2020 *Nano-Micro Lett.* **12**:164
[7] Bhati V S, Hojamberdiev M and Kumar M 2020 *Energy Reports* **6** 46–62
[8] Fan S-W, Srivastava A K and Dravid V P 2009 *Appl. Phys. Lett.* **95** 142106
[9] Dilova T, Atanasova G, Dikovska A Og and Nedyalkov N N 2020 *Appl. Surf. Sci.* **505** 144625
[10] Atanasova G, Dilova T, Dikovska A Og and Nedyalkov N N 2020 *J. Phys.: Conf. Ser.* **1492** 012051
[11] Chaiyo P, Makhachan C, Nutariya J, Thiabgoh O, Sumran S and Pukird S 2019 *J. Phys.: Conf. Ser.* **1259** 012005
[12] Gui Y, Li S, Xu J and Li C 2008 *Microelectronics Journal* **39** 1120–1125
[13] Wongchoosuk C, Subannajui K, Wang C, Yang Y, G¨uder F, Kerdcharoen T, Cimalla V and Zacharias M 2014 *RSC Adv.* **4** 35084
[14] Dikovska A, Gancheva M, Nikov Ru, Avdeev G, Iordanova R and Nedyalkov N 2021 *J. Phys.: Conf. Ser.* **1859** 012005
[15] Atanasova G, Dikovska A Og, Dilova T, Georgieva B, Avdeev G V, Stefanov P and Nedyalkov N N 2019 *Appl. Surf. Sci.* **470** 861
[16] Nikov Ru G, Dikovska A Og, Nedyalkov N N, Atanasov P A, Atanasova G, Hirsh D and Raushenbach B 2017 *Appl. Phys.* A **123** 657
[17] Nedyalkov N, Nikolov A, Atanasov P, Alexandrov M, Terakawa M and Shimizu H 2014 *Opt. Laser Technol.* **64** 41
[18] Chinh N D, Hien T T, Van Do L, Hieu N M, Quang N D, Lee S M, Kim C and Kim D, 2019 *Sens. Actuat. B Chem.* **281** 262–272
[19] Wang J, Shen H, Xia Y and Komarneni S 2021 *Ceramics Internationl* **47** 7353–7368
[20] Zhang C, Boudiba A, De Marco P, Snyders R, Olivier M G and Debliquy M 2013 *Sens. Actuat. B Chem.* **181** 395–401
[21] Yin Z, Wang X, Sun F, Tong X, Zhu C, Lv Q, Ye D, Wang S, Luo W and Huang Y 2017 *Sci. Rep.* **7** 12206

Journal of Physics: Conference Series **2240** (2022) 012017 doi:10.1088/1742-6596/2240/1/012017

Quartz crystal microbalance sensor for NO_2 detection based on electrospun amorphous titanium oxide fibers

K Starbova[1,4], V Georgieva[1], N Starbov[1], P Stefanov[2], B Georgieva[3] and J Lazarov[1]

[1]Institute of Solid State Physics, Bulgarian Academy of Sciences, 72 Tsarigradsko Chaussee, 1784 Sofia, Bulgaria
[2]Institute of General and Inorganic Chemistry, Bulgarian Academy of Science, Acad. Georgi Bonchev Str., bld. 11, 1113 Sofia, Bulgaria
[3]Institute of Electronics, Bulgarian Academy of Sciences, 72 Tsarigradsko Chaussee Blvd., 1784, Sofia, Bulgaria

E-mail: ilka_05@yahoo.com

Abstract. The sensitivity to various concentrations of NO_2 is studied of a 16-MHz quartz crystal microbalance (QCMs) coated by immobilized electrospun amorphous titanium oxide fibers. The morphology, phase and chemical composition of the sensing material are determined by scanning and high-resolution transmission electron microscopy (SEM, HRTEM), transmission high-energy electron diffraction (THEED) and X-ray photoelectron spectroscopy (XPS). Resonance frequency shifts are established of the QCM-TiO_2 system within a broad interval of NO_2 concentrations. Experimental evidence is found for a reversible gas sorption at the lowest tested gas concentration of 50 ppm. Additional sensor response is established to gradually increasing NO_2 concentrations up to 5000 ppm. It is demonstrated that electrospun amorphous titanium oxide fibers are promising sensing materials for NO_2 detection.

1. Introduction

In addition to the broad spectrum of applications in the fields of photocatalysis, medical implants, etc., solid-state titanium-oxide materials are successfully used as a functional medium in various gas and vapor sensors [1-3]. The thin-film form is usually preferred as a sorption medium of various volatile agents [4]. The easily controlled reversible physisorption allows one to monitor the agent concentration/sensing property dependence even at small amounts of pollutants.

In the last two decades, a relatively new class of inorganic nanomaterials has been synthesized via electrospinning. This inexpensive, simple, exceptionally versatile and refined fabrication technique has been successfully applied for fabrication of meso-, micro- and nanostructured ceramic fibers with a high length-to-diameter ratio, specific surface area and porosity [5, 6, 7]. The pioneering studies have shown the sensing ability of TiO_2 electrospun fibers on a Pt electrode for both oxidizing and reducing gases [8, 9]. A fast reversible response at operating temperatures of 300 °C and 350 °C has been established. Further efforts have been focused on electrospun TiO_2-based composite and hybrid nano-heterostructured media [10]. High sensitivity has been achieved toward various gas environments under ambient conditions.

[4] To whom any correspondence should be addressed.

Content from this work may be used under the terms of the Creative Commons Attribution 3.0 licence. Any further distribution of this work must maintain attribution to the author(s) and the title of the work, journal citation and DOI.
Published under licence by IOP Publishing Ltd

At the same time, a large number of novel techniques for environmental monitoring of toxic gases have been developed, thus expanding the variety of engineered sensors. In this respect, the quartz crystal microbalance (QCM) has been proved to be a promising tool for real-time monitoring of the adsorption of nano-amounts of various materials in solid and gaseous state under ambient conditions, together with a good sensitivity and reversibility of the corresponding process [11, 12]. The general relation between the sensitivity and the specific surface area [13] is a reasonable stimulus for further research on the sensing behavior of ceramic materials in fibrous state synthesized via electrospinning.

Bearing in mind the great variety of chemical and phase compositions and morphologies of pure, hybrid and composite electrospun titania fibrous mats [14], the research field of their application seems to be permanently open to developments of novel gas sensors. This is why we combined a QCM with a fibrous sensing medium of ceramic immobilized mats [15]. The latter can be prepared with proven efficiency via optimizing the spinning solutions and the electrospinning conditions.

The present experimental study is a first attempt to fabricate and characterize a QCM sensor for monitoring NO_2 with an electrospun titanium oxide mat as the sorption medium.

2. Experimental

2.1. Test sensor configuration and measurement process

The test sensor device was based on a QCM with polished AT-cut quartz plates with gold electrodes and a resonance frequency of 16 MHz. The sorption properties of the electrospun TiO_2 mats were followed by measuring the resonance frequency shift of the QCM-TiO_2 structures under different NO_2 concentrations. The adsorbed mass was calculated using Sauerbrey's equation [16].

A specially developed set up was used for the measurements, the latter including the following consecutive stages: Purging the chamber with the QCM-TiO_2 structures by dry synthetic air; introducing a certain concentration of NO_2 within the chamber; following the gas sorption process until reaching sorption saturation at relatively constant frequency values; purging again with synthetic dry air thus restoring the sensor and preparing it for the next measurement. In the experiments presented, we used 10 000 ppm nitrogen dioxide in synthetic air atmosphere. The NO_2 concentration and the diluting gas flow within the test chamber were controlled by mass flow controllers. The experiments were carried out at a temperature of 26 °C maintained with an accuracy of ±0.2 °C. The QCM frequency was registered by a Hameg 8123 frequency counter and the data obtained were recorded on a computer [17, 18].

2.2. TiO_2 fibers synthesis

[$Ti(i\text{-}OC_3H_7)_4$] modified to [$Ti(C_3H_7OOH)_4\,(CH_3COO)]_n$ oligomers via glacial CH_3COOH was used as a Ti^{+4} precursor solution [19]. High molecular weight (800 KD) polyethylene oxide (PEO) served as a partner polymer; it could be easily removed from the hybrid fibers via thermal degradation [20]. The chemicals used were products of Sigma Aldrich (≥(99 %). We used 3 wt% of PEO (800 KD) dissolved in 50/50 vol % $CHCl_3/C_2H_5OH$ (98 % Sigma Aldrich). The blend solution with excellent viscoelastic behavior under electrospinning condition was in a weight ratio of 3.0 wt% PEO/Ti precursor solution of 5:1. A syringe with a stainless-steel needle was filled with this solution. The electrospinning conditions of 0.6 kV/cm applied field strength, flow rate of 0.15 ml/hour and ambient temperature of 21 °C resulted in depositing hybrid fibers with good adhesion to the quartz/electrode surface. Finally, the solvents used and the PEO were removed by applying a thermal procedure developed by us [21] resulting in fiber mats immobilized on the quartz/electrode surface. The necessity to preserve the Au electrodes' good adhesion to the quartz crystal carrier restricted the applied maximal temperature of the thermal treatment to 250 °C.

2.3. TiO_2 fibers characterization

The morphology of the electrospun fibers was displayed by SEM (JEOL 6390). The samples' phase composition was evaluated by transmission high-energy electron imaging (THEEM) and electron diffraction (THEED) recorded by a JEOL 200 CX high-resolution transmission electron microscope (HRTEM).

The chemical composition of the fibers was determined by an Ulvac-Phi Scanning ESCA Microprobe Quantum 2000 XPS system.

2.4. TiO_2 fibers characterization

The morphology of the electrospun fibers was displayed by SEM (JEOL 6390). The samples' phase composition was evaluated by transmission high-energy electron imaging (THEEM) and electron diffraction (THEED) recorded under JEOL 200 CX high-resolution transmission electron microscope (HRTEM).

The chemical composition of the fibers was determined by an Ulvac-Phi Scanning ESCA Microprobe Quantum 2000 XPS system.

3. Results and Discussion

Figure 1 shows scanning (a) and bright-field transmission electron micrographs and a diffraction pattern (b) of the surface morphology and phase composition of POE-free electrospun mat of Ti-O fibers deposited on the quartz surface.

The scanning electron micrographs clearly show electrospun fibers uniformly distributed over the substrate surface. They are homogeneous along their length, with a diameter within the range 100 – 1000 nm. The morphology displayed represents a highly developed surface as necessary for the purpose of the present work.

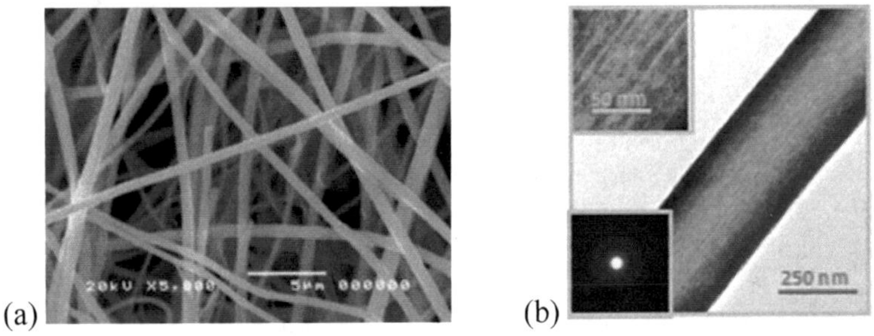

Figure 1. Scanning (a) and bright-field transmission electron micrographs and diffraction pattern (b) of a PEO-free Ti-O electrospun mat.

The transmission electron micrographs and the diffraction pattern unambiguously show typical hollow fibers with amorphous structure. The latter is not surprising bearing in mind that the maximal temperature for the thermal post processing of the as-spun hybrid fibers was restricted to 250 °C. Also, resolvable groove-like features are seen aligned along the fiber length.

The results from the recorded high-resolution XPS spectra of the fibers show a $Ti2p_{3/2}$ doublet and $Ti2p_{1/2}$ peaks at 458.8 eV and 464.5 eV, evidencing a valence state of Ti in the fibers of the order of +4. The O1s spectra could be fitted by peaks around 530.2 eV attributed to oxygen species in the TiO_2.

On this basis, we conclude that the electrospun and thermally-treated fibers were built up of amorphous titania (a-TiO_2).

3.1. Parameters of the QCM-TiO_2 system

The QCM-TiO_2 resonators were characterized by measuring the series resonant frequency (F_s) and the dynamic resistance (R_q), from which the quality factor Q was subsequently calculated. The parameters of the unloaded QCM were F_s = 15.96 MHz, R_q = 17.18 Ω and Q = 25131. The measured and calculated values for the QCM structure loaded with an electrospun TiO_2 fibrous mat were F_s = 15.96 MHz, R_q = 22.10 Ω and Q = 17 810.

Loading the resonator by an electrospun TiO_2 fibrous mat resulted in increasing R_q by 4.92 Ω and lowering the Q-factor by 7 321 compared to the values of the unloaded QCM, where R_q and the Q-factor were 17.18 Ω and 25 131, respectively. Therefore, the parameters deteriorated slightly as a result of the TiO_2 fibrous mat deposition. Further, the multiple exposures of the QCM-TiO_2 samples to various NO_2 concentrations was accompanied by a minor increase of the dynamic resistance, namely, by 2.0 Ω (R_q = 24.1 Ω) resulting in a reduction of the Q-factor value to 16 467. Following multiple exposures to NO_2 and subsequent recovery of the sensor (i.e., NO_2 desorption), negligible deviations in the virgin series frequency were only measured.

The stability of the QCM-TiO_2 resonators' parameters under all test conditions studied led us to the conclusion that they are capable of being used repeatedly as QCM-TiO_2 sensors.

3.2. Sorption properties of the TiO_2-QCM structure

The QCM-TiO_2 structures were tested for NO_2 sensing in a dynamic mode with the gas concentration ranging from 50 ppm to 5000 ppm. Figure 2 presents a typical frequency-time characteristic (FTC) of a QCM coated with TiO_2 during injection of 50 ppm NO_2.

As seen, a dynamic process takes place consisting of several consecutive steps. Initially, after introduction of the gas in the carrying gas flow, the resonance frequency decreases as a result of mass-loading. This rapid response, taking place within 192 seconds with a mean sorption rate of 2.19 Hz/min is related to NO_2 sorption on the fiber surface. Further, the frequency reaches a constant value thus evidencing that a dynamic equilibrium is reached. Once the exposure to NO_2 ceases, its desorption from the QCM-TiO_2 structure begins at a mean rate of 1.2 Hz/min. It is also evident that this process continues until the system fully recovers reaching finally its initial frequency.

The QCM-TiO_2 structure studied was able to detect a concentration of about 50 ppm NO_2 with a minimal frequency change. In this case, the test gas sorption process occurred almost twice as fast as the desorption one. Similar FTCs were measured for several NO_2 concentrations up to 5000 ppm.

As seen in figure 3, increasing the NO_2 concentration tenfold to 500 ppm leads to a change in the sorption mechanism. The FTC clearly outlines fast and slow sorption areas. Initially, the frequency of the QCM-TiO_2 structure decreases by 13 Hz for 350 seconds, corresponding to a sorption rate of 2.24 Hz/min. Then, as time progresses, a reduction to 1.02 Hz/min is seen until a dynamic equilibrium is reached for the NO_2 concentration tested. The total sorption rate measured of 1.68 Hz/min at this concentration is significantly lower than that at 50 ppm (2.19 Hz/min). Obviously, establishing a dynamic equilibrium requires a longer period of time, meaning that the sorption rate decreases as the NO_2 concentration is raised.

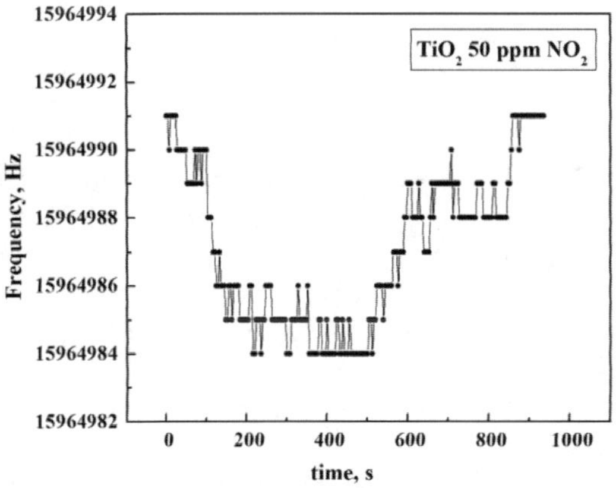

Figure 2. Frequency-time characteristic of a QCM-TiO_2 structure at 50 ppm NO_2.

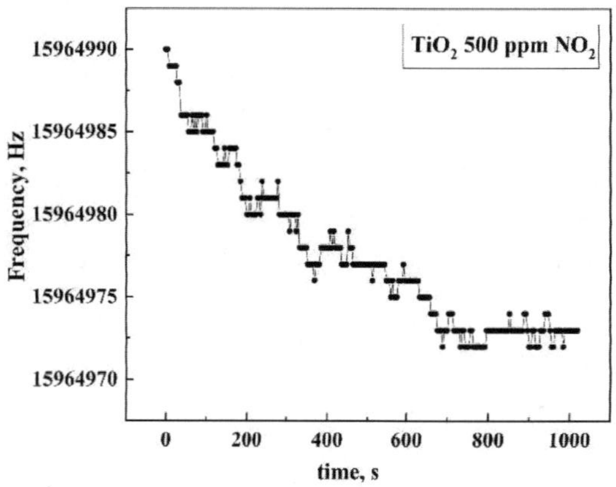

Figure 3. Frequency-time characteristic of QCM-TiO_2 structure at 500 ppm NO_2.

Further, the time evolution of the resonance frequency during exposure to various NO_2 concentrations was recorded. With this purpose, the NO_2 concentration in the test chamber was raised once a dynamic equilibrium was reached at the preceding concentration; the results are shown in figure 4.

The FTCs of the QCM-TiO_2 structure were measured sequentially for each concentration of NO_2 until dynamic equilibrium was established. The results show that the QCM-TiO_2 structure detects the NO_2 concentration variation in the whole range studied. The data clearly demonstrates the presence of a fast and a slow course of the frequency/time dependence after introducing 500 ppm and 2500 ppm NO_2, with the frequency change being lower for the higher concentration. Due to the reduced TiO_2 fibers sorption ability, a negligible frequency change was measured in the interval from 2500 ppm to 5000 ppm.

A frequency variation of 17 Hz to 43 Hz was measured for 500 ppm and 5000 ppm, respectively.

Based on the frequency offset measured for each concentration, the corresponding mass of the sorbed NO_2 was calculated according to Sauerbrey's equation (figure 5). A sorbed mass ranging from 1.53 ng to 9.39 ng was calculated, respectively for concentrations from 50 ppm to 5000 ppm. The resonators used have Au electrodes with a diameter of 0.4 cm; thus, the areal sensitivity varied from 12.18 ng/cm^2 to 74.76 ng/cm^2. A 100-fold increase in the NO_2 concentration resulted in a six-fold increase in the sorbed mass.

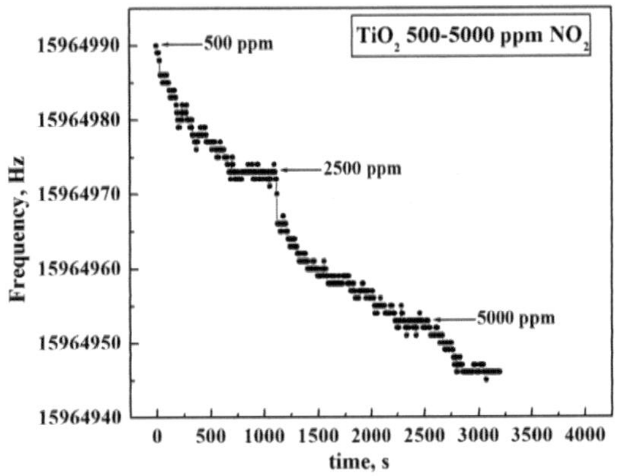

Figure 4. Dependence of the frequency-time characteristic on the NO_2 concentration in the 500 – 5000 ppm range of a QCM coated with electrospun TiO_2 fibers.

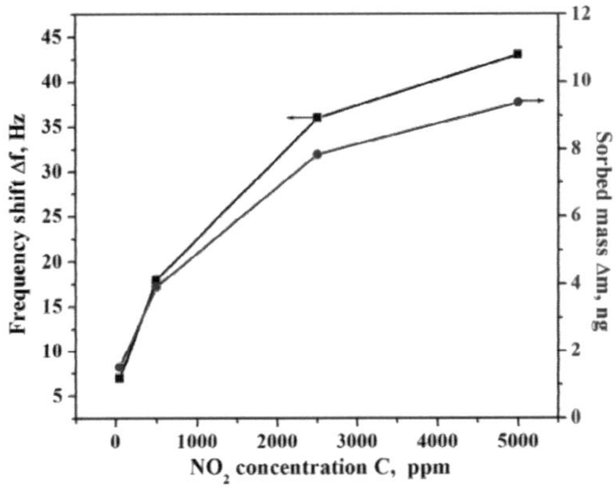

Figure 5. Resonance frequency shift (Δf) and sorbed mass (Δm) of the QCM-TiO_2 structure *vs* NO_2 concentration.

4. Conclusions

The sorption properties to NO_2 are followed of a quartz crystal microbalance coated by TiO_2 fibers. The sensing TiO_2 fibers are deposited and immobilized via consecutive stages of electrospinning and thermal processing. Data for the morphology, phase and chemical composition of the fiber mats are obtained by applying analytical techniques for characterization of solid-state fiber materials. It is thus demonstrated that the immobilized electrospun fibers are amorphous and hollow with a diameter in the submicron range.

Further, sensor structures based on electrospun amorphous titania fibers immobilized on a QCM are fabricated and investigated for detection of NO_2 under ambient conditions.

It is shown that TiO_2 fibers can be used as a sensitive medium for NO_2 detection within the 50 – 5000 ppm concentration range. The formed QCM-fibrous TiO_2 structures exhibit a fast response, which is a substantial prerequisite for gas detection applications.

Based on the measured frequency-time characteristics of the QCM-TiO_2 fibers structure, the maximum Δf response is specified for each nitrogen dioxide concentration. Applying Sauerbrey's equation for the whole concentration interval studied, the evaluated NO_2 sorbed mass is in the range between 1.53 and 9.39 ng.

The results obtained demonstrated the possibility of using electrospun a-TiO_2 structures in a QCM as sensors for NO_2 detection in air.

Acknowledgments
The authors wish to thank Dr. D. Nihtianova for the phase analysis.

References

[1] Madou M J and Morrison S R 1989 *Chemical Sensing with Solid State Devices* (New York: Academic Press, Inc.,) 128
[2] Ferroni M, Guidi V, Martinelli G, Faglia G, Nelli P and Sberveglieri G 1996 *Nanostruct. Mater.* **7** 709–718
[3] Tang H, Prasad K, Sanjinés R and Lévy F 1995 *Sens. Actuators B: Chem.* **26** 71–75
[4] Yordanov R, Boyadjiev S and Georgieva V 2014 *Dig. J. Nanomater. Biostructures* **9** 467–474
[5] Sigmund W, Yuh J, Park H, Maneeratana, Pyrgiotakis G, Daga A, Taylor K and Nino J 2006 *J. Am. Ceram. Soc.* **89** 395
[6] Reneker D and Chun I 1996 *Nanotechnology* **7** 216
[7] Formo E, Camargo P, Lim B, Jiang M and Xia Y 2009 *Chem. Phys. Lett.* **476** 56
[8] Ultra-sensitive metal oxide gas sensor and fabrication method thereof **Patent number:** 7640789
[9] Kim Il-Doo, Rothschild A, Tuller H L, Kim D Y and Jo S M 2006 *Proceedings of the 2006 NSTI Nanotechnology Conference and Trade Show* vol 3
[10] Wang Y et al 2016 *Nanotechnology* **27** 425503
[11] Cunningham A 1998 *Introduction to Bioanalytical Sensors* (New York: Wiley–Interscience)
[12] Mirmohseni A and Oladegaragoze A 2004 *Sens. Actuators B: Chem.* **98** 28 – 36
[13] Sigmund W, Yuh J, Park H, Maneeratana V, Pyrgiotakis G, Daga A, Taylor K and Nino J 2006 *J. Am. Ceram. Soc.* **89** 395
[14] Korotcenkov Gh 2021 *Nanomaterials* **11** 1555
[15] Starbova K, Petrov D, Starbov N and Lovchinov V 2012 *Ceram. International* **36** 4645
[16] Sauerbrey G 1959 *Z. Physik* **155** 206–222
[17] Lazarova V, Spassov L, Gueorguiev, Andreev S, Manolov E and Popova L 1996 *Vacuum* **47**(12) 1423–1425
[18] Georgieva V, Spassov L, Spassov D, Donkov N and Petkov P 2007 *J. Optoelectron. Adv. Mater.* **9**(2) 252–255
[19] Doeuff S, Henry M, Sanchez C and Livage J 1987 *J. Non-Crystall. Solids* **89** (No 1 - 2) 206–216
[20] Fong H, Chun I and Reneker D H 1999 *Polymer* **40** 4585–4592
[21] Starbova K, Spasova M, Yordanova V, Starbov N and Rashkov I 2007 *J. Optoel. AdvMater. Rapid Communications.* **1** 629–635

Study of biogenic iron oxyhydroxide for application in electronics and biotechnology

R Angelova[1,6], L Slavov[1,2], B Blagoev[3], Ch Ghelev[1], D Kovacheva[4], M Iliev[5], V Groudeva[5] and I Nedkov[1]

[1]Institute of Electronics, Bulgarian Academy of Sciences, 72 Tsarigradsko Chaussee, 1784 Sofia, Bulgaria
[2]Frank Laboratory of Neutron Physics, Joint Institute for Nuclear Research, Dubna, Moscow Distr., 141980, Russia
[3]Institute of Solid State Physics, Bulgarian Academy of Sciences, 72 Tsarigradsko Chaussee, 1784 Sofia, Bulgaria Sofia, Bulgaria
[4]Institute of General and Inorganic Chemistry, Bulgarian Academy of Sciences, Acad. Georgi Bonchev Str., bld. 11, 1113 Sofia, Bulgaria
[5]Faculty of Biology, St. Kliment Ohridski University of Sofia, 8 Dragan Tsankov Blvd., 1164 Sofia, Bulgaria

E-mail: raly_angelova@abv.bg

Abstract. The object of this study is a biogenic material obtained from neutrophilic sheath-forming iron bacteria of the genus *Leptothrix* cultivated in a specific medium. Light microscopy and scanning electron microscopy were used to characterize the biogenic product. The X-ray diffraction spectrum of the control sample (without inoculated bacteria) is presented to demonstrate the important role of the bacteria in the Fe^{2+} transformation. Unlike the biogenic product, where nanoparticles of lepidocrocite are present, no traces of stable single-phase iron oxide or oxyhydroxide were detected in the control sample. The study of the biogenic sample showed a lack of significant differences between the naturally obtained and the artificially produced sheaths. Magnetostriction effect measurements were conducted to probe how the different magnetic behavior at temperatures below and above the Neel temperature (50 – 77 K) could be used in possible applications of the material in actuating and sensing devices.

1. Introduction

The need for reliable information on the structural and functional characteristics of nanosized materials produced by *Leptothrix* bacteria has grown in recent years [1-3]. The microaerophilic iron-oxidizing β-proteobacteria from the genus *Leptothrix* are of great interest because during their growth they produce extracellular formations containing iron oxides and oxyhydroxides. These bacteria grow as chains of cells into filaments of 0.4 – 7 μm in width, while it is known that *Leptothrix ochracea* create mostly hollow niches [4]. The filaments (sheaths) may appear yellow to dark-brown because of the deposition of iron and manganese oxides. In a natural environment, the process takes place in Fe-rich waters [5], such as free-flowing water streams [6]. The microbial biomass in such habitats is abundant; however,

[6] To whom any correspondence should be addressed.

Content from this work may be used under the terms of the Creative Commons Attribution 3.0 licence. Any further distribution of this work must maintain attribution to the author(s) and the title of the work, journal citation and DOI.
Published under licence by IOP Publishing Ltd

cultivating these bacteria under laboratory conditions does not usually result in forming large quantities of iron hydroxides, which still constitutes a major research problem.

Biogenic iron oxides and oxyhydroxides are important eco-friendly materials, which already have many biotechnological, biomedical and environmental applications [7-9]. They can be used as innovative red pigments, combustion catalysts, functional materials for targeted drug delivery and contrast agents for magnetic resonance imaging, indicators for organic pollution of water, precursors for synthesis of hybrid battery-supercapacitor systems [7, 10, 11]. Magnetized sheath structures are capable of adsorbing various substances that endanger the environment, e.g., dyes, Fe^{2+} ions, heavy metals, etc. [12, 13].

This study is focused on acquiring new physical data on possible application in electronics and biotechnology of a biogenic material obtained via the *Leptothrix* sp. metabolism under laboratory conditions. Light microscopy and scanning electron microscopy (SEM) were used to characterize the biogenic products formed in the nutrition medium selected for cultivation. X-ray diffraction (XRD) measurements were conducted on both a biogenic and a control sample without bacteria inoculation. The spectrum of the latter one is presented, revealing the role of the bacteria in the iron transformation. The magnetostriction effect in the biogenic material was measured at two different temperatures in order to test for possible applications in actuators and sensors.

2. Materials and Methods

Considering the specific environmental requirements of the *Leptothrix* bacteria, a small water source containing deposits with the typical fluffy texture and reddish color located in Vitosha Mountain was chosen as the area of sampling [14]. To imitate the natural conditions for iron bacteria cultivation, we used the silicon iron glucose peptone (SIGP) medium [15] and a pure culture isolated from the area of the sampling [14]. The cultivation vessels used (Erlenmeyer flasks of 500 ml) provide high volume and surface for aeration of the culture fluid in order to allow the bacteria to grow. The flasks were inoculated (10% v/v) and cultivated at a temperature of 20 °C under dynamic conditions (70 rpm).

Following 14 days of cultivation, the material from the cultivation and control vessels was filtered and dried at room temperature. Its morphology was studied by light microscopy and scanning electron microscopy (SEM) (JEOL JSN-5510, JEOL, Japan). The samples for SEM image manipulations were prepared by dropping bacterial suspension on cover glasses for light microscopy followed by drying at room temperature. This way of preparation was chosen because of the fragility of the bacterial sheath structures. The structural characterization was carried out by X-ray diffraction using a Bruker D8 diffractometer in the Bragg-Brentano reflection geometry with CuKα radiation ($\lambda = 1.5418$ Å). A superconducting magnet installation (Oxford Instruments) was used for magnetostriction measurements.

3. Results and Discussion

The characteristic cell morphology and shape of the *Leptothrix* genus resulting from the bacterial metabolism in the SIGP medium can be seen in figure 1 (a). The typical hollow sheath structures and *Leptothrix* cells outside the sheaths (figure 1 (b)) were observed by light microscopy (figure 1 (c)) and SEM (figure 1 (d)); the sheaths approximate size after a fourteen-day period of cultivation in the SIGP medium was around 10 μ with an average diameter of about 1 μm. Their diameters and lengths vary probably depending on their maturity [16]. The measurements of these structures length may not be reliable because the sheaths of *Leptothrix* sp. are brittle and easily broken into shorter fragments [17].

The resulting SIGP biomaterial and a control sample without inoculated bacteria in it but in the same nutrition medium were characterized by X-ray diffraction (XRD). The spectrum for the biogenic material can be found in [17]. The analysis of the biogenic XRD spectra [17] revealed the presence of a single-phase iron-containing phase, i.e., lepidocrocite – γ-FeOOH, with 8-nm average size of the oxyhydroxide particles.

Unlike in its biotic counterpart, no traces of stable iron oxide or oxyhydroxide were detected in the control sample (figure 2). The two broad maxima around 34° and 61° seen in the spectrum are a benchmark of metastable 2-line ferrihydrite (Fh) [18]. Because of its low stability, it easily transforms

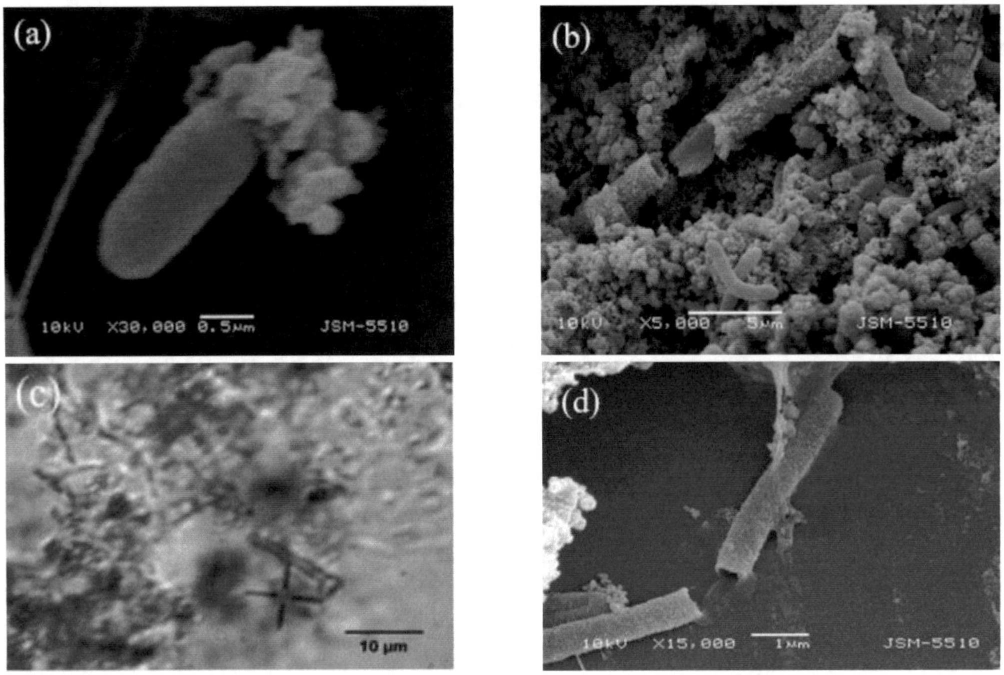

Figure 1. (a) and (b) SEM of bacteria from the genus *Leptothrix* cultivated in an SIGP medium; (c) light and (d) SEM images of sheath structures of the *Leptothrix* sp. cultivated in an SIGP medium.

into goethite and hematite [18], but none of their characteristic peaks are observed. Thus, it seems that in such abiotic conditions, the 14-day period is not sufficient for a transformation of iron to a more stable phase. The other phase with iron in its composition is amorphous iron hydrogen phosphate hydrate ($H_3Fe(PO_4)_2.2.5H_2O$). This compound could be regarded as an intermediate phase to a well-crystallized form of iron hydrogen phosphate hydrate, $FePO_4.H_3PO_4.4H_2O$. The well-crystallized form can emerge in phosphorus-rich environment prior to or simultaneously with rust [19]. It has also been reported [20] that its orthorhombic phase expected to evolve (characteristic peaks at 26°, 29° and 31°) can emerge after slow phosphatation from β-FeOOH (akaganite). It is highly probable that iron oxyhydroxides would form abiotically from a $FeSO_4$ source, but in the case of absence of bacteria, they are still in the phase of formation and transformation.

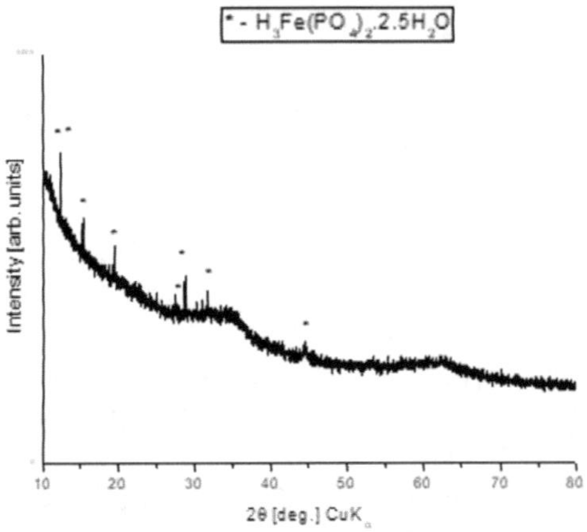

Figure 2. XRD spectrum of the control sample, *without* inoculated *Leptothrix* bacteria and the same SIGP nutrition medium.

The clear difference in the composition of the biotically involved material and the product obtained in abiotic conditions confirm the role of the bacteria in the quicker transformation of Fe^{2+} into a stable Fe^{3+} oxyhydroxide as an end product. Within the same time period (14 days) and under the same medium conditions, stable oxyhydroxide (lepidocrocite) was already formed in the biotic conditions [17], while in abiotic conditions no traces of stable iron oxide or oxyhydroxide were detected, although expected as an end product.

The single-phase lepidocrocite obtained biotically using an SIGP nutrition medium for *Leptothrix* bacteria is in principle antiferromagnetic. However, on a nanometer scale, a second magnetic character (superpara or ferri) starts to affect the bulk magnetism of the material [21]. The effect is detectable via magnetic measurements and is more pronounced at low temperatures. We conducted magnetostriction effect measurements in order to probe how the different magnetic behavior at temperatures below and above the Neel temperature range (50 – 77 K) could affect the material's possible applications.

A change in the shape and size of the material during its magnetization is measurable in many ferromagnetic materials [22]. Thanks to the appearance of this magneto-mechanical effect, such materials find application in devices for vibration control, actuation and sensing, drives, electric motors etc. [23, 24]. Figure 3 shows the results of the measurements of the SIGP biogenic lepidocrocite sample.

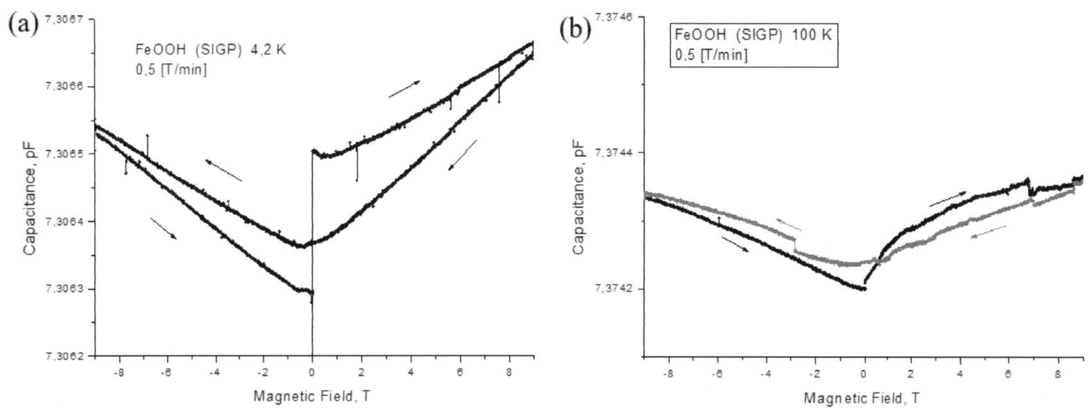

Figure 3. Magnetostriction effect in biogenic lepidocrocite (a) at 4.2 K and (b) at 100 K.

The measurements of the nanosized particles were performed at 4.2 K and 100 K and magnetic flux density of up to 9 T. A positive magnetostriction effect in the range of 10^{-5} was found; the change in the sample's dimension continued to flux density of 9 T. The measurements also demonstrated a stronger magnetostriction effect at 4.2 K indicating that the second magnetic component, more pronounced at low temperatures, favors a stronger effect.

4. Conclusions

Biogenic iron oxyhydroxide and sheath structures were produced by *Leptothrix* sp. under laboratory conditions. Such stable iron oxyhydroxide was not detected by XRD in a control sample not containing bacteria. The biotically obtained nanosized single-phase lepidocrocite was studied in view of the existence of a magnetostriction effect at 4.2 K and 100 K. A positive magnetostriction effect in the range of 10^{-5} was found in the material, with the effect being stronger at the lower temperature. Understanding the structural and magnetic properties of the non-toxic nanosized iron oxyhydroxide particles obtained by biological means is important for fundamental research, but also could suggest the development of practical applications in electronics and biotechnology.

Acknowledgments

This work was supported by the Bulgarian Ministry of Education and Science under the National Research Program "Young scientists and postdoctoral students" approved by DCM # 577 / 17.08.2018 and JINR project No 04-4-1142-2021/2025. The authors acknowledge the contribution of Krzysztof P. Filar (Institute for High Magnetic Fields and Low Temperatures, Polish Academy of Sciences) for the magnetostriction measurements.

References

[1] Brooks C and Field E 2020 *Mbio* 2020 **11** 02720
[2] Tamura K, Kunoh T, Nakanishi M, Kusano Y and Takada J 2020 *ACS Omega* **5** 27287
[3] Tamura K, Kunoh T, Nagaoka N and Takada J 2020 ACS *Appl. Bio Mater.* **3** 5699
[4] Bergey D and Holt J 1994 *Sheathed bacteria* (Baltimore: The Williams & Wilkins Co) p 477
[5] Singh V K, Singh A L, Singh R and Kumar A 2018 *Environmental Sustainability* **1** 221
[6] Matýsek D and Jirásek J 2021 *Bull. Mineral. Petrol.* **29** 115
[7] Kappler U 2021 *J. Inorg. Biochem.* **221** 111488
[8] Gareev K, Grouzdev D, Kharitonskii P, Kosterov A, Koziaeva V, Sergienko E and Shevtsov M 2021 *Magnetochemistry* **7** 86
[9] Emerson D 2019 *Front. Mar. Sci.* **6** 22
[10] Hashimoto H 2011 *Characteristics of biogenous iron oxide microtubes formed by iron-oxidizing bacteria, Leptothrix ochracea* (Singapore: Jenny Stanford Publishing) p 155
[11] Veleva S, Angelova R, Stoyanov L, Groudeva V, Kovacheva D, Mladenov M and Raicheff R 2014 *Nanosci. Nanotechnol.* **14** 50
[12] Safarik I, Angelova R, Baldikova E, Pospiskova K and Safarikova M 2017 *Mater. Sci. Eng. C* **71** 1342
[13] Whitaker A, Peña J, Amor M and Duckworth O 2018 *Environ. Sci. Process. Impacts.* **20** 1056
[14] Angelova R, Slavov L, Iliev M, Mitova M, Blagoev B, Nedkov I and Groudeva V 2015 in Annuaire de l'Université de Sofia "St. Kliment Ohridski" **100** 231
[15] Sawayama M, Suzuki T, Hashimoto H, Kasai T, Furutani M, Miyata N, Kunoh H and Takada J 2011 *Curr. Microbiol.* **63** 173
[16] Spring S 2006 *The genera Leptothrix and Sphaerotilus* (New York: Springer) p 758
[17] Angelova R, Iliev M, Slavov L, Koutzarova T, Closset R, Nedkov I and Groudeva V 2015 *J. BioSci. Biotech.* 165
[18] Drits V, Sakharov B, Salyn A and Manceau A 1993 *Clay Miner.* **28** 185
[19] Balasubramaniam R 2003 *Current Science* **84** 534
[20] Lin R, Amrute A, Krumeich F, Lázár K, Hauert R, Yulikov M and Pérez-Ramírez J 2016 *Cryst. Eng. Comm.* **18** 3174
[21] Guyodo Y, Bonville P, Till J, Ona-Nguema G, Lagroix F and Menguy N 2016 *Front. Earth Sci.* **4** 28
[22] Ekreem N, Olabi A, Prescott T, Rafferty A and Hashmi M 2007 *J. Mater. Process. Technol.* **191** 96
[23] Shi P 2021 *J. Magn. Magn. Mater.* **537** 168212
[24] Burdin D, Ekonomov N, Gordeev S and Fetisov Y 2021 *J. Magn. Magn. Mater.* **521** 167530

Silver nanoparticles synthesis and their effect on the SOPC lipid structure

Z Slavkova[1,4], J Genova[1], H Chamati[1], V Boev[2] and D Yancheva[3]

[1]Institute of Solid State Physics, Bulgarian Academy of Sciences, 72 Tsarigradsko Chaussee, 1784 Sofia, Bulgaria
[2]Acad. Evgeni Budevski Institute of Electrochemistry and Energy Systems, Bulgarian Academy of Sciences, Acad. Georgi Bonchev Str., bld. 10, 1113 Sofia, Bulgaria
[3]Institute of Organic Chemistry with Center of Phytochemistry, Bulgarian Academy of Sciences, Acad. G. Bonchev Str., bld. 9, 1113 Sofia, Bulgaria

E-mail: zslavkova@issp.bas.bg

Abstract. We focus our attention on the influence of hydrophobic silver nanoparticles (Ag NPs) on the 1-Stearoyl-2-oleoylsn-glycero-3-phosphocholine (SOPC) model system. Results obtained by differential scanning calorimetry (DSC) and infrared (IR) spectroscopy were compared to their counterpart for lipid systems with incorporated hydrophobic gold nanoparticles (AuNPs) and pure SOPC lipid in a water environment. The results show a strong effect of hindering the gel-to-liquid crystalline phase transition for both types of noble metal hydrophobic NPs. The effect of the phase transitions shift to the lower temperatures is more pronounced for the silver particles for the concentrations studied. According to the IR spectral analyses, no negative effect is observed of the NPs on the lipid hydration for the concentration studied, and the plasmon effect of the Ag NPs appears to be more pronounced.

1. Introduction

Nanotechnology has a broad spectrum of applications in a great variety of technology and industry sectors, like information processing, homeland security, medicine, transportation, energy, food safety, environmental science and many others. Biosensors with nano-scaled inorganic inclusions are being extensively investigated [1], in particular, noble-metal-based sensors have a well-established role [2].

One of the most essential benefits of Ag NPs is their potential use in medicine, since they possess antifungal [3] and antibacterial [4, 5] properties. They have been proven effective in diagnosis [6, 7] and treatment [6, 8-10] of cancer. The combination of Ag and Au has shown very promising results [9, 10]. When used together, the ensuing systems have shown better chemical stability and photothermal efficiency [11]. Researchers have found that different Au/Ag ratios of the NPs are suitable for different treatment approaches. The combination Ag:Au = 3:1 has exhibited superior antitumor efficiency as "chemotherapeutic agents", while the Ag:Au = 1:3 one is the least toxic (in the proportions studied) making it appropriate for use in *in vivo* methods (targeted therapy) [12]. Studies of Ag-Au NPs together with tryptophan (TRP) have shown similar results. $Ag_3Au_1Trp_{1:2}$ NPs exhibited *in vivo* anti-tumor and anti-metastatic action and, thanks to the enhancement and retention effect, reach the tumor site and cause selective apoptotic extinction of cancer cells through the extrinsic TRAIL-dependent pathway [13].

[4] To whom any correspondence should be addressed.

Content from this work may be used under the terms of the Creative Commons Attribution 3.0 licence. Any further distribution of this work must maintain attribution to the author(s) and the title of the work, journal citation and DOI.
Published under licence by IOP Publishing Ltd

Bearing the above in mind, which is but a small excerpt of all available research and innovations in the field, we believe that conducting fundamental research on the influence of NPs on model lipid systems is essential in advancing our knowledge on the lipids behavior in the presence of inorganic NPs. So far, scientists have mostly focused their interest on Ag NPs and lipid interaction from a medical point of view [3-5, 6-10, 14]. Our long-term goal is studying the effect of different NPs on the physical properties of SOPC model lipid systems. In this paper, the emphasis is on the phase transition from gel to a liquid crystalline state. The liquid crystalline state is of great importance because in this state living cells are able to perform all their vital functions [15, 16].

2. Materials and methods

2.1. Ag NP's synthesis and sample preparation

Colloidal silver was produced following the reverse micelles approach described in [17]. The Ag NPs were covered by AOT to provide them with hydrophobic surface properties. We performed UV-Vis spectroscopy analysis on the Ag NPs solution in order to ascertain the presence of colloidal silver. The results obtained are shown in figure 1. The UV/Vis absorption spectra were registered on a Perkin Elmer Lambda 25 UV/Vis spectrometer in the range 200 – 900 nm. Spectroscopy grade n-hexane was used as a solvent. The spectrum is consistent with that of Ag NPs [17]. According to reference [18], gold NPs prepared by the reverse micelles method and coated by an AOT film do not alter after evaporation of the solution. Following our previous work [19], we assume that Ag NPs are functionalized and proceed with their incorporation in the lipid structure. Samples for both DSC and IR analyses containing 0.5 wt% Ag NPs were prepared using the protocol described in our paper on the influence of hydrophobic gold NPs on SOPC lipid [20]. The necessary quantities of SOPC and Ag NPs solutions were mixed in flasks in order to produce specimens with 0.5 wt% Ag NPs concentration. The solutions obtained were dried under vacuum, then 10 wt% of double distilled water was added.

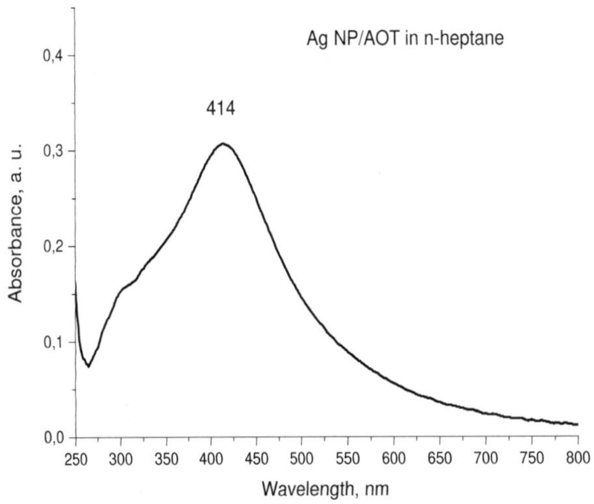

Figure 1. UV-Vis spectrum of Ag NPs/AOT.

2.2. Differential scanning calorimetry (DSC)

This technique of thermal analysis was developed in 1960 and was introduced to the public in 1963 [20]. It has become a valuable tool in various research on biological materials [21]. DSC is capable of estimating the enthalpy, entropy, and heat capacity of the studied material and determine whether it undergoes a phase transition. The thermograms are obtained by taking into account the difference between the heat-fluxes required to maintain the reference pot and the sample pot at the same temperature. When this excess heat is being absorbed by the sample, the transformation of the system is endothermic and vice versa, released excess heat indicates an exothermic reaction. The DSC equipment used was Discovery 250 (TA Instruments, USA). The required quantity of the sample varied between 10 – 30 mg. The heating and cooling rates of the experiment were 5 °C.

2.3. Infrared (IR) spectroscopy

IR radiation induces atoms' oscillations around their bonds and, simultaneously, the whole molecule is in a continuous translational or rotational motion [22]. These oscillations are unique for each chemical bond (functional group), and therefore, this technique is highly valuable for the examination of structural changes in biologically relevant molecules. The IR spectroscopy has been proven very useful in the research of proteins, lipid bilayers and membranes [23–26]. Attenuated total reflection (FTIR) is a big improvement of IR spectroscopy. It facilitates the sample preparation and avoids the short path-length of a liquid cell necessary for transmittance mode when studying water solutions. Since the specimens studied are in water, this approach would be suitable. The spectrometer used was a Bruker Tensor 27 provided with deuterated triglycinesulphate detector. The sample was directly placed on the attenuated total reflectance (ATR) element (diamond crystal). The spectrum was collected in the region of 4000 – 900 cm^{-1} with 64 scans at a 2 cm^{-1} resolution and air as background.

3. Results and discussion

The results of the calorimetric measurements of the studied systems are presented in figure 2. The thermograms obtained reveal a substantial and similar effect of both types of NPs on the phase transition from gel to a liquid crystalline state. They tend to shift it towards the lower temperatures. However, this effect seems to be more pronounced for the Ag NPs, relocating it from 3.91 °C [27] to 0.7 °C. The transition for SOPC-Au is notably hindered and smeared out and harder to trace, resulting in a lower enthalpy value of about one order of magnitude compared to the sample containing Ag NPs. It seems that the Au NPs are hindering the phase transformation more intensively, at least for the 0.5 wt% concentration. The thermograms obtained unveil the conversion of the phase transition from a simultaneous to a non-simultaneous with the addition of NPs and the emergence of an intermediate pre-transition state, similar to the effect of adding cholesterol [25].

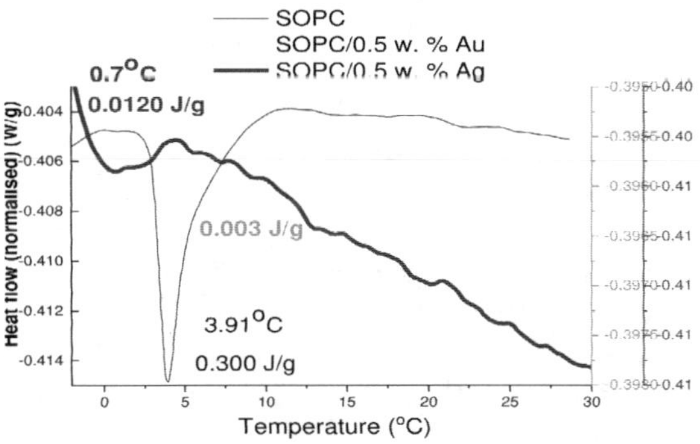

Figure 2. DSC thermograms of the SOPC lipid system containing 0.5 wt% of Ag NPs and Au NPs.

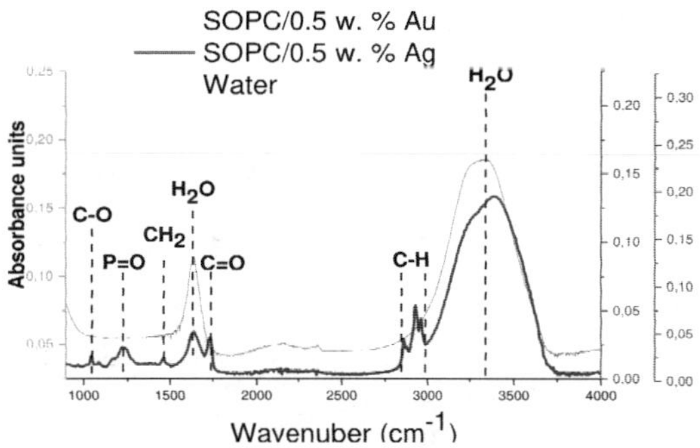

Figure 3. Infrared absorption spectra of the SOPC sample with 0.5 wt% Ag NPs and Au NPs.

In figure 3 we present the IR spectrum of the aqueous solution studied of SOPC with Ag NPs along with that of SOPC-Au system investigated in reference [19]. The strong IR absorption of water is visible in the IR spectra of both samples around 3340 cm^{-1} and 1640 cm^{-1}. The IR bands of the SOPC molecules in the SOPC-Ag system were observed at the same positions as those of SOPC-Au [19]: 2962 cm^{-1}, 2930 cm^{-1}, 2877 cm^{-1}, 2860 cm^{-1} (C-H stretching vibrations of the methyl and methylene groups of the alkyl chains), 1463 cm^{-1}, 1414 cm^{-1}, 1394 cm^{-1}, 1381 cm^{-1}, 1360 cm^{-1} (C-H bending vibrations), 1217 cm^{-1} (asymmetric stretching vibrations of the phosphate groups), and 1046 cm^{-1} (C-O stretching vibrations). Only the band for the C=O stretching vibrations of the lipid ester's carbonyl groups appeared slightly shifted – from 1730 cm^{-1} (SOPC-Au) to 1735 cm^{-1} (SOPC-Ag), but still corresponding to the position reported for fully hydrated diacyl phosphatidylcholines [26]. As previously discussed for the SOPC-Au systems [19], the position of the band for the asymmetric stretching vibration of the phosphate groups-at 1217 cm^{-1} is also a sign for effective hydration of the phosphate groups, obviously not influenced by the presence of metal NPs. On the other hand, in the SOPC-Ag system the relative intensity of the bands for the SOPC molecules compared to those of water is much higher than in the SOPC-Au system with the same concentration of the metal NPs (0.5 wt %), as it could be seen in figure 3. This effect is most probably due to the stronger plasmon effect of the Ag NPs compared to Au NPs.

4. Conclusions

Since both Au NPs and Ag NPs have hydrophobic properties, they are expected to be incorporated in between the tails of the SOPC lipid molecules. Even a small addition of 0.5 wt% of these NPs have a strong hindering effect on the gel to liquid crystalline phase transition. It could be concluded that, at least for this concentration, Au NPs have a far more substantial smearing effect on the examined phase transformation when compared to Ag NPs, resulting in about one order of magnitude lower enthalpy value. On the other hand, Ag NPs tend to shift the transition of the SOPC lipid towards even lower temperatures (0.7 °C). According to the IR spectral analyses, there seems to be no negative effect of the NPs on the lipid hydration for the studied concentration; also, the plasmon effect of the Ag NPs appears to be more pronounced. We plan to carry on further with our work on the matter, including other concentrations and assessing the bending elasticity moduli of the systems.

Acknowledgments

Z. Slavkova is supported by the Bulgarian Ministry of Education and Science under the National Research Program "Young scientists and postdoctoral researchers" approved by DCM No 577/17.08.2018. J. Genova and H. Chamati acknowledge the financial support of the Joint Institute for Nuclear Research (JINR), Dubna, Russian Federation – THEME04-4-1133-2018/2023, and Grant No D01-378/18.12.2020 of the Ministry of Education and Science of Bulgaria. Research equipment of the INFRAMAT Distributed Research Infrastructure, part of the Bulgarian National Roadmap for Research Infrastructures and supported by Bulgarian Ministry of Education and Science, was used in a part of this investigation.

References

[1] Su H, Li S, Jin Y, Xian Z, Yang D, Zhou W, Mangaran F, Leung F, Sithamparanathan G and Kerman K 2017 *Adv. Health Care Technol.* **3** 19–29

[2] Malekzad H, Zangabad P S, Mirshekari H, Karimi M and Hamblin M R 2017 *Nanotechnol. Rev.* **6** 301

[3] Radhakrishnan V S, Dwivedi S P, Siddiqui M H and Prasad T 2018 *Int. J. Nanomedicine* **13** 91

[4] Ran X, Du Y, Wang Z, Wang H, Pu F, Ren J and Qu X 2017 *ACS Appl. Mater. Interfaces* **9** 19717–19724

[5] Park H H, Park S, Ko G and Woo K 2013 *J. Mater. Chem.* B **1** 2701–2709

[6] Shi J, Wang L, Zhang J, Ma R, Gao J, Liu Y, Zhang C and Zhang Z 2014 *Biomaterials* **35** 5847–5861

[7] Darabdhara G, Das M R, Singh S P, Rengan A K, Szunerits S and Boukherroub R 2019 *Adv. Colloid Interface Sci.* **271** 101991

[8] Xu R et al. 2009 *Cell Res.* **19** 1031–4

[9] Rajasekharreddy P and Rani P U 2014 *Mater. Sci. Eng.* C **39** 203–212

[10] Sangour M H, Ali I M, Atwan Z W and Al Ali A A A L A 2021 *Egypt. J. Med. Hum. Genet.* **22** 9

[11] Jiang T, Song J, Zhang W, Wang H, Li X, Xia R, Zhu L and Xu X 2015 *ACS Appl. Mater. Interfaces* **7** 21985–21994

[12] Mukha I, Vityuk N, Grodzyuk G, Shcherbakov S, Lyberopoulou A, Efstathopoulos E and Gazouli M 2017 *J. Nanosci. Nanotechnol.* **17** 8987–8994
[13] Katifelis H et al. 2020 *Int. J. Nanomedicine* **15** 6019
[14] Darabdhara G, Das M R, Singh S P, Rengan A K, Szunerits S and Boukherroub R 2019 *Adv. Colloid Interface Sci.* **271** 101991
[15] Bouligand Y 2008 *Comptes Rendus Chim.* **11** 281–96
[16] Alfsen A and Tatischeff I 2014 *J. Biomater. Nanobiotechnology* **05** 105–115
[17] Singha D, Barman N and Sahu K 2014 *J. Colloid Interface Sci.* **413** 37–42
[18] Longo A, Calandra P, Casaletto M P, Giordano C, Venezia A M and Liveri V 2006 *Mater. Chem. Phys.* **96** 66–72
[19] Slavkova Z, Genova J, Chamati H, Koroleva M and Yancheva D 2020 *Colloids Surf. Physicochem. Eng. Asp.* **603** 125090
[20] Akash M S H and Rehman K 2020 Differential Scanning Calorimetry *Essentials of Pharmaceutical Analysis* ed M S H Akash and K Rehman (Singapore: Springer) pp 199–206
[21] Sun W Q 2021 DSC Analysis of Thermophysical Properties for Biomaterials and Formulations *Cryopreservation and Freeze-Drying Protocols* Methods in Molecular Biology ed W F Wolkers and H Oldenhof (New York, NY: Springer US) pp 285–302
[22] El-Azazy M 2019 *Infrared Spectroscopy: Principles, Advances, and Applications* (BoD – Books on Demand)
[23] Barth A 2007 *Biochim. Biophys. Acta* **1767** 1073–101
[24] Tatulian S A 2019 FTIR *Methods Mol. Biol. Clifton NJ* **2003** 281–325
[25] Genova J, Chamati H and Petrov M 2020 Physico-chemical characterizations of lipid membranes in presence of cholesterol (*Advances in Biomembranes and Lipid Self-Assembly* vol 31) ed A Iglič, M Rappolt and A J García-SÁez (Academic Press) chapter 1 pp 1–42
[26] Tamm L K and Tatukian S A 1997 *Q. Rev. Biophys.* **30** 365–429
[27] Genova J, Slavkova Z, Chamati H and Petrov M 2019 *Phase Transit.* **92** 323–33

Double-discharge plasma system for deposition of carbon nanostructures

S Marinov[1], V Vachkov, I Ivanov and Zh Kiss'ovski[2]

Faculty of Physics, St. Kl. Ohridski University of Sofia, 5 J Bourchier Blvd., 1164 Sofia, Bulgaria

E-mail: smar@phys.uni-sofia.bg, kissov@phys.uni-sofia.bg

Abstract. Deposition of carbon nanostructures from ethanol precursor at atmospheric pressure can be enhanced via the use of a double-discharge system – a microwave and a DC discharge. We used a surface-wave discharge at a frequency of 2.45 GHz in a gas mixture of Ar, H_2 and C_2H_5OH to create a plasma column in a small-diameter ceramic tube and a plasma plume above it. The plume spectra in the UV/Vis range were measured by an optical system with an iHR550 spectrometer; the gas temperature and the electron temperature and density were calculated from the emission lines. The high gas temperature T_g ~3000 K and electron density (~6×10^{20} m^3) in the microwave plasma plume cause an effective dissociation of the ethanol even at low levels of absorbed power (10 W) and a strong carbon Swan band is registered in the spectrum. Transporting the created neutral and charged particles to the substrate at atmospheric pressure is difficult due to their short mean-free-path; we, therefore, applied an additional DC discharge. This two-discharge system allows more reactive species and ions from the ethanol decomposition to reach the substrate, thus speeding up the deposition of carbon nanostructures. Such structures were deposited on Ni foil and Ni foam substrates under fixed plasma parameters and controlled substrate temperature. The morphology of the produced graphene structures was studied using SEM and Raman spectroscopy.

1. Introduction

Carbon nanostructures (CNS) are an excellent choice for various applications because of their unique electrical, mechanical and optical properties. They have high mechanical strength and flexibility, high electrical and thermal conductivity and good structural properties, which depend on the diameter, length, atomic arrangement and nanostructure morphology. The CNS can be classified via their dimensions in several categories: 0D (nanoparticles), 1D (nanowires, nanotubes), 2D (nanowalls, nanosheets) or 3D structures. The carbon nanowires are wires with a very small diameter, sometimes as small as one nanometer, while a nanotube is a sheet of carbon atoms rolled into a cylinder. The carbon nanowalls (CNW) are structures consisting of a few graphene sheets with an individual sheet thickness of several to tens of nanometers vertically aligned on the substrate. They are of great interest due to their specific properties and morphology (a high aspect ratio and a large surface area). CNWs have potential applications in electronics as sensors, field emitters, battery electrodes and thin-film supercapacitors.

[1] To whom any correspondence should be addressed.
[2] To whom any correspondence should be addressed.

Content from this work may be used under the terms of the Creative Commons Attribution 3.0 licence. Any further distribution of this work must maintain attribution to the author(s) and the title of the work, journal citation and DOI.
Published under licence by IOP Publishing Ltd

In the past, the synthesis of CNW was mainly conducted in vacuum or at low pressure [1]. However, the most promising method for graphene production in recent years is the plasma-enhanced chemical vapor deposition (PECVD), which can vary depending on the type of gas discharge (DC, capacitively-coupled or inductively-coupled RF, helicon or MW).

In the present study, a double-discharge system was used for CNW deposition at atmospheric gas pressure under controlled conditions. The collisions of electrons with gas atoms and molecules in microwave discharges generate dense plasma. Furthermore, in comparison to the RF discharges, the microwave discharge at 2.45 GHz produces plasma with higher electron density and gas temperature resulting in higher concentrations of atomic hydrogen and C_2 radicals in the dissociation processes. In addition, the microwave PECVD, being an electrode-less process, avoids contamination of the films on the substrate due to electrode erosion.

2. Experimental setup and diagnostics

The experimental setup (figure 1) presented in [2] has been upgraded to provide an easier to reach optimal operating mode, a more homogeneous deposition of CNS and a better reproducibility of the results.

A low-power microwave discharge in a ceramic tube is used to create plasma in a gas mixture of Ar, CH_4 and additional H_2 to improve the deposition process. The option to use an alternative carbon precursor is added. It is produced by bubbling Ar through a tank containing liquid C_2H_5OH. The ethanol molecules decompose more easily even at low power, which facilitates and speeds up reaching the desired operating regime compared to using CH_4 [2]. An MPG-4M generator is used to feed a microwave plasma exciter through a coaxial line. The microwave coaxial exciter includes alumina ceramic as a surface-wave discharge tube. The source and the substrate (Ni foam or Ni plate) are placed in a quartz vessel with metal flanges filled with the gas mixture. The discharge is ignited by the MW exciter and an additional DC voltage (~ 600 V) is applied between the plume (coaxial exciter) and the substrate holder, which is the cathode. The positive column of the resultant DC discharge ensures the transport of reactive neutral and charged particles from the plasma plume to the substrate for quick deposition of the CNS. This hybrid structure of the discharge results in a higher concentration of the hydrocarbon species reaching the substrate and more homogeneously deposited films.

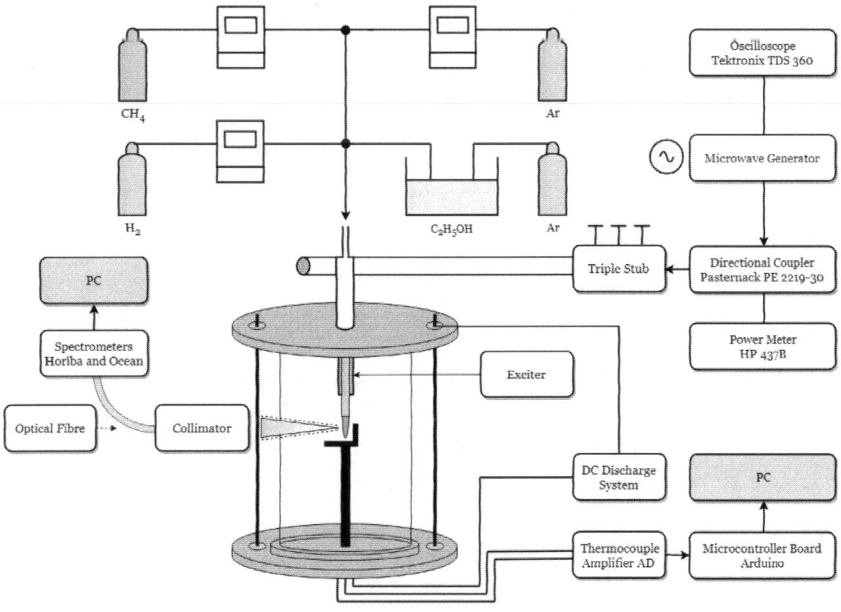

Figure 1. Experimental setup.

The substrate is heated by both the plasma plume and an additional cartridge heater mounted in the substrate holder. We control the heating by controlling the power fed to the heater. The substrate temperature is monitored precisely by a thermocouple system keeping the temperature within the desired range (600 – 700 °C) [3]. The reflected and forward microwave power is measured by an HP 437B power meter through a Pasternack PE 2219-30 directional coupler; the data received are used to better match the plasma source by means of a Maury 1878C triple stub. The emission spectra of the plasma plume are measured by both an HR Ocean Optics spectrometer and a Horiba iHR550 imaging spectrometer using an optical cable connected to a collimator. The data obtained is used to determine the plasma rotational temperature using the LIFBASE program [4] to fit and compare simulated OH and CN-bands to the experimental ones. The electron temperature and electron density are investigated using the line-ratio method [5] and the H_β-line broadening [6]. Also, the presence and strengths of certain bands (Swan band, CN, CH, C-line 247.8 nm, H_α, H_β) are used to determine that a successful production takes place of the reactive species usually associated with the deposition process.

3. Results

Using LIFBASE, we fitted the experimental spectra to simulated ones to determine the gas temperature (T_g) of the plasma [4]. The bands used for this were the OH (A) band ($A^2\Sigma+ \rightarrow X^2\Pi$) and the CN (B) band, more specifically, the portion of the spectra between the (1,1) and (0,0) band heads [7]. Due to the necessity of a controlled environment, the processing chamber was filled with an Ar/H_2 gas mixture. Over time, the oxygen was pushed out of the chamber by the gas mixture; the OH band became less intense and, after adding CH_4 or C_2H_5OH to the mixture, we used the CN band only. Argon is the working gas, H_2 serves as a catalyst and the hydrocarbon carrier gas is decomposed in the plasma column in order to serve as a source of carbon and hydrocarbons radicals and H_2.

Without the additional DC discharge, the transport of particles takes place via diffusion, which at atmospheric pressure allows transport over small distances only. This leads to a slow deposition process and a small area of graphene structures on the substrate. Adding the DC discharge enabled us to transport more ions to the substrate surface allowing for a longer distance between the microwave plasma plume and the substrate and providing a more homogenous coverage of the Ni substrate with CNS. The DC discharge current was measured using the voltage drop on a resistor (29 kΩ) connected in series to the DC power supply. The current was kept in the range 1 – 20 mA by varying the applied voltage up to 1000 V at different plume-substrate distances.

The spectra were used to estimate the rotational temperature at different microwave power levels (6 – 14 W) fed to the plasma exciter. The results show that as the power and the plasma density were increased, the gas temperature varied in the range 3200 – 3900 K.

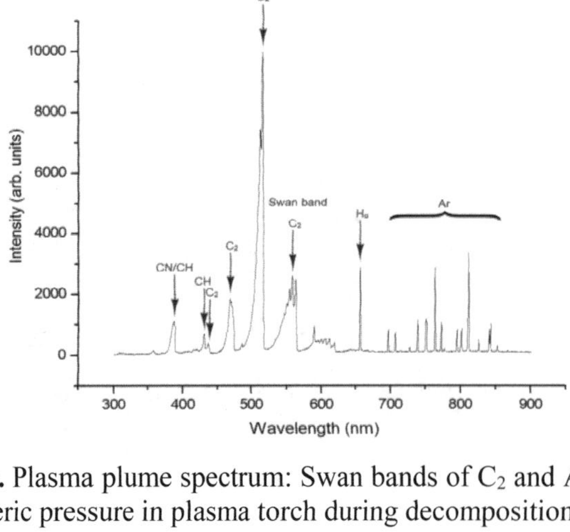

Figure 2. Plasma plume spectrum: Swan bands of C_2 and Ar lines at atmospheric pressure in plasma torch during decomposition.

Journal of Physics: Conference Series **2240** (2022) 012020 doi:10.1088/1742-6596/2240/1/012020

The presence in the measured spectra of a Swan band with a high strength in comparison with the Ar lines at gas temperatures above 3000 K confirms the effective decomposition of the hydrocarbon gas and the production of C_2 radicals (figure 2).

The argon lines in the spectra were used to evaluate the electron temperature (T_e) and density (n_e) by means of the line-ratio method [5] and the H_β-line broadening [6]. During the deposition of graphene nanostructures under the already defined conditions, the plasma parameters were determined to be in the range $n_e \approx 4 - 6\times10^{20}$ m^{-3} and $T_e \approx 1 - 1.1$ eV.

SEM imaging was used to observe the CNS morphology. When the deposition time was too short (less than 6 min), a thin carbon layer is seen (figure 3 a). It is a necessary first step in the process, which forms the foundation of the growing nanostructures. The additional DC discharge (~10 mA) allows for the growth of structures for only 5 – 6 min; exceeding this time results in a thick carbon layer with no visible nanostructures. At lower substrate temperatures (600 – 650 °C) and microwave power (~8 W), the CNS are predominantly carbon nanowires and nanotubes (figure 3 b). The higher temperatures (700 – 750 °C) and power (~12 W) lead to the formation of carbon nanosheets (both horizontal and vertical (figure 3 c), d). In methane, the growth rate was estimated at ~ 100 nm/min, while in ethanol it was ~ 150nm/min.

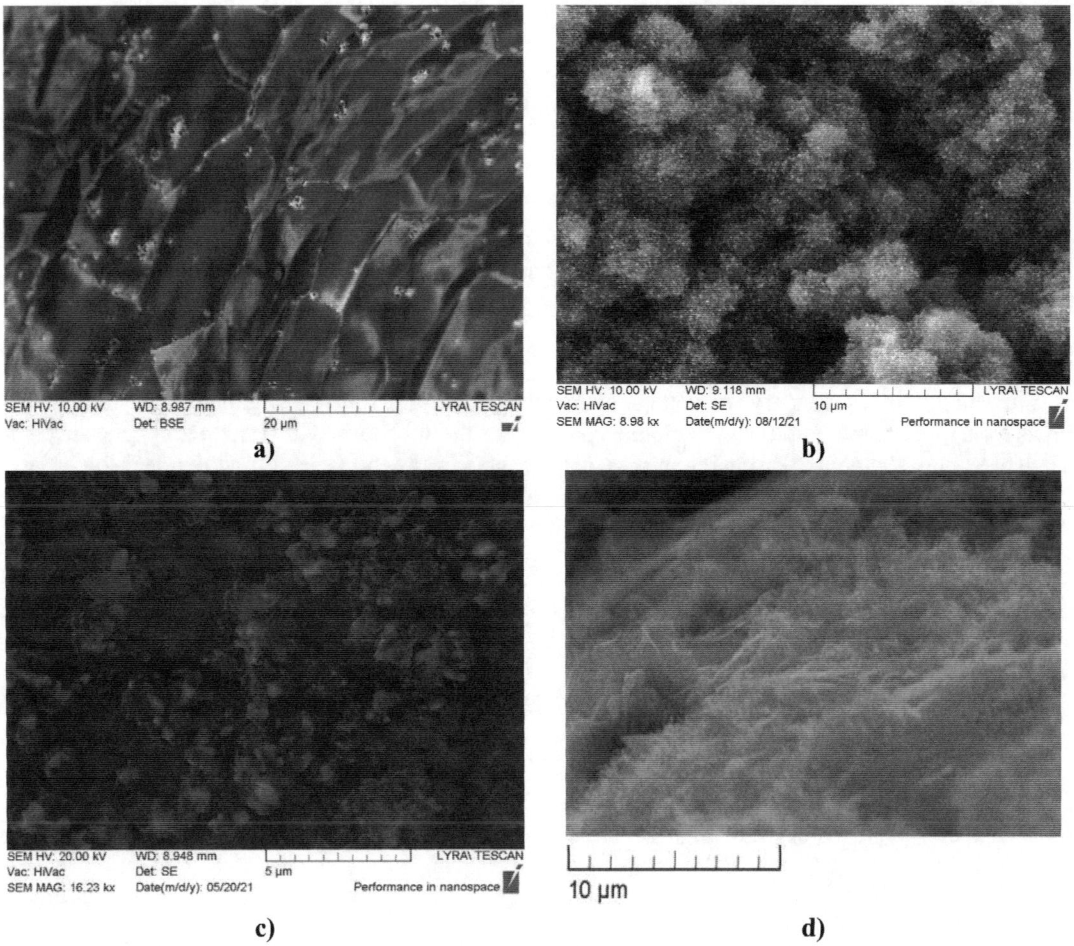

Figure 3. a) SEM image of a thin carbon layer; **b)** SEM image of carbon nanotubes deposited at low power and low substrate temperature; **c)** SEM image from the top of horizontal carbon nanosheets deposited at high power and temperatures above 700 °C; **d)** SEM image of vertical carbon nanosheets and nanowires.

Both types of structures required a sufficiently high substrate temperature, which we maintained by means of a cartridge heater. The carbon nanostructures deposited on Ni foam and Ni plate (thin layer or vertical sheets with height of 500 nm – 1 μm) were studied by Raman spectroscopy (figure 4); it showed that the bands observed conformed with the expectations for vertical graphene structures on the substrate [8, 9].

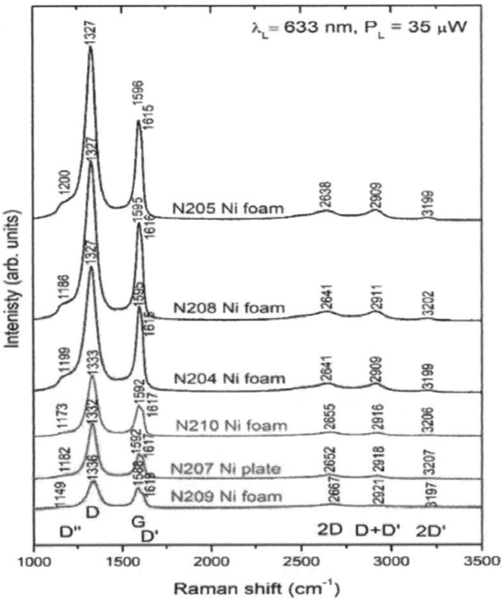

Figure 4. Results from Raman spectroscopy of the deposited nanostructures.

4. Conclusions

The introduction of a secondary DC voltage source contributing to transporting material to the substrate allows for a more homogenous and complete coverage of the substrate with vertical graphene sheets for less than 8 min. The double discharge improves the system's flexibility by allowing operation at larger distances between the source and the substrate. Also, using C_2H_5OH facilitates reaching the desired operating regime in less time compared to using methane.

Acknowledgments

This work was supported by the European research project PEGASUS, H2020, NIS-SU N3257.

References

[1] Hiramatsu M and Hori M 2010 *Carbon Nanowalls Synthesis and Emerging Applications*; (NewYork: Springer, Wien)
[2] Marinov S, Vachkov V and Kiss'ovski Zh 2020 *J. Phys. Conf. Series* **1492** 012032
[3] Vachkov V and Kissovski Zh 2020 *J. Phys. Conf. Series* **1492** 012061
[4] Luque J and Crosley D R 1999 *LIFBASE: Database and Spectral Simulation Program*; SRI International Report MP, 99
[5] Kiss'ovski Zh, Ivanov A, Iordanova S and Koleva I 2011 *Journal of Physics* D*: Applied Physics* **44** (20) 205203
[6] Qian M, Ren C, Wang D, Zhang J and Wei G 2010 *J. Appl. Phys.* **107** (6) 063303
[7] Nassar H 2012 *J. Phys. Conf. Series* **370** 012050
[8] Kurita S, Yoshimura A, Kawamoto H, Uchida T, Kojima K and Tachibana M 2005 *J. Appl. Phys.* **97** (10) 104320
[9] Ni Z H, Fan H M and Feng Y P 2006 *J. Chem. Phys.* **124** (20) 204703

Extraction of accompanying metals through electron beam refining of copper

V Vassileva, K Vutova[1] and M Markov

Institute of Electronics, Bulgarian Academy of Sciences, 72 Tsarigradsko Chaussee, 1784 Sofia, Bulgaria

E-mail: katia@van-computers.com

Abstract. This work presents a study on applying the electron beam method for refining pyro-refined copper (as initial material) and obtaining multiple metals through electron beam melting. Studies of and results on the refining processes under different technological modes are presented and discussed. The experiments were performed in single-melt operations with different e-beam power inputs and refining times. Evaluation is provided concerning the extraction of some valuable metals (such as Ni, As, Bi, Pb, Sb, Ag) generated in the condensate during their removal from the liquid metal in the e-beam refining process.

1. Introduction

Copper is one of the most important metals in modern life, with up to 50 % of the copper produced being used in the electrical industry for manufacturing cables, heat exchangers, refrigerator parts, vacuum equipment, etc., and about 40 % used for production of alloys with Zn, Sn, Al, Ni, Fe, Mn, Be, Si and other elements [1]. The metal is basic in the development and maintenance of modern electronic mobility, renewable energy, energy and electrical systems and networks, architecture and "green" construction, air conditioning, electronics and telecommunications, transport (air, land and sea), mechanical engineering and equipment, chemical industry, agriculture, healthcare, households, etc.

Chile is the undisputed leader in copper mining, followed by Peru, China, United States, Congo, Australia, etc. [2]. The balance between production and consumption indicates a shortage of refined copper, which is currently compensated by reserves available from previous periods [3].

Vacuum metallurgy has a leading position among the industrial methods used for producing metals and alloys of high purity. As part of the vacuum metallurgy, electron beam melting (EBM) has achieved a significant success in the research practice in the last decades of the 20th century. Following the development of EB technologies as high-performance units, it has also established itself as a promising industrial method for obtaining particularly pure metals and alloys in vacuum in [4-6].

Most of the metal impurities released in the EBM of copper are valuable metals, which in their pure form are needed in a number of industries. The ratio between the evaporated impurities and the base (main) metal and their concentration in the condensate differ depending on the technological processes used for the refining process.

[1] To whom any correspondence should be addressed.

Content from this work may be used under the terms of the Creative Commons Attribution 3.0 licence. Any further distribution of this work must maintain attribution to the author(s) and the title of the work, journal citation and DOI.
Published under licence by IOP Publishing Ltd

This paper presents the results of studies where EBM is applied at a certain stage of the classical metallurgical process for the production of copper – before the electrolysis of fire-refined copper intended for anodes in electrolysis. The refining processes are discussed and the quantities are evaluated of some valuable metals (such as Ni, As, Bi, Pb, etc.) released by evaporation in the technological schemes studied.

2. Refined copper production stages

The classic scheme in the production of refined copper includes obtaining a flotation-processed concentrate – blister (black) copper with a high content of impurities (such as As, Sb, Bi, Ni, Zn, Sn, Pb, Fe, S, Al, Ag, Au, Se, Te) and gases. Blister copper is processed sequentially using pyrometallurgical processes (fire refining) to obtain copper with a purity of 99.5 % and electrolytic refining to obtain cathode copper with a purity of 99.95 % – 99.98(9) %.

In fire refining of blister copper, the separation of impurities is incomplete; for some of them, such as Au, Ag, Se and Te, it is even impossible. Due to their lower affinity to oxygen compared to that to copper, the precious metals do not oxidize and do not pass into the oxidized slag. The oxygen content in the obtained fire-refined (pyro-refined) copper is as high as 0.1% [7].

The electrolysis of fire-refined copper is a mandatory next step in the overall refining process of copper, which allows the extraction of precious metals and rare elements (such as Au, Ag, Se, Te, Pd, Pt, etc.). The efficiency of the electrolysis process strongly depends on the type and concentration of impurities in the anode (fire-refined copper). The presence of the impurities Ni, Zn, Fe with electronegativity greater than that of copper and well soluble in the sulfuric acid electrolyte, even in small quantities, leads to their accumulation in the electrolyte. This hinders and complicates the electrolysis process.

The impurities Sn and Pb, which are also more electronegative than copper, pass into the copper slurry and hamper the extraction of precious metals. The metals with potentials close to that of the copper slurry, such as As, Sb and Bi, are the most harmful to the electrolysis process. They get both into the electrolyte and the slurry, and under certain conditions are deposited on the cathode together with the copper thus contaminating it.

Due to the great flexibility in choosing the parameters and conditions for the processes leading to reducing the volatile impurities in metals, as well as in choosing the type, size, shape and amount of starting material, the EB method can be successfully used in the refining of both enriched and separated natural concentrate obtained after thermal reduction, in the production of particularly pure metals (such as W, Zr, Ti, Mo, etc.) and for processing metal waste from metallurgical production and mechanical treatment of valuable metals, which have been highly oxidized during their previous processing (such as Ti, Hf, Zr, Ta, Co, etc.).

3. Results and discussion

The experiments reported here were performed in the Physical Problems of Electron Beam Technologies Laboratory of the Institute of Electronics, Bulgarian Academy of Sciences, using the ELIT-60 electron beam installation for melting and refining. The maximum beam power is 60 kW and the working vacuum pressure is $3 - 6\times10^{-3}$ Pa [6, 4].

The experiments reported here were conducted with the purpose of EB refining of pyro-refined copper (Cu 99.239), which is the anode in classic electrolysis for the production of electrolytic copper.

The starting (initial) material was subjected to a process of oil removal. The chemical composition variation of the metal samples before and after e-beam refining was determined by using emission spectral analysis.

A series of EBM experiments were performed with three different e-beam melting powers (P_b) of 7.5 kW (T = 1400 K), 10 kW (T = 1500 K) and 15 kW (T = 1700 K) for different time of refining (τ) – 3 min, 10 min, 15 min, 20 min and 30 min. Each metal sample (specimen of about 1260 g) was placed on a water-cooled copper crucible and was refined by a single procedure.

For each EB processing regime, the changes were monitored in the chemical composition and weight of the metal ingots (material losses ΔG). The degree of refining (removal efficiency) was evaluated for each of the controlled impurities (η_i) and the overall degree of refining (η_{tot}) was calculated for the regimes studied.

Table 1 presents data on the concentrations of impurities in fire-refined copper (Cu 99.239) and anode copper (Cu 99.45, the anode residue after the electrolysis process). The data in table 1 show that before the electrolysis process the copper of the anode (pyro-refined copper) has a content of Sb, Sn, Pb, Ni, Zn higher than the content in the anode residue after the electrolysis. During the electrolysis process, these impurities get into the copper slurry, into the electrolyte, and even onto the cathode and their presence impedes the full flow of the refining processes and reduces the efficiency of the electrolysis refining.

Table 1. Impurity concentrations (mass %) of pyro-refined copper and anode copper.

impurity	pyro-refined Cu	anode Cu
As	0.05	0.07
Sb	0.05	0.0095
Sn	0.05	0.0034
Bi	0.003	0.015
Pb	0.05	0.021
Ni	0.4	0.1
Ag	0.03	0.05
Zn	0.0282	0.0023
O	0.1	0.2251

In the work reported, the investigated copper material (pyro-refined copper) contained 99.239 % Cu and Sb, Sn, Pb, Ni, and Zn with high concentrations (table 1). The other controlled impurities in this study were arsenic, bismuth, silver, and oxygen.

The values of the removal efficiency (η_i) for each of the controlled impurities, the total (overall) removal efficiency (η_{tot}) (removal efficiency of all the impurities in the sample) for the investigated EB process parameters and the material losses (ΔG) were evaluated (table 2).

Table 2. Removal efficiency of specific impurities (η_i), overall removal efficiency (η_{tot}) and material losses (ΔG) in e-beam processing of pyro-refined copper.

P_b [kW]	τ [min]	η_{As} [%]	η_{Sb} [%]	η_{Sn} [%]	η_{Bi} [%]	η_{Pb} [%]	η_{Ni} [%]	η_{Ag} [%]	η_{Zn} [%]	η_{O} [%]	η_{tot} [%]	ΔG [g]
7.5	**3**	97.6	94.7	97.9	95.0	92.8	98.7	97.3	99.6	70.0	78.5	0.35
	10	97.7	94.8	98.0	94.3	93.0	98.8	97.3	96.4	96.0	97.5	0.76
	15	97.7	95.0	98.0	94.0	93.2	98.8	97.4	94.0	95.0	97.3	1.11
	20	97.8	95.1	98.1	94.0	93.4	98.8	97.4	93.6	94.0	97.2	1.46
	30	97.8	95.3	98.3	93.3	93.8	98.9	97.5	96.1	90.0	96.9	2.14
10	**3**	97.9	96.2	98.9	95.3	94.9	98.8	97.4	95.4	75.0	95.0	1.54
	10	98.0	96.3	99.0	94.7	95.1	98.9	97.4	100	97.0	98.1	4.28
	15	98.0	96.3	99.0	94.3	95.2	98.9	97.4	100	96.0	98.0	6.24
	20	98.0	96.4	99.1	93.7	95.3	98.9	97.4	100	93.0	97.6	8.20
	30	98.1	96.6	99.1	93.0	95.6	98.9	97.5	87.9	88.0	96.6	12.12
15	**3**	98.48	99.12	100	96.00	99.00	98.98	97.70	100	79.00	96.37	4.04
	10	98.48	99.12	100	95.00	99.02	98.98	97.6	98.23	97.00	98.66	11.31
	15	98.56	99.12	100	94.33	99.02	98.98	97.53	97.87	98.00	98.79	16.50
	20	98.48	99.12	100	93.67	99.04	98.98	97.43	97.52	98.9	98.88	21.69
	30	98.50	99.12	100	92.33	99.06	98.99	97.3	66.31	99.00	98.90	32.07

The material losses were estimated using the weight of the initial metal specimens and the obtained refined ingots. The removal efficiency of a specific impurity η_i and the overall removal efficiency η_{tot} were calculated using:

$$\eta_i = \frac{C_{0,i} - C_{r,i}}{C_{0,i}} 100\% \quad (1)$$

$$\eta_{tot} = \frac{C_{0,tot} - C_{r,tot}}{C_{0,tot}} 100\% \quad (2)$$

where $C_{0,i}$ and $C_{r,i}$ are the specific impurity concentrations of the starting material (before EBM) and of the ingot after e-beam refining, respectively; $C_{0,tot}$ and $C_{r,tot}$ are the overall concentrations of the all impurities of the initial material and of the ingot after refining, respectively.

It is seen that the increase of the electron beam power (P_b) leads to an increase in the removal efficiency (i.e., a decrease in the concentration $C_{r,i}$ after refining, equation (1)) of As, Sb, Sn, Pb, and Ni. At the highest beam power (P_b = 15 kW), the refining efficiency of these impurities (As, Sb, Sn, Pb, Ni) is the highest (table 2), i.e., the content of As, Sb, Sn, Pb, Ni is the lowest. At the beam power of 7.5 kW, the degrees of refining of these impurities are the highest for the longest refining time applied (30 min).

The results obtained show that the values of the degree of refining (η_i) of all the controlled impurities and the total removal efficiency (η_{tot}) for all the investigated process regimes are high (70 % – 100 %, table 2) and the impurities were eliminated to low levels (equations (1, 2)). At a certain temperature (beam power), the duration of the refining time did not significantly affect the As, Sb, Sn, Pb, Ni, and Ag removal.

Also, the increase of the operating temperature (beam power) led to an increase in the overall removal efficiency (η_{tot}) and in the material losses (table 2), which are mainly due to evaporation.

We evaluated the process of extraction of some valuable metals (such as Ni, As, Bi, Pb, Sb, Ag, etc.) generated in the condensate during their removal from the liquid metal in the refining process; their concentrations ($C_{i,cond}$) in the condensate for the regimes studied are presented in table 3.

Table 3. Impurities' concentration in EBM of pyro-refined copper.

P_b [kW]	τ [min]	$C_{As,cond}$ [%]	$C_{Sb,cond}$ [%]	$C_{Sn,cond}$ [%]	$C_{Bi,cond}$ [%]	$C_{Pb,cond}$ [%]	$C_{Ni,cond}$ [%]	$C_{Ag,cond}$ [%]	$C_{Zn,cond}$ [%]	C_{Cu} [%]
7.5	**3**	6.00	6.00	6.00	0.28	5.71	48.57	3.71	3.43	99.837
	10	4.74	4.61	4.74	0.26	4.47	38.16	2.76	2.76	99.981
	15	2.88	2.79	2.79	0.16	2.79	23.42	1.71	1.53	99.980
10	**3**	2.34	2.34	2.34	0.13	2.27	18.83	1.43	1.3	99.962
	10	0.7	0.7	0.7	0.05	0.7	5.61	0.42	0.5	99.986
	15	0.59	0.26	0.59	0.02	0.26	3.21	0.24	0.22	99.985
15	**3**	3.46	3.46	3.46	0.2	3.46	28.46	2.1	2.03	99.972
	10	2.21	2.21	2.3	0.13	2.21	17.95	1.33	1.24	99.898
	15	1.52	1.52	1.58	0.13	1.52	12.42	0.91	0.91	99.991

The total amount of evaporated metal increases with increasing the duration of the EB impact on the liquid metal (table 2), but the concentrations of metal impurities in the separated condensate decrease significantly (table 3). Therefore, a high degree of Cu refining from the metal inclusions concentrated in the condensate deposited on the surface of specially placed screens is achieved in EB refining via a short impact of the low-power beam.

The results obtained show that Ni as the impurity with the highest concentration in the source (initial) material and with a high degree of refining (i.e., a low Ni content in the ingots after EBM) for the regimes studied was generated in the largest quantity in the condensate. For a beam power of 7.5 kW and a short refining time (3 min and 10 min), processing of 1 kg of condensate resulted in obtaining nickel quantity as high as 485.7 g and 381.6 g (table 3). The evaluations made showed that up to 60 g of metals (such as As, Sb, Sn, Bi, Pb, Ag and Zn) can also be produced by processing the condensate.

It was further found that after EB refining of pyro-refined Cu (99.239 %) (before electrolysis), the copper purity increased by two orders of magnitude and the best purification of Cu (99.991 %) was achieved at a beam power of 15 kW for a refining time of 15 min (table 3).

The EB refining of pyro-refined Cu (99.239 %) (before electrolysis) raised the copper purity by two orders of magnitude (table 3). For a 15-kW beam power and a refining time of 15 min, the total refining efficiency (η_{tot}) reached 98.79 % (table 2) with a metal purity of 99.991 (table 3). Under these process conditions, the degree of oxygen refining was high: $\eta_O = 98$ % (table 2), resulting in a high reduction in the oxygen content (0.002 %) in the refined material (equation (1)). In this regime, a balance was achieved between the high overall degree of refining and the small weight losses (table 2), resulting in a refined Cu (99.991) with a very high purity (table 3).

4. Conclusions

In the present work, the effect was examined of applying the e-beam method to the processing of pyro-refined Cu and obtaining multiple metals under different melting conditions. The influence was evaluated of the beam power and the refining time on the refining efficiency for each of the controlled impurities, on the overall removal efficiency, on the extraction of some valuable metals (such as Ni, As, Pb, Sb, Ag) generated in the condensate, and on the degree of refining of copper.

The degree of refining of each of the controlled metal impurities (such as As, Sb, Sn, Bi, Pb, Ni, Ag and Zn) was high (92 % – 100 %) – their amounts in the base material was efficiently reduced. The degree of oxygen removal was in the range of 70 % – 99 % for the process parameters applied. The copper purity increased by two orders of magnitude (Cu 99.991 %) after EB refining of pyro-refined copper (Cu 99.239 %).

For an electron beam power of 7.5 kW and refining times of 3 min and 10 min, about 485.7 g and 381.6 g of Ni could be obtained, respectively, from the processed condensate. Under these melting conditions, up to 37 g of silver could also be generated in the condensate, while the purity of refined copper reached Cu 99.98 %.

Using the electron beam refining method for purifying pyro-refined copper leads to the removal of impurities, such as Ni, Zn, Sn, Pb, As, Sb and Bi, that are detrimental to the process of electrolysis, and to obtaining refined copper with a high purity and a low oxygen content

Acknowledgements

The work was supported by the Bulgarian National Science Fund under contract DN17/9.

References

[1] Schlesinger M E, King M J, Sole K C and Davenport W G 2011 *Extractive Metallurgy of Copper*, (Elsevier)
[2] International Copper Study Group 2020 *The World Copper Factbook 2020,* http://www.icsg.org/index.php/component/jdownloads/finish/170/3046
[3] European Copper Institute *Copper: A circular material for a resource efficient Europe* https://copperalliance.eu/benefits-of-copper/recycling/
[4] Vutova K and Vassileva V 2020 *Met Sci Heat Treat* **62** 345
[5] Vutova K, Vassileva V, Stefanova V and Naplatanova M 2020 *in 11th Int. Symposium on High-Temperature Metallurgical Processing. Minerals, Metals and Materials Series,* eds. Z Peng et al. (Springer) pp 941-951 DOI: 10.1007/978-3-030-36540-0_84
[6] Mladenov G, Koleva E, Vutova K and Vassileva V 2011 *in book Practical Aspects and Application of Electron Beam Irradiation*, eds. M Memtanu and M Brasoveanu (Transworld Research Network India) chapter 3 pp 43–93
[7] Vaniukov A V and Zaitzev V J 1993 *Theory of pyrometallurgical processes* (Moscow: Metall. Publ. House, in Russian)

Coherent optical spectroscopy characterization of the magnetic properties of oriented Fe_3O_4 nanoparticles

E Taskova[1,5], E Alipieva[1], S Kolev[1,2], T Koutzarova[1] and D Brazhnikov[3,4]

[1]Academician Emil Djakov Institute of Electronics, Bulgarian Academy of Sciences 72 Tsarigradsko Chaussee, 1784 Sofia, Bulgaria
[2]Neofit Rilski South-Western University, 66 Ivan Mihailov Str., 2700 Blagoevgrad, Bulgaria
[3]Institute of Laser Physics, Siberian Branch, Russian Academy of Sciences, 15B Lavrentyev Ave, 630090 Novosibirsk, Russia
[4]Novosibirsk State University, 1 Pirogov Str., 630090 Novosibirsk, Russia

E-mail: taskova@ie.bas.bg

Abstract. Magnetite (Fe_3O_4) nanoparticles are widely used in microwave components, such as microwave absorbers and anti-reflection coatings. Their magnetic properties are important for these applications. In this work, the weak magnetic field created by the Fe_3O_4 nanoparticles was estimated using a coherent-spectroscopy all-optical method for measurement of weak magnetic fields. Both the magnitude and the direction of the magnetic field were evaluated through the position and amplitude of a magneto-optical resonance obtained in a paraffin-coated sensor cell containing Rb vapors. A pump-probe scheme was used to prepare a high-contrast magneto-optical resonance. Different possible geometries related to this concept are discussed, including sensor miniaturization.

1. Introduction

The all-optical methods for magnetic field measurements have recently attracted the attention and interest of many scientists due to the potential for miniaturizing the magnetic sensors and their ability to detect very weak magnetic fields – in the range of 10^{-9} – 10^{-13} T. The magnetic field (MF) sensors are based on the nonlinear interference effects arising after the interaction of coherent light fields with degenerate atomic energy levels [1, 2].

Our experiment is based on the electromagnetically-induced absorption (EIA) [3, 4] in the so-called "Hanle configuration", where a magneto-optical resonance is observed by addressing the optical transition of the Rb D_1 line using a single-frequency coherent light field. The Hanle-type resonance is detected by scanning an externally applied magnetic field (MF) around its zero value. At zero magnetic field, the Zeeman sub-levels involved degenerate and the coherent light field creates alignment/orientation of the atomic magnetic moments. This alignment vanishes rapidly when the degeneracy is removed by the applied external MF. As a result, a resonance structure with a nearly Lorentzian shape is observed by detecting the laser beam transmitted through the cell with the position of its extremum situated exactly at zero MF. Any additional magnetic field created along with the

[5] To whom any correspondence should be addressed.

Content from this work may be used under the terms of the Creative Commons Attribution 3.0 licence. Any further distribution of this work must maintain attribution to the author(s) and the title of the work, journal citation and DOI.
Published under licence by IOP Publishing Ltd

scanned MF will shift the position of the resonance on the magnetic field scale – its value can thus be measured [5].

The aim of this work is to investigate the magnetic properties of nanocomposite Fe_3O_4 bulk samples. To achieve a high-contrast magneto-optical resonance, the two-beam pump-probe scheme proposed in [6] was used.

2. Experiment

In our study, we prepared composite samples with 30-nm Fe_3O_4 particles as a filler. The ferrite powders were dispersed homogeneously in the polymer matrix (epoxy resin). During the process of polymerization, the sample was placed in a magnetic field with a strength of 0.4 T. The monodomain particles were thus oriented along the magnetic force lines of the external magnet [7, 8]. Figure 2 (b) presents a photograph of the sample having a diameter (D) of 26 mm and a thickness (t) of 2 mm used in this test experiment. To characterize its magnetic properties, we used the experimental set-up shown in figure 1 (a).

The MF sensor is a paraffin-coated ^{87}Rb cell, which is 2 cm in diameter (d_{cell}) and 2.5 cm long (L_{cell}) and heated by a bifilar coil to maintain a temperature of 50 °C. The cell is placed in a 30-cm long solenoid providing a homogeneous dc magnetic field along the cell axis. Both are enclosed in a two-layer magnetic shield that isolates the laboratory magnetic field. A single-mode diode laser (DL), oscillating at $\lambda = 794.76$ nm, two polarization beam splitters (PBS) and a system of mirrors are used to form two laser beams with orthogonal polarizations counter-propagating through the cell and perfectly overlapping in space. A $\lambda/2$ plate is used to redistribute the optical power between the probe and pump waves and to control the laser polarization. The laser frequency is controlled by monitoring the fluorescence from a second (reference) Rb vapor cell. The sample with oriented Fe_3O_4 nanoparticles is placed next to the cell. At each position, the magnetic field created by the sample was measured at two directions of the nanoparticle orientation – along the scanned magnetic field, (||) $\mathbf{B}_{scan}$, and perpendicularly to it, (⊥) $\mathbf{B}_{scan}$.

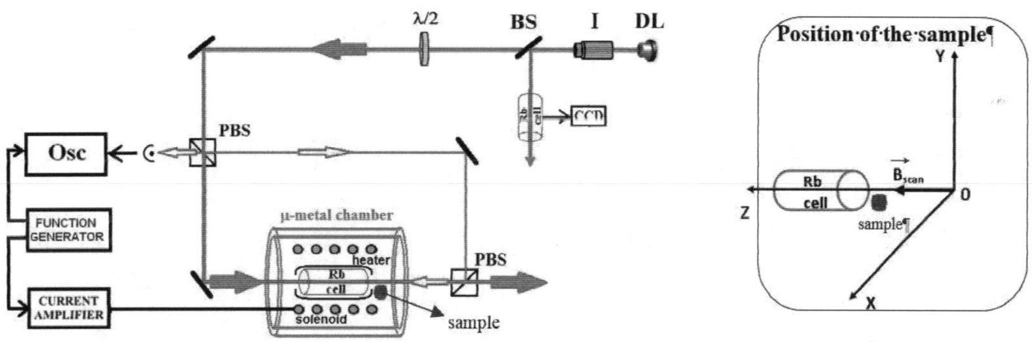

Figure 1. (a) Experimental set-up. **DL** – diode laser, **I** – Faraday optical isolator, **PBS** – polarizing beam-splitter; filled arrows mark the pump beam path; the empty ones show the probe beam path. **(b)** Sample position.

Our experiment was performed at the $5^2S_{1/2} \rightarrow 5^2P_{1/2}$ transition ($F_g = 2 \rightarrow F_e = 1$) of the ^{87}Rb D_1 line. The coherent field of the pump beam aligns the atomic media and the probe beam absorption detects this alignment. The absorption of the probe beam was registered depending on the dc magnetic field $\mathbf{B}_{scan}$ scanned at a frequency 0.33 Hz around zero value with amplitude ± 40 mG and ± 400 mG.

3. Experimental results and discussion

The EIA resonances were registered for a pump beam power $P_{pump} = 500$ μW and a probe beam power $P_{probe} = 8$ μW. The beam diameters (d_{beam}) were in the order of 4 mm. For the magnetic field measurement, we used the distinguishing feature of the Hanle-type EIA resonances: they arise always with a maximum situated at the point where the MF is zero. An additional magnetic field parallel to the laser beam propagation direction does not disturb the resonance shape, but only moves its position on the magnetic field scale. The MF values are deduced by determining the position of the coherent resonance on the magnetic field scale. The difference between the actual position of the resonance and the position of zero MF created by the solenoid (determined when the current through the solenoid is zero) gives us the value of the MF component, as shown in figure 2 (a). In this way, not only the value but the sign of the MF can be determined as well.

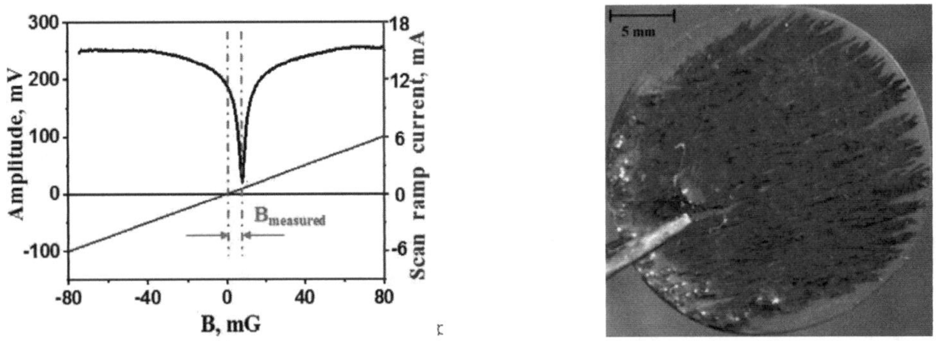

(a) (b)

Figure 2. (a) Method for weak MF measurement. **(b)** Photograph of the sample: *D* 26 mm; *t* 2 mm.

The sample containing Fe_3O_4 nanoparticles with oriented magnetic moments (figure 2 (b)) was placed close to the Rb cell, consecutively at three positions (in the planes XY – upright, facing the cell's window; YZ – upright, transverse, and ZX – lying in front of the cell). We measured the magnetic field

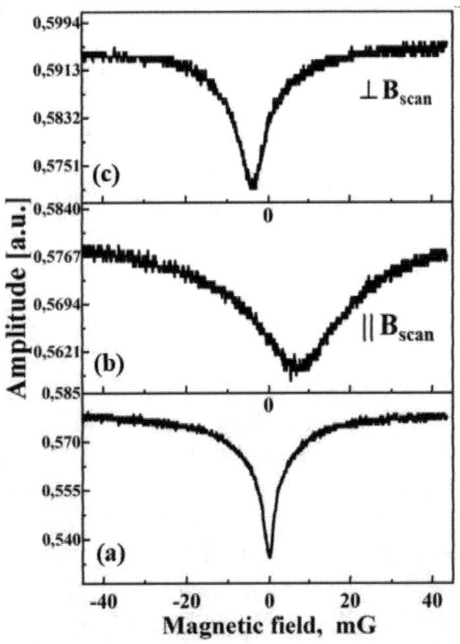

Figure 3. (a) The reference resonance position. The EIA resonance shift when the oriented magnetic moment is: **(b)** || and **(c)** ⊥ to the $\mathbf{B}_{scan}$. The sample is lying in plane ZX.

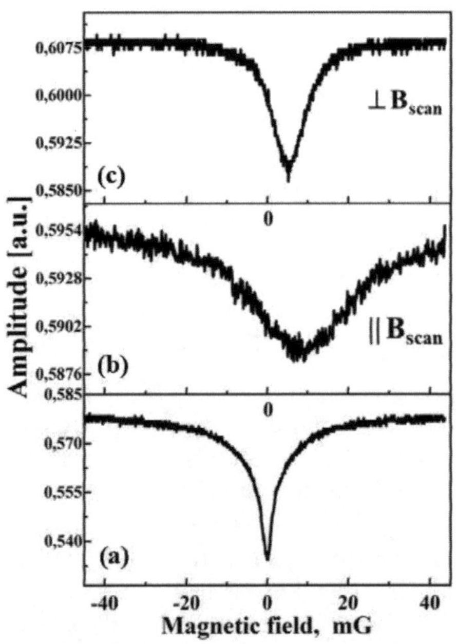

Figure 4. (a) The reference resonance position. The EIA resonance shift when the oriented magnetic moment is: **(b)** || and **(c)** ⊥ to the $\mathbf{B}_{scan}$. The sample upright, in plane YZ.

created by the sample in the volume where the laser beam interacts with Rb atoms. Figures 3, 4 and 5 represent the measured probe-beam transmission for the three positions of the sample, respectively.

Figures 3 (a) and 4 (a) show the reference position of the EIA resonance, without the sample. The MF created by the sample shifts it on the MF scale. The nanoparticles magnetic moments orientation in figure 3 (b) and figure 4 (b) is similar along **0**Z and, as a result, the response of the medium is the same. The width and the shift of the resonances are equal. The measured magnetic field created by the sample at these positions are (6.55$\pm$0.57) mG. The cylindrical geometry of the cell in the present experiment results in the identity of the two of these three positions.

The MF created by the nanoparticles magnetic moments orientation in positions orthogonal to $\mathbf{B}_{\mathbf{scan}}$, figures 3 (c) and 4 (c), is of the same order – ~ 5 mG, but with opposite signs. The MF direction depends on the polarity orientation of the nanoparticles magnetic moments. In figure 3 (c), this orientation is along **0**X; in figure 4 (c), along **0**Y.

The response of the Rb medium when the sample is placed in the XY plane – facing the cell's window is shown in figure 5. The particles orientation is along **0**Y. The MF created at this position is (7.26 ± 1.21) mG, but the width of the resonance is around ten times larger. This is because the axial dimension of the volume of Rb vapor illuminated by the laser beam (L_{cell} = 20 mm) is much larger than its radial dimension (d ~ 4 mm = d_{beam}). As the recorded EIA signal results from the integral contribution of all Rb atoms inside the illuminated volume, it integrates the existing MF gradients in this volume. The XY-position of the sample creates a much larger MF gradient along the **0**Z-axis of the illuminated volume, compared to the gradient created in the other two cases. When the sample is situated in the XZ- and the YZ-planes, it causes a rather homogeneous MF along **0**Z and a far smaller gradient along the **0**X and **0**Y-axis. The difference of the resonance widths is also due to the variation of the radial field created in the three sample positions [9, 10].

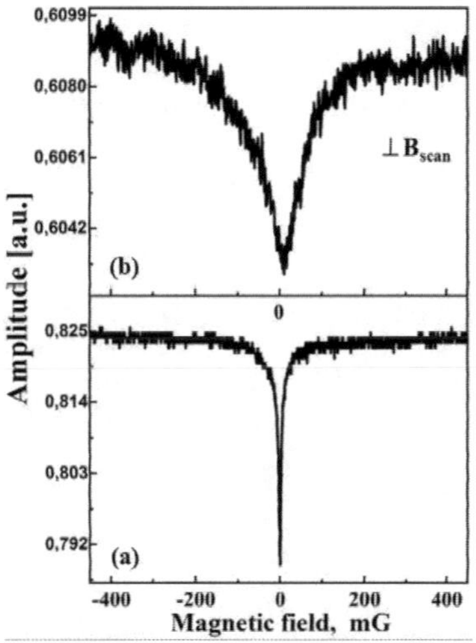

Figure 5. (a) The reference resonance position. **(b)** The EIA resonance shift when the oriented magnetic moment is ⊥ to the $\mathbf{B}_{\mathbf{scan}}$. The sample is in the XY-plane.

When using miniaturized cells with alkali metal vapors, the situation changes and the recorded values of the magnetic field will characterize the sample locally. With this size ratio, very detailed data can be obtained on the distribution of the magnetic field along the studied object. We plan to explore such an arrangement in the future.

4. Conclusions

In the present work, we applied an all-optical method based on coherent spectroscopy to characterize the magnetic field created by Fe_3O_4 nanoparticles with oriented magnetic moments. The magnitude of the magnetic field generated by the sample when it was located at three different positions in front of the sensor was measured to be of the same order (~ 5 – 7 mG). The sensitivity of the measurement method strongly depends on the ability to accurately determine the resonance maximum position, which is closely related to the resonance width. For the case where the sample has dimensions smaller than or comparable to the dimension of the sensor Rb-cell, we demonstrated that the least favorable sensor's position is the one creating the larger MF gradient in the illuminated volume of the Rb-cell. To measure more precisely the MF of a similar sample, a miniaturized Rb-cell sensor with dimensions smaller than the ones of the sample will be more appropriate. In such a case, the magnetic field gradients will be reduced significantly, leading to an increased sensitivity to the magnetic field. It will also allow creating maps of the MF around the sample.

Acknowledgments
This work was supported by the Bulgarian National Science Fund under contracts: (i) bilateral project Bulgaria-Russia, "Investigation of new magneto-optical effects in alkali-metal atoms prepared in long-lived coherent states for application to quantum magnetometry" (BNSF Grant no. KP-06-Russia/11, RFBR Grant no. 20-52-18004) and (ii) KP-06-N48/5 "Multifunctional composite structures based on ferrites (including magneto-electrics) and carbon materials".

References
[1] Alexandrov E, Chaika M and Hvostenko G 1991 *Interference of Atomic States* (Berlin: Springer) (https://doi.org/10.1007/978-3-642-84442-3)
[2] Arimondo E 1996 *Prog. Opt.* **35** 257–354
[3] Akulshin A M, Barreiro S and Lezama A 1998 *Phys. Rev.* A **57** 2996–3002
[4] Dancheva Y, Alzetta G, Cartaleva S, Taslakov M and Andreeva C 2000 *Opt. Commun.* **178** 103–110
[5] Alipieva E, Gateva S and Taskova E 2005 *IEEE Trans. Instrum. Meas.* **54** (2) 738–741
[6] Brazhnikov D V, Taichenachev A V, Tumaikin A M and Yudin V I 2014 *Laser Phys. Lett.* **11** 125702
[7] Kolev S, Yanev A and Nedkov I 2006 *Phys. Stat. Sol.* (c) **3** 1308–1315
[8] Koutzarova T, Kolev S, Ghelev Ch, Paneva D and Nedkov I 2006 *Phys. Stat. Sol.* (c) **3** 1302–1307
[9] Taskova E, Alipieva E and Todorov G 2018 *J. Phys. B: At. Mol. Opt. Phys.* **51** 035005
[10] Brazhnikov D, Ignatovich S, Mesenzova I, Novokreshchenov A and Goncharov A 2020 *Opt. Lett.* **45** 3309–3312

Effect of cation substitutions in Y-type $Ba_{0.5}Sr_{1.5}Me_2Fe_{12}O_{22}$ hexaferrites on the magnetic phase transitions

B Georgieva[1,6], S Kolev[1,2], Ch Ghelev[1], K Krezhov[1], D Kovacheva[3], B Vertruyen[4], R Closset[4], L-M Tran[5], M Babij[5], A Zaleski[5] and T Koutzarova[1]

[1]Institute of Electronics, Bulgarian Academy of Sciences, 72 Tsarigradsko Chaussee, 1784 Sofia, Bulgaria

[2]Neofit Rilski South-Western University, 66 Ivan Mihailov Str., 2700 Blagoevgrad, Bulgaria

[3]Institute of General and Inorganic Chemistry, Bulgarian Academy of Sciences, Acad. Georgi Bonchev Str., bld. 11, 1113 Sofia, Bulgaria

[4]Greenmat, Chemistry Department, University of Liege, 11 allée du 6 août, 4000 Liège, Belgium

[5]Institute of Low Temperature and Structure Research, Polish Academy of Sciences, ul. Okólna 2, 50-422 Wrocław, Poland

E-mail: b.georgiewa@abv.bg

Abstract. We investigated the magnetic properties and magnetic phase transition in Y-type $Ba_{0.5}Sr_{1.5}NiMgFe_{12}O_{22}$ hexaferrite powder prepared by citrate sol-gel spontaneous combustion. The saturation magnetisation value of 32 emu/g at 4.2 K was lowered to 24 emu/g at 300 K. The magnetisations curves did not saturate even at a magnetic field of 50 kOe for both temperatures – 4.2 K and 300 K. A step-like behaviour appeared in the initial magnetisation curve at 4.2 K. A magnetic phase transformation from a spiral magnetic ordering to a conical spin one was observed at 40 K.

1. Introduction

The studies of the magnetic properties and, in particular, of the magnetic phase transitions in Y-type hexaferrites during the last decade were provoked by the observation of a distinct magnetoelectric effect close to room temperature. This gave rise to opportunities of broadening the field of Y-type hexaferrites applications: multi-ferroic gyrators, electrical-field-controlled spintronic devices, ac/dc magnetic and electric field sensors in, e.g., biomedicine, power generators, electrical-field-controlled microwave components, e.g., filters and switches for wireless technologies, among many others [1-3]. The magnetoelectric effect in hexaferrites was first observed by T. Kimura et al. [4] in a $Ba_{0.5}Sr_{1.5}Zn_2Fe_{12}O_{22}$ single crystal in a magnetic field of about 1 T near room temperature, with the effect being significant at $T < 130$ K [5]. $Ba_{0.5}Sr_{1.5}Zn_2Fe_{12}O_{22}$ is a Y-type hexaferrite ($A_2Me_2Fe_{12}O_{22}$), with A = Ba or Sr and Me, a divalent cation. It is a dielectric ferromagnetic material with a spiral spin order stabilised beneath its Neel temperature ($T_N \approx 326$ K); it goes through several thermally-induced magnetic phase transitions [4, 6]. In general, the hexaferrites $Ba_2Me_2Fe_{12}O_{22}$ assume a $R\bar{3}m$ space group structure. The Me^{2+} and

[6] To whom any correspondence should be addressed.

Content from this work may be used under the terms of the Creative Commons Attribution 3.0 licence. Any further distribution of this work must maintain attribution to the author(s) and the title of the work, journal citation and DOI.
Published under licence by IOP Publishing Ltd

Fe^{3+} ions are distributed over six distinct crystallographic sites, both tetrahaedral and octahaedral. The type of these divalent cations and the sites they occupy within the lattice cell can give rise to significant changes in these materials′ structural and magnetic properties. Their unit cell comprises three formula units and is constructed by a sequential arrangement of the S and the T-blocks, namely, TST′ST″S″, the primes denoting a 120° rotation around the *c*-axis; the easy magnetisation axis is in a plane perpendicular to the *c* axis [7]. The magnetic configuration can be presented by L and S blocks lumped alternately in parallel with the [001] axis with oppositely directed large and small magnetisation *M*, respectively, thus forming the magnetic sublattice.

The studies on the magnetic properties and magnetoelectric effect have been mainly focused on the $Ba_{0.5}Sr_{1.5}Zn_2Fe_{12}O_{22}$ hexaferrite and on the effect of substituting Zn with other cations. The results of our previous works [8, 9] on replacing the non-magnetic Zn^{2+} with the magnetic Ni^{2+} cations in $Ba_{0.5}Sr_{1.5}Zn_{2-x}Ni_xFe_{12}O_{22}$ demonstrated that a half substitution of Zn^{2+} with Ni^{2+} displaces the helical spin magnetic order to higher temperatures. Also, the electrical polarisation caused by a weak magnetic field in $Ba_2Mg_2Fe_{12}O_{22}$ has to do with the Mg^{2+} cations reducing the magnetic anisotropy [10]. Zhang et al. [11] found an increase of the temperature of transformation from a proper screw spin ordering to a collinear ferrimagnetic spin ordering in $Ba_{0.5}Sr_{1.5}Zn_{2-x}Mg_xFe_{12}O_{22}$ due to the Mg doping. Bearing this in mind, in this paper we present a study on the effect of combining the magnetic Ni^{2+} and the non-magnetic Mg^{2+} on the magnetic parameters and magnetic-phase transitions in the hexaferrite $Ba_{0.5}Sr_{1.5}NiMgFe_{12}O_{22}$ powder.

2. Experimental

The $Ba_{0.5}Sr_{1.5}NiMgFe_{12}O_{22}$ powder was prepared by citrate sol-gel spontaneous combustion. The procedure started with mixing the respective metal nitrates; solution of citric acid ($C_6H_8O_7$) was carefully introduced to the mixture as a chelating agent, thus forming a homogenous solution with the metal cations. At 120 °C by slow dehydration, it was converted to a dark brown bulk, which self-ignited afterward and burned. The spontaneously combusted powders were annealed at 600 °C. The precursor powders thus produced were homogenized and retained in an oven at 1170 °C for five hours in air to obtain the material of desired composition $Ba_{0.5}Sr_{1.5}NiMgFe_{12}O_{22}$.

The structure and phase content of the sintered powder material were explored by X-ray diffraction (XRD) on a Brucker D8 instrument (40 kV, 30 mA, Cu-K_α radiation, 1.5418 Å) operated by DIFFRACTPLUS software in Bragg-Brentano geometry. The magnetic characteristics were determined by a PPMS magnetometer (Quantum Design). The hysteresis measurements were conducted at 4.2 K and at room temperature. The measurements of the magnetisation-*vs*-temperature dependence (4.2 – 300 K) were carried out by first cooling the sample from 300 K down to 4.2 K with no magnetic field applied (zero-field cooling, ZFC); the magnetisation was then measured as the temperature was raised from 4.2 K to 300 K at a rate of 3 K/min in a 50-Oe magnetic field. The field cooling (FC) plot was registered by cooling the sample from 300 K to 4.2 K in the same magnetic field.

3. Result and discussion

The XRD pattern (figure 1) at room temperature of the synthesized $Ba_{0.5}Sr_{1.5}NiMgFe_{12}O_{22}$ powder was indexed successfully in the centrosymmetric $R\overline{3}m$ space group; it exhibits peaks characteristic of the Y-type hexaferrite structure studied as the major phase. The additional peaks indicate a minor phase amounting to 2.4 wt% and identified as belonging to the spinel $(MgNi)Fe_2O_4$.

As is known, Y-type hexaferrites are complex oxides, so that a number of obstacles are encountered in attempting to produce single-phase samples. The synthesis of Y-type hexaferrites is commonly accompanied by the formation of different oxides, usually spinels, mainly because the temperature range for synthesising the Y-phase is rather narrow. However, in what concerns the magnetic phase transitions, our previous studies [9, 12, 13] on other Y-hexaferrite compositions have shown that the trace amount of spinel ferrites in the material studied does not affect the magnetic-phase transition temperatures.

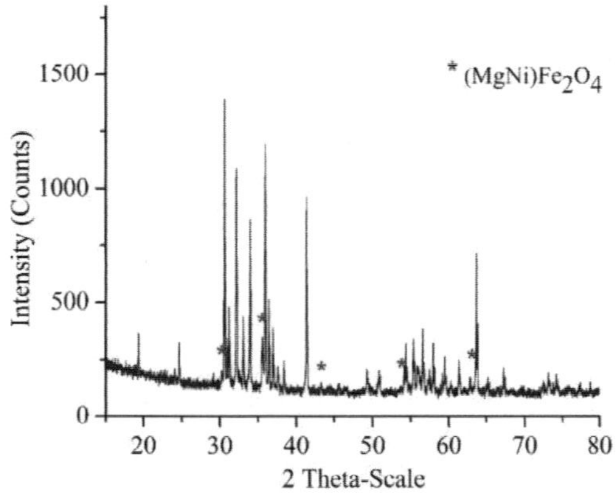

Figure 1. XRD pattern of $Ba_{0.5}Sr_{1.5}NiMgFe_{12}O_{22}$ powder.

In figure 2, we display the initial magnetisation and the hysteresis curves of the $Ba_{0.5}Sr_{1.5}NiMgFe_{12}O_{22}$ powder at 300 K and 4.2 K. As seen, the magnetisation curves for both temperatures do not reach saturation at the highest magnetic field used of 50 kOe, the highest magnetisation values being 32 emu/g and 24 emu/g respectively at 4.2 K and 300 K. The magnetisation values determined are typical for Y-type hexaferrites. At 300 K, the magnetisation grows sharply up to $H \approx 1$ kOe, then adheres to the behaviour typical for ferrimagnetic compounds. A stepwise behaviour is seen in the initial magnetisation plot at 4.2 K. Analogous behaviour has been described for a $Ba_{0.5}Sr_{1.5}Ni_2Fe_{12}O_{22}$ single crystal at 20 K, as well as for other Y-type hexagonal ferrites with distinct magnetoelectric coupling, e.g., $Ba_{0.5}Sr_{1.5}Co_2Fe_{12}O_{22}$ [14]; it points to transformations taking place between various magnetic orders effected by the magnetic field [14, 15]. Such a behaviour is usually associated with a spiral magnetic order and the presence of metamagnetic or intermediate phases [16]. In our case, three different magnetic phases are present depending of the applied magnetic field. The first one appears up to a magnetic field of 2.6 kOe; the second magnetic phase exists in the range 2.6 – 3.5 kOe. The magnetisation curves have a typical ferromagnetic behaviour for magnetic field values higher than 3.5 kOe.

The hysteresis curve at 300 K is very narrow with a coercive field H_c of 35 Oe. This value is representative for hexaferrites with a planar magneto-crystalline anisotropy. The H_c value is 44 Oe at 4.2 K. The remanent magnetisation is very low. The hysteresis curve behaviour at 4.2 K is similar to that of $Ba_{0.5}Sr_{1.5}Zn_{2-x}Mg_xFe_{12}O_{22}$ [11] pointing to the existence of two orderings both of ferrimagnetic type with differing magnetisation values.

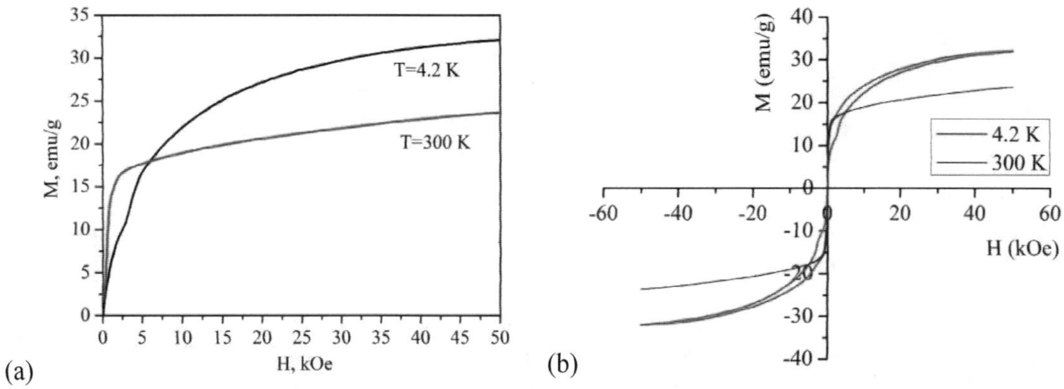

Figure 2. Magnetic field dependence of the initial magnetisation (a) and hysteresis curves (b) at 4.2 K and 300 K.

Figure 3 presents the ZFC and FC magnetisation plots of the $Ba_{0.5}Sr_{1.5}NiMgFe_{12}O_{22}$ powder in a 50-Oe magnetic field as depending on the temperature in the 4.2 – 300 K range. The data presented in the figure bear some comment: In both bulk and nanostructured ferrites of the system $Mg_xNi_{1-x}Fe_2O_4$, the reduction of the saturation magnetisation and magnetisation with the increasing magnesium content has been convincingly evidenced. According to literature data [17] and our previous studies [18], no magnetic phase transitions take place at low temperatures (e.g. below 100 K for $x = 0$) in these spinel ferrites. Moreover, in the ZFC and FC magnetisation graphs presented here, no behaviour is seen characteristic of a blocking temperature, as is commonly the case in spinel ferrites. Consequently, no contribution of the spinel phase (about 2.4 wt%) could be clearly distinguished in the material's magnetic response. The FC magnetisation curve begins decreasing gradually at a temperature of about 38 K, but around 150 K the trend changes to a smooth increase. Conversely, the ZFC magnetisation clearly increases as the temperature is raised. The maximum observed at 70 K is related to the intermediate magnetic phase transition. Another maximum at 40 K is also seen – it arises from the magnetic phase transformation from a spiral magnetic order to a conical spin order at low temperatures. Such a maximum was not observed in our earlier study on the $Ba_{0.5}Sr_{1.5}ZnNiFe_{12}O_{22}$ powder [9]; this indicates that replacing the Zn^{2+} ionic species with the smaller Mg^{2+} cations [19] leads to important changes in the paths of magnetic interactions leading to a conical spin structure arising at low temperatures. The ZFC and FC curves showing a stable magnetisation growth at temperatures approaching 300 K means that the temperature of transformation to a paramagnetic state exceeds this temperature, so that this transition could not be registered in the experiments reported here. In accordance with Hiraoka's works [14, 15] for $Ba_{0.5}Sr_{1.5}Ni_2Fe_{12}O_{22}$, one might expect that the transition from a screw-type magnetic ordering to a ferrimagnetic state above room temperature should take place via consecutive magnetic field-caused (metamagnetic) transformations related to the development of magnetic structures having to do with the magnetoelectric effect.

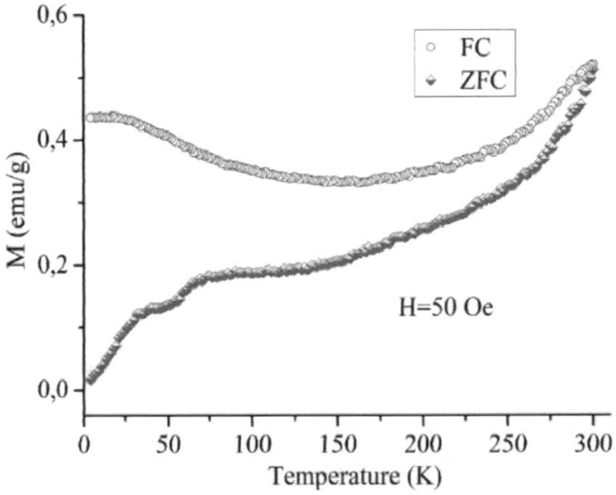

Figure 3. ZFC- and FC-magnetisation *vs* temperature of $Ba_{0.5}Sr_{1.5}MgNiFe_{12}O_{22}$ powder in a 50-Oe magnetic field.

4. Conclusions

$Ba_{0.5}Sr_{1.5}NiMgFe_{12}O_{22}$ polycrystalline powder was prepared by citrate sol-gel spontaneous combustion. The magnetisation loop taken at 4.2 K showed a behaviour pointing to the existence of two types of ferromagnetic orderings with differing magnetisation values. The transition to conical spin order at 40 K in a magnetic field of 50 Oe is known to bring about the appearance of strong magnetoelectric correlation. The combination of the magnetic Ni^{2+} and the non-magnetic Mg^{2+} instead of Zn^{2+} in $Ba_{0.5}Sr_{1.5}Zn_2Fe_{12}O_{22}$ shifts the transition temperature from a screw magnetic state to a ferrimagnetic arrangement at temperatures exceeding 300 K.

Acknowledgements
This study is financed by the BNSF, contract KP-06-N48/5 "Multifunctional composite structures based on ferrites (including magnetoelectrics) and carbon materials". B. Georgieva would like to acknowledge the assistance of the Bulgarian Ministry of Education and Science under the National Research Program "Young scientists and postdoctoral students" approved by DCM # 577/17.08.2018. The research is carried out in the framework of a joint project between the Bulgarian Academy of Sciences and WBI, Belgium, and a joint project between the Bulgarian Academy of Sciences and W. Trzebiatowski Institute of Low Temperature and Structure Research, Polish Academy of Sciences.

References
[1] Scott J F 2007 *Nat. Mater.* **6** 25
[2] Rao C N R, Sundaresan A and Saha R 2012 *J. Phys. Chem. Lett.* **3** 2237–2246
[3] Petrov R V, Murthy D V B, Sreenivasulu G and Srinivasan G 2013 *Microwave Opt. Technol. Lett.* **55** 533
[4] Kimura T, Lawes G and Ramirez A P 2005 *Phys. Rev. Lett.* **94** 137201
[5] Utsumi S, Yoshiba D and Momozawa N 2007 *J. Phys. Soc. Jpn.* **76** 034704
[6] Chai Y S, Chun S H, Haam S H, Oh Y S, Kim I and Kim K H 2009 *New J. Phys.* **11** 073030
[7] Pullar R C 2012 *Prog. Mater. Sci.* **57** 1191–1334
[8] Koutzarova T et al 2020 *J. Magn. Magn. Mater.* **505** 166725
[9] Georgieva B, Koutzarova T, Kolev S, Krezhov K, Kovacheva D, Ghelev Ch, Vertruyen B, Tran L M and Zaleski A 2019 *AIP Conference Proceedings* **2075** 160032
[10] Wang Y, Zhang S, Zhu W K, Ling L, Zhang L, Qu Z, Pi L, Tong W and Tian M 2019 *J. Mater. Chem.* C **7** 340–5
[11] Zhang M, Dai J, Yin L, Kong X, Liu Q, Zi Z and Sun Y 2016 *J. Alloys Compd.* **689** 75–80
[12] Koutzarova T *et al.* 2019 *Materials* **12** 1414
[13] Georgieva B *et al.* 2019 *J. Magn. Magn. Mater.* **477** 131–5
[14] Hiraoka Y, Tanaka Y, Kojima T, Takata Y, Oura M, Senba Y, Ohashi H, Wakabayashi Y, Shin S and Kimura T 2011 *Phys. Rev.* B **84** 064418
[15] Hiraoka Y, Nakamura H, Soda M, Wakabayashi Y and Kimura T 2011 *J. Appl. Phys.* **110** 033920
[16] Wang F, Zou T, Yan L-Q, Liu Y and Sun Y 2012 *Appl. Phys. Lett.* **100** 122901
[17] Özçelik B, Özçelik S, Amaveda H, Santos H, Borrell C J, Sáez-Puche R, Fuente G F de la and Angurel L A 2020 *J. Materiomics* **6** 661–70
[18] Ghelev Ch *et al.* 2012 *J. Phys. Conf. Ser.* **356** 012048
[19] Shannon R D and Prewitt C T 1969 *Acta Cryst.* B **25** 925–946

Design of a plasmonic sensor based on a nanosized structure for biochemical application

Hocine Ben Salah[1,2,6], Hocine Bahri[3], Abdesselam Hocini[1], Imane Zegaar[4], Sven Ingebrandt[5] and Vivek Pachauri[5]

[1]Laboratoire d'Analyse des Signaux et Systèmes, Department of Electronics, University of M'Sila, BP.166, Route Ichebilia, M'sila 28000, Algeria
[2]Université Yahia fares Médéa, Médéa, Algeria
[3]Laboratory of Electronics and New Technologies (LENT), University Oum El Bouaghi, Algeria
[4]LIST Laboratory, University of M'Hamed Bougara Boumerdes, Avenue of Independence, 35000 Boumerdes, Algeria
[5]Institute of Materials in Electrical Engineering 1, RWTH Aachen University, Sommerfeldstraße 24, D-52074 Aachen, Germany

E-mail: hocine.bensalah@univ-msila.dz

Abstract. A novel design of an integrated ring cavity consisting of two merged and opposite triangles formed on a metal-insulator-metal (MIM) waveguide is suggested and analyzed for refractive index sensing application. The cavity design can be optimized provide the best sensing performance. In this work, we simulated numerically the device design by utilizing the finite-difference-time-domain (FDTD) technique in a two-dimensional (2D) structure. The detection of the refractive index changes was numerically simulated and analyzed using RSoft™. Small variations in the geometric parameters can enhance the sensitivity achieved, which we found to be 3575 nm/RIU. We believe that the sensor can achieve a resolution of 2.79×10^{-6}. The structure proposed has a simple design for easy and compact realization, paving the way of detecting rare biochemical analytes and for finding applications in optical filters.

1. Introduction

A surface plasmon polariton (SPPs) is electromagnetic excitation propagating along the interface between a metal and a dielectric medium whose amplitude decays exponentially with increasing the distance from the interface into each medium [1]. They exhibit interesting and valuable properties, such as energy asymptote in dispersion curves, resonant modes, field localization and enhancement, high sensitivity, and subwavelength confinement. The SPPs have thus offered vast potential uses in medical diagnostics and environmental monitoring, as DNA [2] and glucose concentration [3]. Further, due to the possibility of a high degree of integration and miniaturization, of overcoming the diffraction limits and of localizing light in subwavelength regions, the surface plasmon resonance (SPR) has attracted strong attention within the scientific community; in particular, the potential to guide, concentrate, and scatter light on a nanoscale has been studied experimentally and theoretically [4]. Moreover, the metal-insulator-metal (MIM) waveguide structure supports the propagation of SPPs and has high transmission

[6] To whom any correspondence should be addressed.

Content from this work may be used under the terms of the Creative Commons Attribution 3.0 licence. Any further distribution of this work must maintain attribution to the author(s) and the title of the work, journal citation and DOI.
Published under licence by IOP Publishing Ltd

losses (that can be neglected on a nanoscale) and has been used in many fields as sensors, filters [6], switches [5]. Over the last few years, we have witnessed several unique cavity designs improving the sensing performance of the plasmonic sensors based on the MIM waveguide, such as disk resonators [4] and rectangular resonators [7]. The efforts of multiple research teams have been directed to enhancing the transmission spectrum and increasing the sensitivity by choosing unique and properly selected structural dimensions. For example, the design has been explored extensively of on-chip optical biosensors with miniature and sharp geometric feature using a plasmonic waveguide.

In this work, we applied the FDTD simulation method to study a plasmonic biosensor using a double triangle as a ring cavity. Varying the structure's geometric parameters allowed us to control and manipulate the resonance wavelengths. The studies of the sensor proposed can be extended to a broad range of medical applications.

2. Theory and Simulation

The MIM waveguide structure is illustrated in figure 1. We propose a highly sensitive refractive index (RI) sensor consisting of a double ring triangular cavity (hereinafter denoted as DRT). Additionally, the device performance is compared with a standard triangular cavity, with the RSOFT CAD software used to calculate the transmission characteristics. We designed a waveguide with a width of $W = 50$ nm coupled with two opposite triangular rings with $n = 1$ and dimensions $a = 400$ nm, $b = 320$ nm and $g = 10$ nm. Further, the transmittance is defined as $T = P_{out}/P_{in}$ – its value is determined by partial reflection and partial coupling with the DRT; the transmitted fraction is measured as P_{out}. The gray and white areas represent the silver layer and the dielectric (air) material, respectively. Using silver is the best choice as the metal in the MIM structure at optical frequencies; it improves the absorption by trapping the incident light waves thus generating more SPRs and has the smallest damping constant. The metal also offers a framework through which the fluid test substance could flow. The Drude-Lorentz model allows the calculation of the relative dielectric constant of silver [8]:

$$\varepsilon_m = \varepsilon_\infty - \frac{\omega_p^2}{\omega^2 - i\omega\gamma} \quad (1)$$

Here, ε_∞, γ, and ω_p are the dielectric constant at infinite frequency, the electron collision frequency and the bulk plasma frequency, respectively. The parameters are set to be $\varepsilon_\infty = 3.7$, $\omega_p = 1.38\times10^{16}$ Hz, $\gamma = 2.73\times10^{13}$ Hz; ω is the incident light angular frequency. Since the wavelength of the incident light is greater than the waveguide's width, only a single propagation mode TM_0 can exist in the structure and the resonance wavelength of the ring resonator can be approximately given by the following:

$$\lambda_m = \frac{2Re(\eta_{eff})L_{eff}}{m}, \; m=1,2,3.... \quad (2)$$

$Re(\eta_{eff})$ is the real part of the effective RI of the SPP, L_{eff} is the effective resonance length.

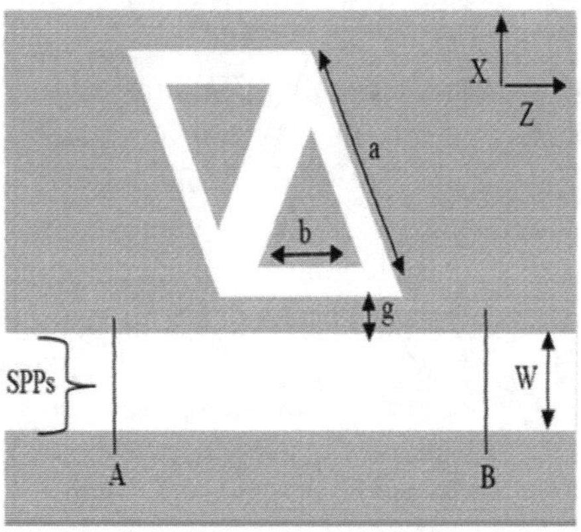

Figure 1. Schematic diagram of the proposed MIM-based plasmonic structure.

3. Numerical results and discussion

A two-dimensional FDTD method is used; a perfectly matched layer (PML) is assumed as an absorbing boundary condition for the computational window's boundaries, with the grid sizes in the x and z directions being $\Delta x = \Delta z = 3$ nm; the time step derived by the Courant condition, is:

$$\Delta t = 0.95/c\sqrt{(\Delta x)^2 + (\Delta z)^2}$$

to ensure that all the outside waves are emitted. The first step in the simulations assumes air as the insulator ($n = 1$), as shown in figure 1.

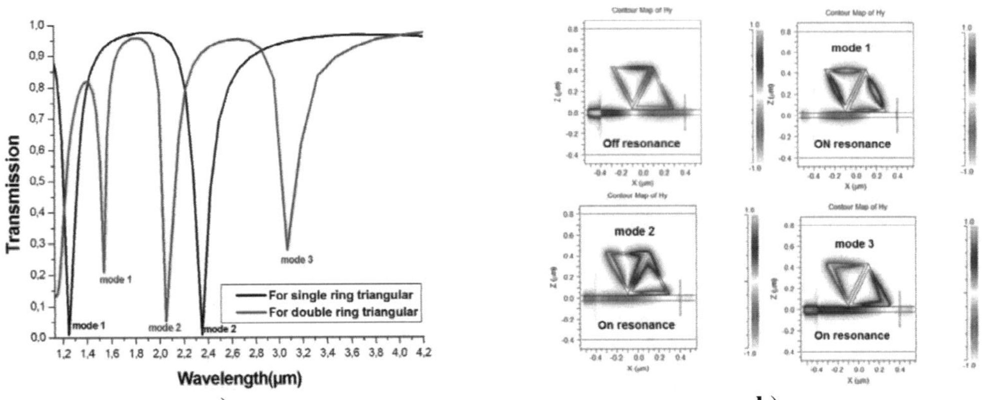

Figure 2. a) Transmission spectra of MIM plasmonic waveguides coupled with a single ring triangular and a double ring triangular resonator with $n = 1$, $a = 400$ nm, $b = 320$ nm and $g = 10$ nm, **b)** magnetic field (H_y) distributions at the resonance wavelengths.

To justify our choosing this structure, we first conducted calculations for the case of a design with a single triangular ring. As shown in figure 2 a), we achieve two transmission peaks at the resonance wavelengths of 1.2485 μm for mode 1 and 2.3502 μm for mode 2 for a single ring triangular resonator; and 1.5366 μm for mode 1, 2.05 μm for mode 2 and 3.0733 μm for mode 3 for the DTR resonator. It is clear that adding a second triangular ring result in shifting the peaks positions and in the appearance of a new sharp peak in the mid-infrared (mid-IR) range with a higher transmitted power at $\lambda = 3.0733$ μm. The mid-IR range is particularly well-suited for bio-sensing as it encompasses the molecular vibrations that uniquely identify the biochemical building blocks of life.

Moreover, to provide a deeper understanding of the change in the power transmission mechanism, figure 2 b) depicts the field distributions of H_y at the resonance wavelength of the peaks and at off-resonance wavelengths of the MIM plasmonic waveguides coupled with a DRT resonator. It is seen that the incident light at the wavelength of 1.5366 μm, 2.05 μm and 3.0733 μm couples to the DRT resonator and is prevented from passing through the waveguide, while the rest of the wavelengths are transmitted to the output.

Further, we studied the refractive index (RI) impact on the proposed design performance by varying the RI from 1 to 1.08 at a step of 0.02, as shown in figure 3 a). Clearly, as the RI increases, the resonance wavelengths of the transmitted spectra are red-shifted, with the largest shift occurring for mode 3. Figure 3 b) shows the linear relationships between the resonance wavelengths and the refractive indices for the single ring triangular and for the DRT resonator. As mentioned above, the most important parameter for the sensors is their sensitivity defined as $S = \Delta\lambda/\Delta n$. Thus, the sensitivity for mode 1 is around 1506 nm/RIU, for mode 2 it is around 2890 nm/RIU and for mode 3, around 3202 nm/RIU in the case of an MIM-based plasmonic sensors with DRT resonators; and 1370 nm/RIU for mode 1 and

2365 nm/RIU for mode 2 in the case of a single ring triangular resonator. It is clearly seen that the presence of the DRT structure results in an increase in the device's sensitivity by 10% for mode 1 and 22.2% for mode 2 and reaches ultra-high sensitivity for mode 3.

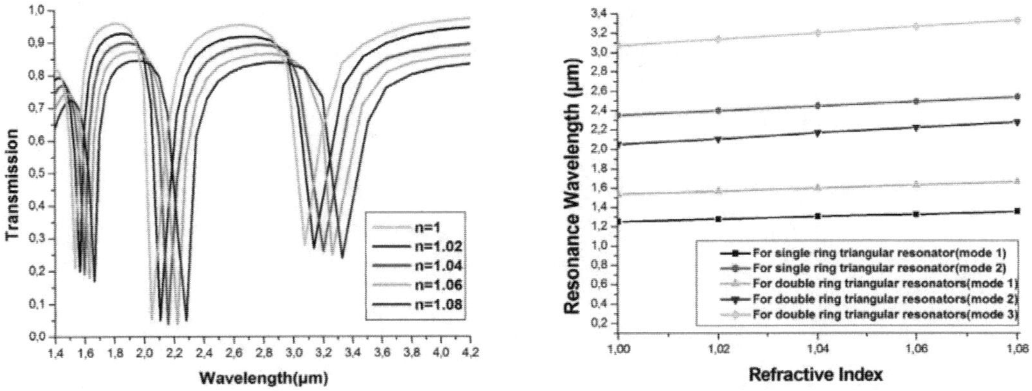

a) **b)**

Figure 3. **a)** Transmission spectra of MIM plasmonic waveguides coupled with DRT for different refractive indices n for $a = 400$ nm, $b = 320$ nm and $g = 10$ nm. **b)** The resonance wavelength versus the refractive index n of the material under sensing.

Furthermore, to optimize these plasmonic structures' characteristic parameters, figure 4 illustrates the effect of varying the side length b of the triangle; as seen, increasing b from 310 nm to 340 nm at a step of 10 nm is associated with an increase in the resonance wavelength. The linear relationship noted between the wavelength resonances and the refractive indices is due to the relationship between the effective refractive index and the resonance wavelength in equation (2) and illustrated in figure 4 a), despite the increase in the value of the side length b of the triangle. In addition, as figure 4 b) demonstrates, increasing the value of b from 310 nm to 340 nm enhances the sensitivity of the sensor. The maximum sensitivity is 3575 nm/RIU, with an improvement of 380 nm/RIU and a sensing resolution of 2.79×10^{-6} RIU achieved for $b = 340$ nm. This improvement in the sensitivity can be explained as being due to the length of the surface plasmon (SP) – as we increase b, we increase the interaction between the surrounding medium and the SP.

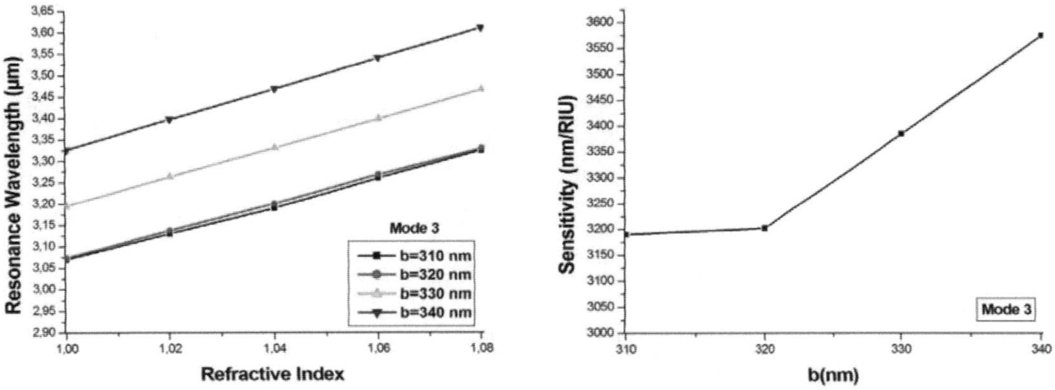

a) **b)**

Figure 4. Sensing properties as functions of b. **a)** Resonance wavelength *versus* the refractive index for b varying from 310 nm to 340 nm. **b)** Plasmonic sensor sensitivity for b varying from 310 nm to 340 nm.

4. Conclusions

The proposed structure is of simple design based on an MIM triangular ring cavity. Based on simulations using the FDTD method, the results show that adding an additional triangle can enhance the sensitivity and achieve a peak in the mid-IR with a linear relationship between the resonance wavelength and the RI. Increasing the length of the ring cavity enhances the structure's sensitivity reaching 3575 nm/RIU with high transmission. Thus, the sensitivity value of this structure is higher than those reported by other authors (table 1). Because of its easy fabrication, simple configuration, and compact structure, this sensor is a promising candidate for wide-ranging applications in bio-sensing and on-chip optical nano-sensors.

Table 1. Sensitivity comparison of different sensor structures.

Reference/ Application	Sensitivity (nm/RIU)	year
[9]	2162	2020
[10]	2365	2021
[11]	3400	2021
[12]	3010	2021
In this work	3575	2021

References

[1] Ritchie R H 1957 *Phys Rev* **106** 874–881
[2] Filion-Côté S, Roche P J R, Foudeh A M, Tabrizian M and Kirk A G 2014 *Rev. Sci. Instrum.* **85** 093107
[3] Bahri H, Hocini A, Mouetsi S and Ben Salah H 2021 *ECS J Solid State Sci Technol* **10** (7) 071015
[4] Rakhshani M R, Tavousi A and Mansouri-Birjandi M A 2018 *Appl Opt* **57** (27) 7798
[5] Ben Salah H, Hocini A, Bahri H and Melouki N 2021 *ECS J. Solid State Sci. Technol* **10** (8) 081001
[6] Hocini A, Ben salah H, Khedrouche D and Melouki N 2020 *Opt Quantum Electron* **52** (7) 336
[7] Rahmatiyar M, Afsahi M and Danaie M 2020 *Plasmonics* **15** 2169–2176
[8] Lee T W and Gray S 2005 *Opt Express* **13** (24) 9652–9659
[9] Xu D, Yan S, Yang X, Su H, Wu X and Hua E 2021 *IEEE Sens J.* **21** (4) 4442–4450
[10] Zegaar I, Hocini A and Ben salah H 2021 *J. Phys. Conf. Ser.* **1859** 012024
[11] Zubaidah binti Haji Jumat S, Chou Chao C T, Chou Chau Y F, Hanif Mahadi A, Raziq Rahimi Kooh M, Kumara N T R N and Chiang H P 2021 *Chinese J Phys* **71** 286–299
[12] Bahri H, Mouetsi S, Hocini A and Ben Salah H 2021 *Opt Quantum Electron* **53** (6) 332

Plasmonic stop-band filter based on an MIM waveguide coupled with cavity resonators

Imane Zegaar[1,5], Abdesselam Hocini[2], Djamel Khedrouche[2], Hocine Ben salah[3] and Hocine bahri[4]

[1]LIST Laboratory, University of M'Hamed Bougara Boumerdes, Avenue of Independence, 35000 Boumerdes, Algeria
[2]Laboratoire d'Analyse des Signaux et Systèmes, Department of Electronics, University of M'Sila BP.166, Route Ichebilia, M'Sila, 28000, Algeria
[3]Université Yahia fares Médéa, Algeria
[4]Laboratory of Electronics and New Technologies (LENT), University Oum El Bouaghi. Algeria

E-mail: i.zeggar@univ-boumerdes.dz

Abstract. In this study, we propose and simulate in the NIR region a tunable stop-band filter based on an MIM waveguide coupled to a triangular cavity. The transmission spectra of the filter are obtained using the R-Soft CAD software based on the two-dimension (2D) finite-difference time-domain (FDTD) method, which uses perfectly matched layers (PML) to absorb the outgoing energy fields. A stop-band plasmonic filter with a large bandwidth and high band-pass transmission is achieved by adding another triangular cavity resonator. One can also control the range filtered and the shift of the central wavelength by tuning the second triangular cavity parameters. This structure has important applications in highly integrated photonic circuits.

1. Introduction

Surface plasmon polaritons (SPPs) are electromagnetic waves that propagate along a metal-dielectric interface. The SPPs can overcome the refraction limit of light to transmit energy and information. This property is important for the realization of highly integrated optical circuits at the nanoscale. Due to the confinement of the SPPs energy to a small region along the interface, a sub-wavelength-size optical elements can be realized [1, 2, 3]. Because of their properties, such as high speed and sensitivity, the SPPs find numerous applications, the most promising of which are in photonic circuits [4] and biological sensing [5, 6].

The SPPs waveguides are of two types, namely, based on Insulator-Metal-Insulator (IMI) or Metal-Insulator-Metal (MIM) structures. Recently, the MIM ones have piqued the interest of researchers due to the distinct potential of light control at nanoscale dimensions, which could enable the development of various types of plasmonic devices for filtering [7, 8], demultiplexing [9, 10], sensing [11, 12], switching [13].

The optical filters have attracted more attention among the applications mentioned because of their critical role in optical communication systems and wavelength selection techniques [14]. Filter devices

[5] To whom any correspondence should be addressed.

Content from this work may be used under the terms of the Creative Commons Attribution 3.0 licence. Any further distribution of this work must maintain attribution to the author(s) and the title of the work, journal citation and DOI.
Published under licence by IOP Publishing Ltd

based on the SPPs phenomenon have been proposed and designed with different characteristics of the cavity coupled with the MIM waveguide, such as nanodisk resonators [15], rectangular resonators [16] etc.

This paper proposes a tunable band-stop plasmonic filter with a two-dimensional (2D) structure composed of an MIM waveguide and a triangular cavity resonator. We obtained and analyzed the transmission spectrum properties of the filter by using the finite-difference time-domain (FDTD) method using the R-Soft CAD software with perfectly matched layers (PML). Moreover, the filtering range width of the plasmonic filter can be controlled and manipulated by tuning the cavity resonator's parameters. This structure may have significant applications in highly integrated photonic circuits.

2. Structure design and analysis method

Figure 1 presents the basic MIM plasmonic filter schematic. It is composed of a waveguide connected to a triangular cavity with a height *H1* and a base *t*; the width of the waveguide is denoted by *w*. Grey and white, respectively, represent the metal (silver) and dielectric (air) sections. We chose silver as the metal for this filter due to its low ohmic loss and permittivity that can be described by the Drude model [17].

$$\varepsilon_m = \varepsilon_\infty - \frac{\omega_p^2}{\omega^2 - i\omega\gamma} \quad (1)$$

Here, $\varepsilon_\infty = 3.7$ represents the dielectric constant at infinite frequency; ω_p and γ are the bulk plasma and electron collision frequencies with values 1.38×10^{16} Hz and 2.73×10^{13} Hz, respectively; ω is the incident light angular frequency. By applying the FDTD method using the R-Soft CAD software with perfectly matched layers (PML), we simulated the transmission spectrum of the proposed structure. The perfect matched layer absorbs the outgoing waves without reflections. Thus, the grid size (Δx, Δz) in the *x* and *z* directions are equal to 5 nm; smaller grids are used to ensure accurate results. The source type in the left-hand side of the MIM waveguide is a Gaussian light beam; the only mode that can be excited to propagate through the MIM waveguides is a single transverse magnetic (TM_0) mode, due to the width of the waveguide being very small compared to the incident wavelength. The resonance wavelength of the triangular cavity for the TM_0 mode can be obtained as follows [18]:

$$\lambda_m = \frac{2R_e(n_{eff})l_{eff}}{m}, m = 1,2,3 \ldots$$

where $R_e(n_{eff})$ is the real part of the effective refractive index of the SPP, l_{eff} is the effective resonance length of the resonator.

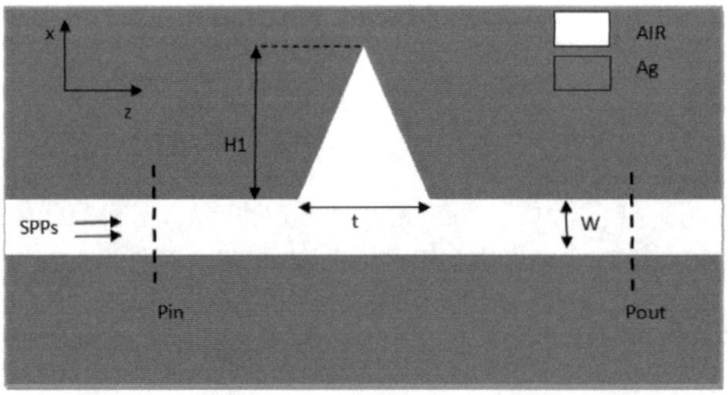

Figure 1. 2D Schematic diagram of plasmonic filter structure.

3. Numerical results and discussion

We begin with studying the influence of the triangular resonator's height on the transmission spectrum, figure 2 a), where the other geometrical parameters of the proposed device are kept at *t*= 80 nm and *w* = 100 nm. As seen, the transmission peaks clearly increase as *H1* is increased from 450 nm to

600 nm in 50-nm steps, while the resonance wavelength shifts to the longer wavelengths. The value $H1 = 500$ nm results in an optimal transmission spectrum. The magnetic field distribution| in the proposed structure at the resonance wavelengths 1750 nm, 2654 nm and 4000 nm are illustrated in figure 2 b), respectively. As shown, the structure resonates at 2654 nm and the magnetic field is confined to the triangular resonator. For incident light with wavelengths of 1750 nm and 4000 nm, the triangular cavity is unexcited and the light propagates through the waveguide.

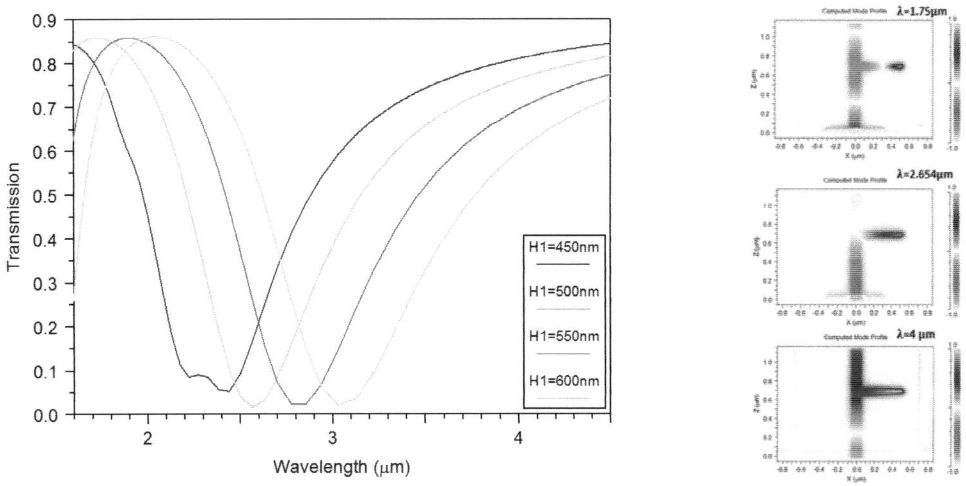

a) b)

Figure 2. a) Transmission spectra of the proposed filter for different *H1*. **b)** Magnetic field contours of the proposed structure at the different resonance wavelengths.

Further in our study, another triangular cavity was added to the structure with the same geometrical parameters as the first one; figure 3 depicts a stop-band plasmonic filter with two triangular cavities.

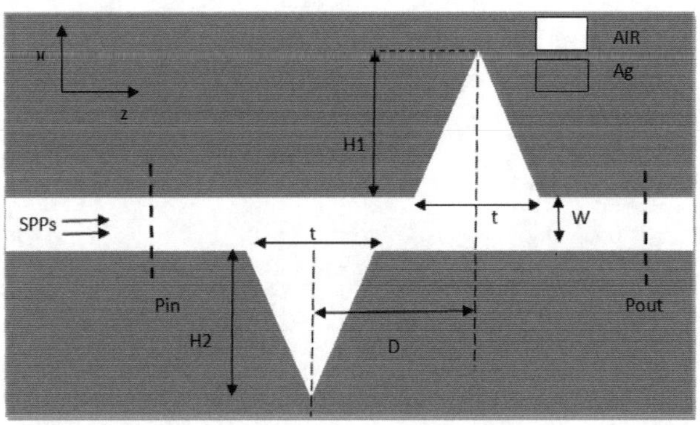

Figure 3. 2D Schematic diagram of plasmonic top-band filter coupled with two triangular cavities.

We studied the influence of the distance *D* between the triangular cavities on the transmission spectrum. Figure 4 a) demonstrates that the bandwidth of the plasmonic filter with two triangular cavities (the solid line) is broader than the bandwidth of the plasmonic filter with only one (the dashed line). Figure 4 b) depicts the influence of varying the distance *D* on the transmission spectrum. It is seen that increasing *D* from 0 to 440 nm in steps of 80 nm broadens the bandwidth of the filter. As a result, the plasmonic filter's band-pass transmission increases from 58% to 80% on the right-hand side of the band, resulting in a U-shaped transmittance. Thus, in the next analysis, the length *D* was fixed at the best value of 440 nm, and the effect was analyzed of varying *H2* on the plasmonic filter properties (figure 5 a), b)). Transmission spectra for values of *H2* varying from 450 nm to 550 nm with a step of 50 nm are presented in figure 5 a). Increasing the value of *H2* broadens the bandwidth of the stop-band filter in the NIR region from 230 nm to 350 nm (figure 5 b)), with a slight effect on the band-pass transmission. It becomes 77%.

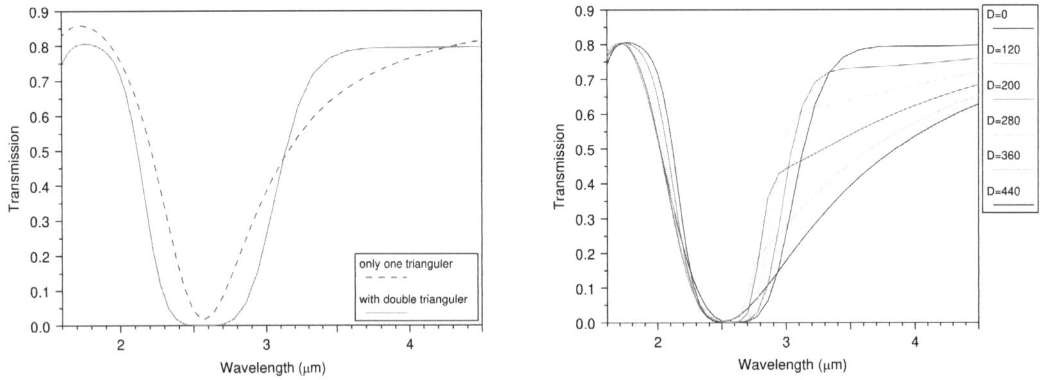

a) **b)**

Figure 4. a) Transmission spectra of a plasmonic double triangular cavity stop-band filter for $D = 440$ nm. **b)** Transmission spectra of the plasmonic double triangular cavity stop-band filter for different *D*.

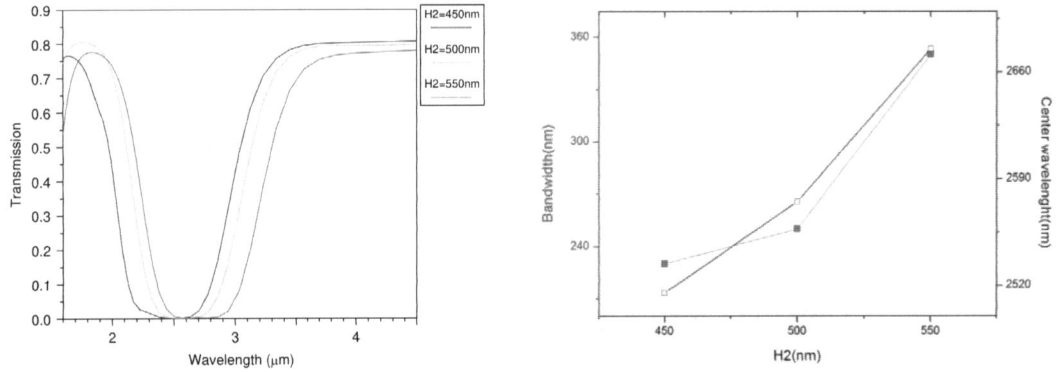

a) **b)**

Figure 5. a) Transmission spectrum of the system with different *H2*. **b)** Bandwidth and central wavelength of the band-stop filter for different *H2*.

Furthermore, the central wavelength is shifted to the longer wavelengths. The filtered wavelengths for different values of *H2* are listed in Table 1.

Table1. The filtered wavelengths for different values of H_2.

H2 (nm)	Filtered Wavelengths
450	$[\lambda1 = 2400\text{nm} - \lambda2 = 2630\text{nm}]$
500	$[\lambda1 = 2450\text{nm} - \lambda2 = 2700\text{nm}]$
550	$[\lambda1 = 2500\text{nm} - \lambda2 = 2850\text{nm}]$

4. Conclusions

A tunable band-stop filter based on an MIM waveguide coupled to one and two triangular cavity resonators is proposed and simulated in the near-infrared (NIR). Using the R-Soft CAD software and perfectly matched layers, we obtained the filter's properties using the finite-difference-time-domain (FDTD) method. The simulations indicated that it is possible to design a plasmonic band-stop filter whose width and central wavelength of the filtered range can be controlled by varying the cavity resonators parameters. This structure has a broad range of applications in highly integrated photonic circuits.

References

[1] Wu J, Lang P, Chen X and Zhang R 2016 *J. Mod. Opt.* **63** 219–23
[2] Duan G, Lang P, Wang L, Yu L and Xiao J 2016 *Mod. Phys. Lett.* B **30** 1–8
[3] Dehghan B, Sadeghi M and Adelpour Z 2019 *Optik (Stuttg).* **184** 44–9
[4] Liu L, Han Z and He S 2005 *Opt. Express* **13** 6645
[5] Špačková B, Wrobel P, Bocková M and Homola J 2016 *Proc. IEEE* **104** 2380–408
[6] Yin Z, Guo H, Li Y, Chiu J and Tian L 2020 *ACS Appl. Mater. Interfaces* **12** 35977–85
[7] Lounis A, Hocini A and Ben Salah H 2021 *J. Phys. Conf. Ser.* **1859** 012027
[8] Zheng G, Chen Y, Xu L, Lai M and Liu Y 2013 *Opt. Commun.* **305** 164–9
[9] Nouai A, Akjouj A, Pennec Y, Gillet J N and Djafari-Rouhani B 2009 *New J. Phys.* **11** 0–19
[10] Pav M R, Granpayeh N, Hosseini S P and Rahimzadegan A 2019 *Opt. Commun.* **437** 285–9
[11] Zegaar I, Hocini A and Ben Salah H 2021 *J. Phys. Conf. Ser.* **1859** 012024
[12] Ben salah H, Hocini A, Temmar M N and Khedrouche D 2019 *Chin. J. Phys.* **61** 86–97
[13] Zhu J H, Huang X G and Mei X 2011 *Plasmonics* **6** 605–12
[14] Zavvari M, Taleb Hesami Azar M and Arashmehr A 2017 *J. Mod. Opt.* **64** 2221–7
[15] Wang G, Lu H, Liu X, Mao D and Duan L 2011 *Opt. Express* **19** 3513
[16] Yun B, Hu G and Cui Y 2013 *Plasmonics* **8** 267–75
[17] Yao X 2014 *Mod. Phys. Lett.* B **28** 1450025
[18] Hocini A, Ben salah H, Khedrouche D and Melouki N 2020 *Opt. Quantum Electron.* **52** 1–10

Cirrus-cloud conditioned ratio of multiple-to-single backscatter atmospheric phase functions

Ts Evgenieva[1,3], V Anguelov[2] and L Gurdev[1]

[1]Institute of Electronics, Bulgarian Academy of Sciences, 72 Tsarigradsko Chaussee, 1784 Sofia, Bulgaria
[2]Institute for Nuclear Research and Nuclear Energy, Bulgarian Academy of Sciences, 72 Tsarigradsko Chaussee, 1784 Sofia, Bulgaria

E-mail: tsevgenieva@ie.bas.bg

Abstract. The present work is related to LIDAR remote sensing of the atmosphere in the presence of cirrus clouds (CC) along the LIDAR line of sight (LOS), when the multiple-backscatter component of the LIDAR return may be essential. E. Eloranta has found analytical expressions of the multiple-backscatter components of different orders as containing the corresponding ratios of the multiple-to-single backscatter phase functions of the backward scattering medium. As a main purpose of this work, we estimated the above-mentioned ratios in cases of backscattering from air and CC, using accurate approximations of the single-scattering phase functions obtained by U. Wandinger. The results obtained show that the multiple-to-single backscatter phase function ratios for air are equal to unity, while for CCs they are quite near unity.

1. Introduction

The present investigation is devoted to LIDAR remote sensing of the atmosphere in the presence of optically dense aerosol objects, such as cirrus clouds (CC), along the LIDAR line of sight (LOS). In such cases, the multiple-scattering component of the LIDAR return may be significant and the profiles of the atmospheric extinction and backscattering coefficients along the LOS would not be accurately determinable using the well-known single-scattering LIDAR equations. Then, one should search for ways of building adequate multiple-scattering LIDAR equations that take into account the multiple-scatter contribution to the LIDAR return and can eventually be inverted with respect to the true LOS profiles of the aerosol extinction and backscatter coefficients. An effective approach to such a problem was developed by E. Eloranta [1] who found analytical expressions of the multiple-backscattering components of different orders. An essential moment concerning the calculation of these components is the necessity to know the ratio of the multiple-to-single backscatter phase functions of different orders of the backward scattering medium. The analytical integral expression of this ratio, given in [1], requires the knowledge of the backward branch of the backscattering-medium phase function and the forward diffraction peak width of the secondary (forward) scattering medium that are obtainable on the basis of the microphysical and optical properties of the aerosol ensembles under consideration.

The main purpose of the work reported here is, using accurate approximations of the phase functions of air and cirrus clouds obtained by U. Wandinger [2], to evaluate the above commented

[3] To whom any correspondence should be addressed.

Content from this work may be used under the terms of the Creative Commons Attribution 3.0 licence. Any further distribution of this work must maintain attribution to the author(s) and the title of the work, journal citation and DOI.
Published under licence by IOP Publishing Ltd

ratio in cases of backscattering from air and cirrus clouds (CC). The secondary scattering medium is the icy matter of CCs.

2. Basic relations, parameters and approximations

A general expression obtained by E. Eloranta [1] of the ratio between the detected multiple- and single-backscattering due LIDAR return signals $S_n(\lambda, z)$ and $S(\lambda, z)$, respectively, can be written in the following, slightly modified, form:

$$S_n(\lambda,z)/S(\lambda,z)=\left[P_{n\pi}(\lambda,z)/P_\pi(\lambda,z)\right]\left[1-\exp\left(-\theta_t^2/\theta_l^2\right)\right]^{-1}\cdot$$
$$\left\{\tau^{n-1}(\lambda,z)/(n-1)!-2^{-(n-1)}\int_{-z}^{z}dx_1\mu(\lambda,x_1)\int_{x_1}^{z}dx_2\mu(\lambda,x_2)\ldots\int_{x_{n-2}}^{z}dx_{n-1}\mu(\lambda,x_{n-1})\exp\left[-\frac{\theta_t z^2}{\theta_s(\lambda)\left(x_1^2+x_2^2+\ldots x_{n-1}^2\right)+\theta_l z^2}\right]\right\} \quad (1)$$

where λ is the wavelength of the sensing laser radiation; $z = ct/2$ is the LOS coordinate of the scattering volume contributing to the signal at a moment t after the sensing-pulse emission; n denotes the order of scattering; $P_{n\pi}(\lambda, z)$ and $P_\pi(\lambda, z)$ are, respectively, multiple and single backscatter phase functions of the backscattering medium; θ_t, θ_l and θ_s are, respectively, the angle of view of the LIDAR receiving optics, the angular divergence of the sensing laser beam, and the angular width of the forward peak of the phase function of the secondary (forward) scattering medium (see in [3]); $\mu(\lambda, z)$ is the LOS profile of the scattering coefficient of the secondary scattering medium; $\tau = \int_0^z \mu(\lambda, x)\mathrm{d}x$; and $x, x_1,\ldots, x_{n-1}$ are current integration variables. Equation (1) describes the case when all the scattering components of the atmosphere, including clouds (modeled appropriately [3]), are extended continuously along all the LOS from $z = 0$ to $z = \infty$. Below, we shall consider single backscattering from air and CCs, accompanied by multiple acts of forward scattering by CCs in forward and/or backward directions.

The mean multiple-backscatter phase function $P_{n\pi}(\lambda, z)$ of the backscattering medium of interest is expressible as (see also in [1])

$$P_{n\pi}(\lambda)=-\int_0^\infty P_b(\theta,\lambda)d\exp\left\{-\theta^2/\left[(n-1)\theta_s^2(\lambda)\right]\right\}, \quad (2)$$

where θ_s is assumed independent of z but proportional to λ [1, 4]; $P_b(\theta, \lambda)$ is the backward-peak vicinity approximation of the phase function of the backscattering medium, considered as independent of z and weakly dependent on λ [2]; the angle θ in $P_b(\theta, \lambda)$ is measured with respect to its peak (the backward) direction. The forward and backward peak zones, $P_{fi}(\theta, \lambda)$, and $P_{bi}(\theta, \lambda)$ respectively, of the phase functions of air (i = m) and cirrus clouds (i = cc) are precisely approximated in [2] as sums of Gaussian shapes

$$P_{fi}(\theta,\lambda)=\sum_{j=1}^{5}\alpha_{fij}\left(\pi\theta_{fij}^2\right)^{-1}\exp\left\{-\theta^2/\theta_{fij}^2\right\},$$
$$P_{bi}(\theta,\lambda)=\sum_{j=1}^{5}\alpha_{bij}\left(\pi\theta_{bij}^2\right)^{-1}\exp\left\{-\theta^2/\theta_{bij}^2\right\} \quad (3)$$

where α_{fij} and α_{bij} are weight numbers, and θ_{fij} and θ_{bij} are e^{-1} widths of the corresponding Gaussian components. The values of α_{fij} and α_{bij} and θ_{fij} and θ_{bij}, evaluated in [2], are given in table 1. Further, we approximate the waveform of $P_{fcc}(\lambda, z)$ obtained in [2] by only one Gaussian shape

$$P_{\mathrm{fcc}}(\theta,\lambda)=\sum_{j=1}^{5}\alpha_{\mathrm{fcc}j}\left(\pi\theta_{\mathrm{fcc}j}^{2}\right)^{-1}\exp\left\{-\theta^{2}/\theta_{\mathrm{fcc}j}^{2}\right\}\approx\alpha_{\mathrm{ccs}}\left(\pi\theta_{\mathrm{ccs}}^{2}\right)^{-1}\exp\left(-\theta^{2}/\theta_{\mathrm{ccs}}^{2}\right),\ \mathrm{cc}=\mathrm{hp\ or\ hc}\,, \qquad (4)$$

allowing us to use equation (2) and determine θ_s appearing there – see figure 1 and table 2. Note that two types of CC ice crystal shapes were considered – hexagonal plates (HPs, index hp) and hexagonal columns (HCs, index hc). The data available for $\alpha_{\mathrm{f,b}\,ij}$ and $\theta_{\mathrm{f,b}\,ij}$ concern the wavelengths $\lambda = 308$ nm, in the case of air, and $\lambda = 550$ nm, in the cases of CCs. The results obtained for these wavelengths can be extended to cases of other wavelengths, using the proportionality of θ_s to λ (table 2) and the weak dependence of $P_b(\theta, \lambda)$ on λ.

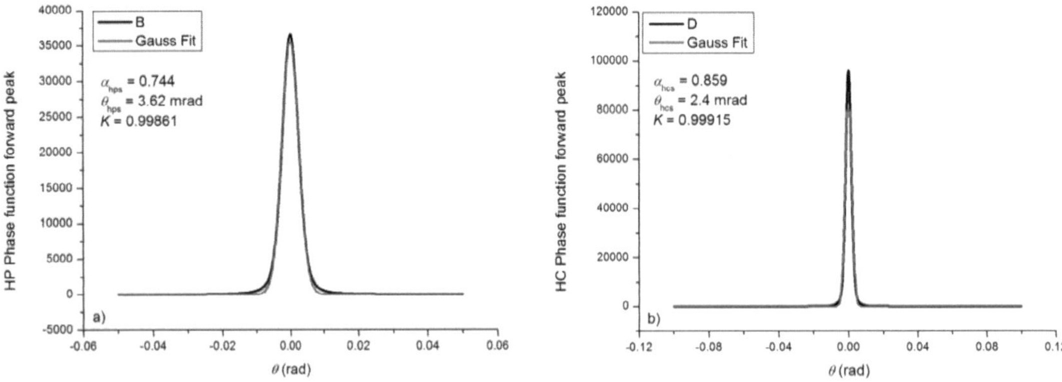

Figure 1. Gaussian fits, $P_{\mathrm{fcc}}(\lambda, \theta) = \alpha_{\mathrm{ccs}}(\pi\theta_{\mathrm{ccs}}^2)^{-1}\exp(-\theta^2/\theta_{\mathrm{ccs}}^2)$, of the forward peaks of the phase function of cirrus clouds built of hexagonal icy plates (a) and hexagonal icy columns (b); K is the correlation coefficient.

Table 1. Phase-function approximation parameters.

	Medium		
Parameters	CC-HP	CC-HC	Air
Wavelength λ (nm)	550	550	308
α_{f1}; θ_{f1} (mrad)	$3.56.10^{-1}$; 3.32	$4.64.10^{-1}$; 2.27	$7.6.10^{-1}$; 1430
α_{f2}; θ_{f2} (mrad)	$1.53.10^{-1}$; 6.28	$1.53.10^{-1}$; 5.39	
α_{f3}; θ_{f3} (mrad)	$7.00.10^{-2}$; 15.9	$3.52.10^{-2}$; 15.9	
α_{f4}; θ_{f4} (mrad)	$3.26.10^{-2}$; 42.9	$1.69.10^{-2}$; 35.8	
α_{f5}; θ_{f5} (mrad)	$5.12.10^{-3}$; 47.8	$1.37.10^{-2}$; 95.8	
α_{b1}; θ_{b1} (mrad)	$3.05.10^{-5}$; 18	$3.32.10^{-5}$; 17.6	$7.6.10^{-1}$; 1430
α_{b2}; θ_{b2} (mrad)	$5.88.10^{-5}$; 26.2	$6.33.10^{-5}$; 26.7	
α_{b3}; θ_{b3} (mrad)	$9.14.10^{-5}$; 36.5	$9.26.10^{-5}$; 36.8	
α_{b4}; θ_{b4} (mrad)	$1.17.10^{-3}$; 97.4	$1.19.10^{-3}$; 97.4	
α_{b5}; θ_{b5} (mrad)	$5.03.10^{-4}$; 567	$2.09.10^{-4}$; 504	

Table 2. Values of angle θ_s calculated for $\lambda = 550$ nm and extrapolated for $\lambda = 514.5$ nm, 532 nm, 700 nm and 1064 nm.

Wavelength λ (nm)	514.5	532	550	700	1064
CC – HP					
θ_s (mrad)	3.39	3.50	3.62	4.61	7.00
CC – HC					
θ_s (mrad)	2.24	2.32	2.40	3.04	4.47

By substituting the expression of equation (3) for $P_{bi}(\theta, \lambda)$ in equation (2) and performing a cascade integration by parts, we obtain an analytical expression of $P_{n\pi i}(\lambda)$ leading to the following expression of the ratio $R_{ni}(\lambda) = P_{n\pi i}(\lambda)/P_{\pi i}(\lambda)$:

$$R_{ni}(\lambda) = \sum_j \left[\alpha_{bij}/\theta_{bij}^2(\lambda)\right]\left[1+(n-1)\theta_{\mathrm{hp,hc\,s}}^2(\lambda)/\theta_{bij}^2(\lambda)\right]^{-1} \Big/ \sum_j \alpha_{bij}/\theta_{bij}^2(\lambda). \qquad (5)$$

3. Results and discussion

Using equation (5) and the data given in tables 1 and 2, we estimated the ratio $R_{ni}(\lambda)$ in cases of single backscattering from air (i = m) and cirrus clouds (i = hp or hc) and additional multiple forward scattering through cirrus clouds (see table 3). The sensing laser wavelengths of concern were $\lambda = 514.5$, 532, 550, 700 and 1064 nm (tables 2 and 3). Calculations (including those of θ_s) based on actual available data were performed for $\lambda = 550$ nm [2]. For the other wavelengths, the angles θ_s were evaluated as proportional to λ. At the same time, the backward branches of the phase functions were considered as practically independent of λ and matching those at $\lambda = 550$ nm. Thus, the results obtained for $R_{ni}(\lambda)$ at $\lambda = 550$ nm are accurate, while those concerning the remaining wavelengths are approximate but useful for rough estimation.

Table 3. Estimated double (d), triple (t) and quadruple (q) – to – single backscatter phase function ratios for backscattering from air (R_m), CC-HP (R_{hp}) and CC-HC (R_{hc}) through CCs.

Wavelength λ (nm)		514.5	532	550	700	1064
$\boldsymbol{R}_{\mathbf{m}}$		1	1	1	1	1
	d	0.986	0.985	0.984	0.974	0.934
$\boldsymbol{R}_{\mathbf{hp}}$	t	0.972	0.970	0.968	0.950	0.897
	q	0.959	0.956	0.954	0.928	0.857
	d	0.993	0.993	0.992	0.988	0.974
$\boldsymbol{R}_{\mathbf{hc}}$	t	0.987	0.986	0.985	0.976	0.950
	q	0.980	0.979	0.977	0.965	0.929

The analysis of the results for $R_{ni}(\lambda)$ given in table 3 shows in general that for lower orders of scattering n and shorter wavelengths λ, the ratios $R_{n\mathrm{hp}}(\lambda)$ and $R_{n\mathrm{hc}}(\lambda)$ are near unity. At $\lambda = 550$ nm and below, they are practically equal to (closely approaching) unity. At longer wavelengths, the difference from unity increases and becomes more essential, which would lead eventually to noticeable systematic errors in the restoration of backscatter coefficient profiles in CCs, using high spectral resolution LIDARs (HSRLs) and single-scattering LIDAR equations [2, 5]. In the case of backscattering from air, the value of the ratio $R_{n\mathrm{m}}(\lambda)$ is equal to unity independently of the wavelength λ of the sensing laser radiation.

4. Conclusions

The results from the calculations performed show that the multiple-to-single backscatter phase function ratio for air is certainly unity for sensing laser radiations within the optical range. In case of backscattering from CCs, this ratio is quite near unity for wavelengths around $\lambda = 550$ nm and below. These results are especially useful, as they show that the backscatter coefficient profiles in CCs are accurately retrievable (without systematic errors) by high spectral resolution LIDARs using the single-scattering LIDAR equations even under multiple-scattering conditions caused by CCs. At wavelengths longer than $\lambda = 550$ nm, the ratio becomes noticeably smaller than unity, which means that the multiple-scattering influence should be taken into account in the retrieval procedures or somehow removed. Further calculations are intended aimed at determining the multiple-to-single backscatter phase function ratios for cases of cumulus and stratus clouds present in the atmosphere. A careful estimation would also be useful of the systematic errors in the restoration of the atmospheric-aerosol backscatter coefficient profiles under multiple-scattering conditions using HSRLs and single-scattering LIDAR equations, depending on the corresponding multiple-to single backscatter phase function ratios.

Acknowledgments

This research was supported partly by the Bulgarian Ministry of Education and Science (support for ACTRIS BG, part of the Bulgarian National Roadmap for Research Infrastructure), the Bulgarian National Science Fund (Grant No. KP-06-N28/10/2018) and the European Union's Horizon 2020 Research and Innovation Program under the ACTRIS IMP project (Grant No. 871115).

References

[1] Eloranta E W 1998 *Appl. Opt.* **37** (12) 2464–72
[2] Wandinger U 1998 *Appl. Opt.* **37** (3) 417–427
[3] Evgenieva Ts, Grigorov V, Anguelov V and Gurdev L 2021 *J. Phys.: Conf. Ser.* **1859** 012029
[4] Whiteman D et al 2001 *J. Geophys. Res.* **106** (D6) 5211–5225
[5] Weitkamp C ed 2005 *Lidar, Range-Resolved Optical Remote Sensing of the Atmosphere* (New York: Springer)

Aerosol types identification during different aerosol events over Sofia, Bulgaria, using sun-photometer and satellite data on the aerosol optical depth and Ångström exponent

Ts Evgenieva[1], L Gurdev, E Toncheva and T Dreischuh

Institute of Electronics, Bulgarian Academy of Sciences, 72 Tsarigradsko Chaussee, 1784 Sofia, Bulgaria

E-mail: tsevgenieva@ie.bas.bg

Abstract. A series of data was analyzed on the aerosol optical depth (AOD) and the Ångström exponent (AE) over Sofia, Bulgaria, obtained using a Cimel CE318-TS9 ground-based sun/sky/lunar photometer and a space-born Moderate Resolution Imaging Spectroradiometer (MODIS). The results show that in cases of Saharan dust passages over the city the values of AOD_{440} at the wavelength of 440 nm fall mainly between 0.15 and 0.50, while those of $AE_{440/870}$ are mainly below 1.1, which is in accordance with the existing AOD- and AE-based classifications of the aerosol types. It is shown as well that a reasonable correlation exists between the data on the AOD provided by the CE318-TS9 Cimel sun photometer and those provided by the MODIS mounted on board of Aqua satellite.

1. Introduction

Atmospheric aerosols have long been known as strongly affecting the global and local climate particularities and, correspondingly, the living conditions on the Earth by impacting the Earth's radiative budget due to their abilities to scatter and/or absorb the incoming solar radiation and to act as cloud condensation nuclei. Aerosol particles, particularly those from the so-called fine mode fraction, have also been found to have adverse effect on human health through causing issues mostly to the respiratory and cardiovascular systems.

Because of their numerous impacts combined with high spatial and temporal variabilities due to their short lifetime, the aerosols are the object of comprehensive global and local monitoring; their sources, type, transport, microphysical and optical properties are being explored on a broad scale or locally using a variety of experimental active and passive ground-based, ship-born, air-born and space-born remote sensing facilities, together with powerful data processing and interpretation program products and models [1-3].

Sofia LIDAR station at the Institute of Electronics, Bulgarian Academy of Sciences, involved in the activities of the European Aerosol Research Lidar Network (EARLINET) [1], performs regular LIDAR monitoring of the aerosol stratification over the city. In May 2020, the station's equipment was enriched with a CE318-TS9 Cimel sun/sky/lunar photometer contributing to the Global Aerosol Robotic Network (AERONET) [2] that provides almost the whole set of column-averaged or integrated aerosol characteristics at different optical wavelengths. Similar publicly available data is

[1] To whom any correspondence should be addressed.

Content from this work may be used under the terms of the Creative Commons Attribution 3.0 licence. Any further distribution of this work must maintain attribution to the author(s) and the title of the work, journal citation and DOI.
Published under licence by IOP Publishing Ltd

provided by the Moderate Resolution Imaging Spectroradiometer (MODIS) mounted on the Aqua and Terra satellites [3]. The present work aims generally at tracking variations in the aerosol optical depth (AOD_{440}) at the wavelength $\lambda = 440$ nm and the Ångström exponent ($AE_{440/870}$) for the wavelength pair 440/870 nm over Sofia, Bulgaria, and their dependence on the aerosol load and type. As another goal, the sun-photometer data are compared with the MODIS-Aqua satellite data concerning the aerosol optical depth respectively at wavelengths $\lambda = 500$ nm (AOD_{500}) and $\lambda = 550$ nm (AOD_{550}). Special attention is paid to aerosol events characterized by a prevailing coarse particle fraction with values of $AE_{440/870} < 0.7 - 0.8$. If, in addition, the AOD_{440} values are between 0.15 and 0.5, according to a classification developed in [4] one may expect the presence of Saharan dust (SD) layers in the atmosphere over the city. Thus, the days of possible desert dust intrusions can be identified. To do it, we used the above-mentioned classification, the National Oceanic and Atmospheric Administration (NOAA) Hybrid Single-Particle Lagrangian Integrated Trajectory model (HYSPLIT) [5,6] and the NMMB-MONARCH model (formerly NMMB/BSC-Dust) [7,8] operating at the Barcelona Supercomputing Center (BSC).

2. Instruments and auxiliary information sources

The CE318-TS9 sun/sky/lunar photometer comprises optical channels at the wavelengths $\lambda = 340$ nm, 380 nm, 440 nm, 500 nm, 675 nm, 870 nm, 936 nm, 1020 nm and 1064 nm and performs measurements of the sun and moon irradiance and the sky radiance on several bands according to predefined scenarios. The corresponding AERONET automatic data processing allows one to retrieve a large set of atmospheric aerosol characteristic parameters, such as AOD, AE, water vapor content, volume size distribution, AOD of the fine (AODf) and coarse (AODc) aerosol fractions at $\lambda = 500$ nm, complex refraction index, phase function, etc.

The Moderate Resolution Imaging Spectroradiometer (MODIS) [3] is an instrument aboard the Aqua and Terra satellites, part of NASA's Earth Observing System (EOS) created to monitor the Earth's atmosphere and surface using 36 visible, NIR, MIR, and thermal spectral channels (bands). Aqua crosses the equator from south to north in the afternoon, while the orbit of Terra around the Earth is adjusted in the opposite direction in the morning. Aqua and Terra are providing a global coverage every 1 to 2 days, acquiring data in spectral bands ranging from 0.4 μm to 14.4 μm. They are covering a swath of 2330 km with a spatial resolution at nadir 0.25 km (bands 1-2), 0.5 km (bands 3-7) and 1 km (bands 8-36). The radiometric resolution is 12 bits per channel. The first seven spectral bands (around 470 nm, 550 nm, 660 nm, 870 nm, 1240 nm, 1640 nm and 2130 nm) provide aerosol-related data by means of several independent algorithms determining the aerosol parameters over ocean and land [9,10].

NOAA HYSPLIT model is used for revealing the origin and route of the air masses reaching Sofia at different heights during days and periods of interest; the NMMB-MONARCH model, operating at BSC, is used to forecast the dust load over Sofia.

3. Results and discussion

The analysis described below of the character, state and dynamics of the aerosol field over Sofia, Bulgaria, during 2020 and 2021 was performed mainly on the basis of AERONET data (level 1.5, https://aeronet.gsfc.nasa.gov/) acquired by the CE318-TS9 sun/sky/lunar photometer measurements in Sofia, Bulgaria (table 1). The concrete aerosol characteristics of interest were the daily-averaged values of the total AOD_{440}, $AE_{440/870}$, $AODf_{500}$ and $AODc_{500}$. Based on the AERONET-provided volume size distributions of the aerosol particles, we consider the sizes of the fine fraction as being in the interval from 0.06 μm to 0.5 – 1.0 μm with a peak around 0.1 – 0.2 μm. The sizes of the coarse fraction are in the interval from 0.5 – 1.0 μm to 10 μm with peak(s) between 2 μm and 8 μm. We directed our attention mainly to the days with $AE_{440/870} < 0.7 - 0.8$ and $0.15 < AOD_{440} < 0.50$, expecting to reveal SD events over the city [4]. In the cases when $AE_{440/870} > 0.7$ and $AOD_{440} < 0.15 - 0.20$, the atmosphere was assumed to contain marine aerosols (MA). Figure 1 is an AOD_{440} vs $AE_{440/870}$ scatter plot characterizing the aerosol environment during the selected days of 2020 and

2021. The SD events are marked by black squares, while the marine and the continental aerosol (CA) events, by red circles and blue triangles, respectively (table 1 and figure 1).

Table 1. AERONET-provided daily-averaged values of the total AOD_{440}, $AE_{440/870}$, $AODf_{500}$ and $AODc_{500}$ over Sofia station during different days of 2020 and 2021. NMMB-MONARCH forecasted SD event occurrence (Y – yes or N – no) and HYSPLIT and AERONET AOD and AE data-based differentiation of the aerosol type – SD and/or MA and/or CA.

Date	AOD_{440}	$AE_{440/870}$	$AODf_{500}$	$AODc_{500}$	NMMB-MONARCH	HYSPLIT	AOD/AE Classification
05.05.2020	0.172	0.780	0.067	0.083	N	MA+CA+SD	●■
11.05.2020	0.168	0.645	0.061	0.089	Y	SD+MA+CA	■
12.05.2020	0.138	0.720	0.051	0.070	Y	MA+SD+CA	●
07.06.2020	0.217	0.837	0.095	0.095	Y	MA+SD	■
09.06.2020	0.197	1.147	0.102	0.067	Y	SD+MA	●▲
31.08.2020	0.188	0.868	0.079	0.083	Y	SD+MA	■
01.09.2020	0.253	0.864	0.109	0.108	Y	SD+MA	■
16.09.2020	0.187	0.936	0.087	0.074	N	SD+MA+CA	■
03.10.2020	0.171	1.041	0.091	0.058	N	SD+MA+CA	●▲
21.10.2020	0.109	1.352	0.068	0.022	N	CA+MA	●
23.12.2020	0.076	1.170	0.038	0.025	N	CA+MA	●
30.12.2020	0.054	0.409	0.015	0.035	Y	SD+MA	●
24.01.2021	0.050	0.890	0.030	0.014	N	MA+SD	●
23.02.2021	0.181	0.786	0.072	0.088	Y	SD+CA+MA	■
24.02.2021	0.483	0.741	0.213	0.223	Y	SD+CA+MA	■
25.02.2021	0.210	0.648	0.085	0.104	N	SD+CA+MA	■
26.02.2021	0.164	0.894	0.069	0.066	N	CA+MA+SD	●■
27.02.2021	0.253	0.689	0.106	0.122	Y	CA+MA+SD	■
05.03.2021	0.152	1.076	0.072	0.056	N	CA+SD+MA	▲●
26.04.2021	0.213	1.572	0.133	0.036	N	CA+MA	▲
28.04.2021	0.301	0.645	0.113	0.157	Y	SD+MA	■
30.04.2021	0.197	0.606	0.071	0.106	Y	SD	■
02.05.2021	0.162	0.425	0.047	0.103	Y	SD+MA	■
03.05.2021	0.126	0.891	0.051	0.057	N	MA+SD	●
17.07.2021	0.225	0.975	0.112	0.083	N	MA+SD	■
26.07.2021	0.415	0.839	0.182	0.180	Y	MA+CA+SD	■
27.07.2021	0.413	0.652	0.161	0.213	Y	MA+CA+SD	■
28.07.2021	0.434	0.579	0.157	0.241	Y	MA+SD	■
29.07.2021	0.363	0.587	0.134	0.198	Y	MA+CA+SD	■
30.07.2021	0.364	0.657	0.140	0.188	Y	MA+CA+SD	■
31.07.2021	0.437	1.092	0.228	0.143	Y	MA+CA+SD	▲
02.08.2021	0.301	0.525	0.101	0.176	Y	SD+MA	■
03.08.2021	0.265	1.039	0.135	0.092	Y	SD+MA	■▲
04.08.2021	0.241	0.972	0.115	0.093	Y	MA+SD	■
05.08.2021	0.426	0.668	0.168	0.218	Y	SD+MA	■
11.08.2021	0.352	1.093	0.184	0.118	Y	SD+CA+MA	▲

To verify our sun-photometer-data-based assumptions regarding the atmospheric-aerosol character, we employed additional information from the NMMB-MONARCH and the HYSPLIT models. The days with SD events forecasted by NMMB-MONARCH are designated in table 1 by "Y" (yes) in the NMMB-MONARCH column. The days without such forecasts are designated there by "N" (no). In the HYSPLIT column of table 1, the aerosol events identified hypothetically on the basis of the HYSPLIT-derived air mass back trajectories are denoted by their above-defined abbreviations SD, MA and CA. In the AOD/AE Classification column, the AERONET-data-based predictions are denoted by the above-described geometric figures. As seen in the table, the SD events forecasted by the NMMB-MONARCH are 24 and those predicted by AERONET are also 24. The coincident predictions are 19, that is, 79 % of both the NMMB-MONARCH and the AERONET predictions. At the same time, all the coincident predictions (Y or N) are 26 of 36, that is, 72 %. HYSPLIT confirms as well the possibility of SD events to occur during all AERONET-predicted days and 8 more; in this

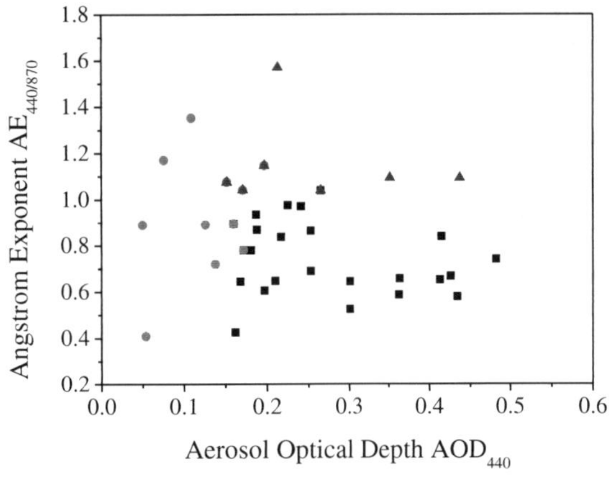

Figure 1. $AOD_{440}/AE_{440/870}$ scatter plot characterizing the aerosol environment over Sofia station during the days of concern of 2020 and 2021. The black squares, red circles and blue triangles indicate SD, MA and CA, respectively.

case, however, HYSPLIT plays an auxiliary role only, because it cannot reveal the prevailing as well the possibility of SD events to occur during all AERONET-predicted days and 8 more; in this case, however, HYSPLIT plays an auxiliary role only, because it cannot reveal the prevailing aerosol component. As a whole, based on the above-obtained estimates one may roughly assume that the AERONET predictions concerning the region of the Sofia station and following the AOD/AE aerosol-type classification rules developed in [4] might be valid with a probability of between 70 – 80%.

Following the data summarized in table 1 (the first three columns from left), the SD and other aerosol events that occurred during the chosen days are presented in figure 1. The scatter plot outlines an AOD/AE area revealing, using the AERONET facilities, the presence of an SD mass over Sofia station. The approximate extent of this area is 0.16 – 0.17 < AOD < 0.48 – 0.50 and AE < 1.06 (see also [4]). In figure 2, we illustrated the aerosol diagnostics results concerning 30 April 2021 (a, d), 09 June 2020 (b, e), and 25 February 2021 (c, f). On 30 April, the values of $AOD_{440} = 0.197$ and $AE_{440/870} = 0.606$ (table 1) indicate possible passages of an SD mass in the atmosphere. Correspondingly, such passages are forecasted by the NMMB-MONARCH (figure 2a) and confirmed by the HYSPLIT back trajectories (figure 2d). On 09 June, $AOD_{440} = 0.197$ and $AE_{440/870} = 1.147$ (table 1), so that one should not expect SD events on that date. Nevertheless, such events are forecasted by the NMMB-MONARCH (figure 2b) and supported by the HYSPLIT back trajectories (figure 2e). On 25 February, the values of $AOD_{440} = 0.210$ and $AE_{440/870} = 0.648$ (table 1) and the HYSPLIT back trajectories (figure 2f) indicate possible SD events. However, NMMB-MONARCH does not predict such events (figure 2c).

The ground-based sun photometer performance allows one to track the local state and dynamics of the aerosol field over Sofia region. A global monitoring of the aerosol field is only feasible, however, by using satellite-born apparatus, such as Terra MODIS and Aqua MODIS. In any case, it is interesting and useful to compare the data about AOD and AE provided by both systems. We compare here the AERONET AOD data obtained by the sun photometer at $\lambda = 500$ nm with the corresponding AOD data obtained by Aqua MODIS at $\lambda = 550$ nm with resolutions of 3 km and 10 km (figure 3). The data concern the daytime period of satellite flyover (relatively) near the region of Sofia station temporally coinciding with the sun photometer measurements. Aqua MODIS data derived from the NASA Worldview application (https://worldview.earthdata.nasa.gov/), part of the NASA Earth Observing System Data and Information System (EOSDIS), were employed in the comparison. It is seen in figure 3 that, while accounting for the difference between the wavelengths $\lambda = 500$ nm and 550 nm, the correspondence between the two sets of AOD values is adequate and reasonable. The lines approximating the dependences (AOD_{550}/AOD_{500}) have relatively small intercepts, and slopes below unity at relatively high values of the adjusted R-square (~ 0.86 and 0.79 for 3 km and 10 km resolution, respectively).

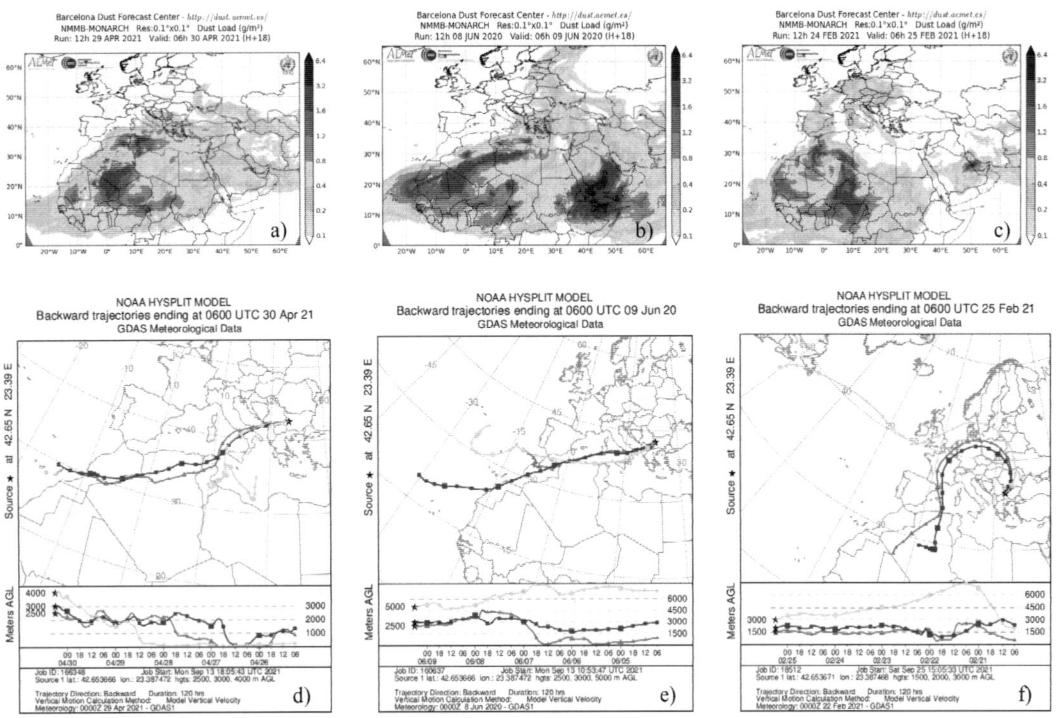

Figure 2. NMMB-MONARCH and HYSPLIT models data on 30 April 2021 (a, d), 9 June 2020 (b, e), and 25 February 2021 (c, f).

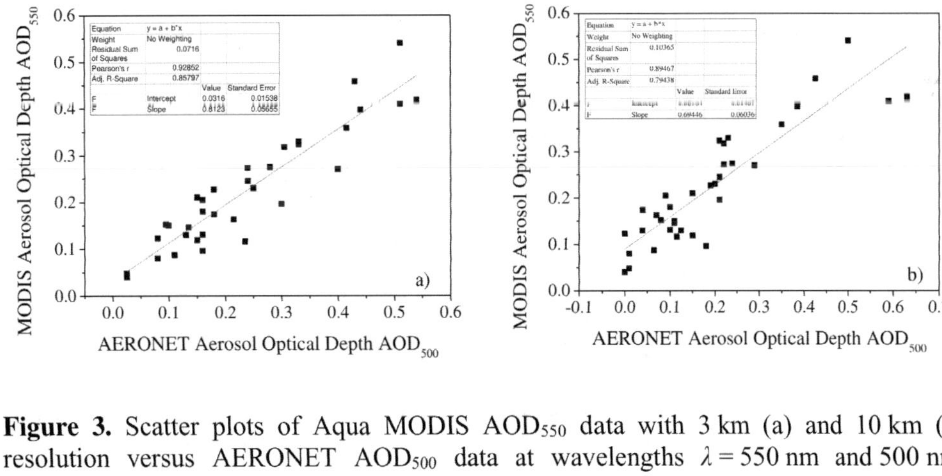

Figure 3. Scatter plots of Aqua MODIS AOD_{550} data with 3 km (a) and 10 km (b) resolution versus AERONET AOD_{500} data at wavelengths $\lambda = 550$ nm and 500 nm, respectively.

4. Conclusions

In the present work, we analyzed AERONET sun photometer data for AOD_{440} and $AE_{440/870}$ of the aerosol load over Sofia, Bulgaria, during different days in 2020 and 2021. Cases were considered when $AE_{440/870}$ was around unity and less, while AOD_{440} varied up to around 0.5. Thus, following the AOD_{440}-vs-$AE_{440/870}$-based aerosol-type recognition approach developed in [4], we expected to distinguish the days with SD events, assuming that then $AE_{440/870} < 0.7 – 0.8$ and $0.15 < AOD_{440} < 0.5$. As additional verifying information, we employed the dust load forecasts of the NMMB-MONARCH and the HYSPLIT-derived back trajectories of the arriving air masses. Among all the days (24) with SD events assumed on the basis of the AOD/AE classification, 19 (79 %) were predicted by the NMMB-MONARCH. This number is also 79 % of all the SD events (24) forecasted by the NMMB-MONARCH. All coincident predictions (with or without SD) are 26 of 36, that is, 72 %. Thus, the AERONET data-based predictions might roughly be considered as valid with a probability of 70 – 80 %. The comparison between AOD_{500} and AOD_{550} determined simultaneously by using the ground-based local (CE318-TS9 sun photometer) and satellite-born (MODIS) instruments, respectively, shows reasonable correlation between the data considered.

Acknowledgments
This research was supported partly by the Ministry of Education and Science of Bulgaria (support for ACTRIS BG, part of the Bulgarian National Roadmap for Research Infrastructure), by the European Commission under the Horizon 2020 – Research and Innovation Framework Program, Grant Agreement number 871115 (ACTRIS IMP), and by the Bulgarian National Science Fund (Grant No. KP-06-N28/10/2018). The authors acknowledge AERONET-Europe, part of ACTRIS IMP project, for providing calibration service. The NOAA Air Resources Laboratory provision is acknowledged of the HYSPLIT transport and dispersion model and the READY website used in this publication. The authors also acknowledge the images from the NMMB-MONARCH, operated by the Barcelona Supercomputing Center (https://dust.aemet.es/forecast), and the use of imagery from the NASA Worldview application (https://worldview.earthdata.nasa.gov/), part of the NASA Earth Observing System Data and Information System (EOSDIS).

References
[1] Pappalardo G et al 2014 *Atmos. Meas. Tech.* **7** 2389–2409
[2] Holben B N et al 1998 *Rem. Sens. Environ.* **66** 1–16
[3] MODIS (Moderate Resolution Imaging Spectroradiometer) https://modis.gsfc.nasa.gov/
[4] Toledano C, Cachorro V E, Sorribas M, Berjón A, de la Morena B A, de Frutos A and Gouloub P 2007 *Q. J. R. Meteorol. Soc.* **133** 795–807
[5] Stein A F et al 2015 *Bull. Am. Meteorol. Soc.* **96** 2059–77
[6] Rolph G, Stein A and Stunder B 2017 *Environ. Modell. Softwares* **95** 210–28
[7] Pérez C et al 2011 *Atmos. Chem. Phys.* **11** 13001–27
[8] Haustein K et al 2012 *Atmos. Chem. Phys.* **12** 2933–58
[9] Remer L A et al 2005 *J. Atmos. Sci.* **62** 947–73
[10] Levy R C, Remer L A, Mattoo S, Vermote E F and Kaufman Y J 2007 *J. Geophys. Res.-Atmos.* **112** D13211

Characterization of the aerosol field over Sofia, Bulgaria, using contact and passive-remote-sensing methods and instruments

Ts Evgenieva[1,3], N Kolev[1,2], P Savov[2], L Gurdev[1], I Grigorov[1] and T Dreischuh[1]

[1]Institute of Electronics, Bulgarian Academy of Sciences, 72 Tsarigradsko Chaussee, 1784 Sofia, Bulgaria

[2]Department of Physics, St. Ivan Rilski University of Mining and Geology, Sofia, Bulgaria

E-mail: tsevgenieva@ie.bas.bg

Abstract. The atmospheric aerosol particles, called also atmospheric particulate matter, that include also fine dust particles (FDP), are an important factor conditioning many physical processes in the atmosphere, affecting the climate and the air quality, and posing serious potential risks to the human health. In this regard, the regular observation of the atmospheric aerosol load and its dynamics is a valuable source of scientific information concerning these processes, while also allowing the relevant institutions to undertake preventive measures in cases of a dangerous FDP concentration increase. This work is devoted to estimating experimentally, by passive optical sensing and contact measurements, some important integral microphysical and optical characteristics and their dynamics of the aerosol ensembles over Sofia, Bulgaria. The results obtained outline the springtime values of the aerosol number and mass concentration in the near-ground urban atmosphere, the depth of the atmospheric boundary layer, the aerosol optical depth at different optical wavelengths, the water vapor content, and the Ångström exponent at the wavelengths of 440/870 nm.

1. Introduction

As it is well known, the aerosol particles (or atmospheric particulate matter, APM), including the fine dust particles (FDP), in the atmosphere strongly influence the energy exchange between the Sun and the Earth's atmosphere and surface. Thus, they impact multiple atmospheric phenomena, the climate, and the living conditions on the Earth. The FDP may be especially harmful to the human health as well. The permanent monitoring and studying the aerosol properties and dynamics contribute to the better understanding of many atmospheric processes, including the climate change, and assist the authorities, especially in urban areas, in being timely informed of dangerous FDP concentration rise above the permissible levels. Correspondingly, the main purpose of the present work is to demonstrate the efficiency of the combined use of passive optical remote-sensing and contact-probing approaches to characterizing the atmospheric aerosol field.

The aerosol characteristics of interest in this work are the aerosol number and mass concentration in the near-ground urban atmosphere, the atmospheric boundary layer depth (BLD), the

[3] To whom any correspondence should be addressed.

Content from this work may be used under the terms of the Creative Commons Attribution 3.0 licence. Any further distribution of this work must maintain attribution to the author(s) and the title of the work, journal citation and DOI.
Published under licence by IOP Publishing Ltd

meteorological parameters influencing the aerosol properties and dynamics, the aerosol optical depth (AOD) at different optical wavelengths, the columnar water vapor content (WVC), and the Ångström exponent (AE) calculated for the 440/870 nm wavelength pair.

2. Experimental sites, instrument specifications and additional information sources

The instruments employed in the experiments were a Cimel sun photometer (CE318-TS9) involved in the global Aerosol Robotic Network (AERONET) [1] and a two-channel laser particle counter (LPC).

The sun/sky/lunar photometer CE318-TS9 (Cimel, France) has optical channels at the wavelengths $\lambda = 340$ nm, 380 nm, 440 nm, 500 nm, 675 nm, 870 nm, 936 nm, 1020 nm and 1064 nm and carries out measurements of the sun and moon irradiance and the sky radiance at several bands according to predefined scenarios. The AERONET automatic data processing allows one to retrieve a large number of atmospheric aerosol parameters, such as AOD, AE, WVC, volume size distribution, complex refraction index, phase function, etc. The photometer is installed on the roof of the building of the Institute of Electronics of the Bulgarian Academy of Sciences (IE-BAS, 42.65N, 23.38E; 621 m ASL) in the southeast part of Sofia near a heavy-traffic thoroughfare.

The laser particle counter BQ20 (Trotec, Germany) comprises two contact measurement channels. The first one, $PM_{2.5}$, selects aerosol particles of sizes from 0 µm to 2.5 µm; the second one, PM_{10}, particles of sizes from 2.5 µm to 10 µm. In both channels, the particle size and the particle number and mass concentration are measured with a temporal resolution of 10 min to 15 min and a sampling rate of 0.9 l/min. The mass concentration measurement range is from 0 µg/m^3 to 2000 µg/m^3 with a resolution of 1 µg/m^3. The contact measurements were performed in a residential area near the above-mentioned thoroughfare at a distance of about 3.5 km from the IE site.

Additional meteorological and modeling information supporting the interpretation and validation of the final results was provided by the Bulgarian National Institute of Meteorology and Hydrology [2], as well as by using the Hybrid Single-Particle Lagrangian Integrated Trajectory (HYSPLIT) [3,4] and the Weather Research and Forecasting Model – Global Data Assimilation System (WRF-GDAS) [4] models developed by the Air Resource Laboratory of the National Oceanic and Atmospheric Administration (NOAA), USA. The first one was used for tracing the origin and pathway of the air masses reaching Sofia at different heights during the different case study experiments; the second one was applied to estimating the BLD.

3. Experimental results and discussion

3.1. Aerosol particle number and mass concentration measured by LPC, and AOD, AE, WVC measured by CE318-TS9 sun photometer

The results from the measurements performed between May 8 and 10, 2020, are illustrated in figures 1 and 2. A maximum is observed in the daily course of $PM_{2.5}$ and PM_{10} number and mass concentrations (figure 1a-f) in the morning due to the intensive traffic during that time period. At all wavelengths used, a noticeable time correlation is seen of the AOD with the number and mass aerosol concentration, and, to a weaker extent, with the WVC.

At the same time, the WVC and the AE measured at the 440/870 nm wavelength pair on May 8 and 9 are obviously anti-correlated, which is most probably due to the increasing aerosol particles size by absorption and condensation of water vapor and coagulation of scattering particles. This assumption is supported by the fact that during the (early-morning) periods of decreasing AE and increasing WVC, the air temperatures were near the corresponding dew points (see table 1). Parallel processes of intensified sedimentation are also likely to have taken place in this case. A moist air mass containing marine aerosol particles might also have been passing (see below). The correlation between AOD and WVC and between AE and WVC on May 10 seems rather positive. On May 8, AOD at $\lambda = 440$ nm and AE vary, respectively, from 0.15 to 0.20 and from 1.6 to 1.1 with readings between 0.7 and 0.9 in the period from 12:50 UTC to 14:10 UTC. On May 9, AOD at the same wavelength varies from 0.10 to 0.17, and AE, from 1.4 to 1.0 with values below 1.0 to 0.7 in the period from 06:30 UTC to

14:10 UTC. On May 10, AOD at $\lambda = 440$ nm varies between 0.15 and 0.17, and AE, between 1.04 and 1.30. These results may be interpreted using the classification developed in [5] of the aerosol types based on AOD and AE. According to it, AOD at $\lambda = 440$ nm and AE concerning the wavelength pair 440/870 nm, in cases of continental aerosols, marine aerosols, biomass burning aerosols, desert dust, and mixed aerosol matter, vary, respectively, within the intervals 0.2 – 0.3 and 1.2 – 1.6; 0.05 – 0.15 and 0.7 – 1.5; 0.30 – 0.65 and 1.5 – 1.7; 0.15 – 0.50 and 0.25 – 0.80; and 0.35 – 0.50 and 1.2 – 1.35.

Table 1. Meteorological parameters on 8, 9, and 10 May 2020.

Time (UTC)	Date	Temperature (°C)	Dew Point (°C)	Relative Humidity (%)	Wind Speed (m/s)	Wind Direction	Visibility (km)
	08.05.2020	3.7	2	89	0	-	25
03:00	09.05.2020	4.9	2.8	86	0	-	20
	10.05.2020	8.8	5.6	80	0	-	20
	08.05.2020	**10.4**	**3.5**	**62**	**0**	**-**	**25**
06:00	**09.05.2020**	**10.8**	**3.8**	**62**	**0**	**-**	**25**
	10.05.2020	**13.5**	**7.1**	**65**	**0**	**-**	**30**
	08.05.2020	16.1	2.8	41	2	WSW	25
09:00	09.05.2020	17.5	6.2	48	0	-	25
	10.05.2020	20.2	8.5	47	0	-	25
	08.05.2020	**18.5**	**4.5**	**40**	**2**	**WSW**	**25**
12:00	**09.05.2020**	**23.4**	**5.2**	**31**	**2**	**WNW**	**25**
	10.05.2020	**25.3**	**10**	**38**	**0**	**-**	**25**
	08.05.2020	19.5	4.9	38	2	W	25
15:00	09.05.2020	24.8	5.2	28	2	NW	30
	10.05.2020	26.2	9.9	36	0	-	25
	08.05.2020	**16**	**3.5**	**43**	**2**	**W**	**25**
18:00	**09.05.2020**	**19.2**	**7.9**	**48**	**0**	**-**	**25**
	10.05.2020	**21.7**	**11.4**	**52**	**0**	**-**	**20**

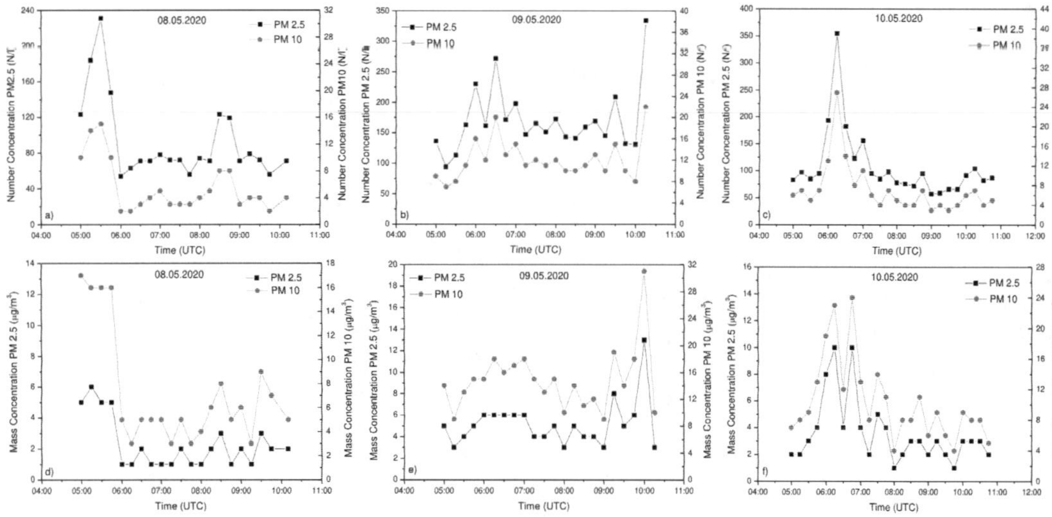

Figure 1. Daily variations in $PM_{2.5}$ and PM_{10} number and mass concentration measured on 08 May (a, d), 09 May (b, e), and 10 May (c, f) 2020.

Therefore, one may assume that the aerosol mass observed on May 8 contained mainly continental particulate matter with a small admixture of marine aerosols. The aerosol mass observed on May 9 appears to be a mixture of continental aerosol and marine or Saharan dust components, while the one on May 10 is rather continental.

Note as well that the optical depths of the fine (AODf) and coarse (AODc) aerosol fractions determined on May 8, 9 and 10 using the CE318-TS9 vary synchronously with the corresponding AEs. Thus, on May 8, AE decreases (figure 2) and the relative contribution of the fine aerosol fraction (with respect to the coarse aerosol fraction) to the scattering process decreases (figure 3).

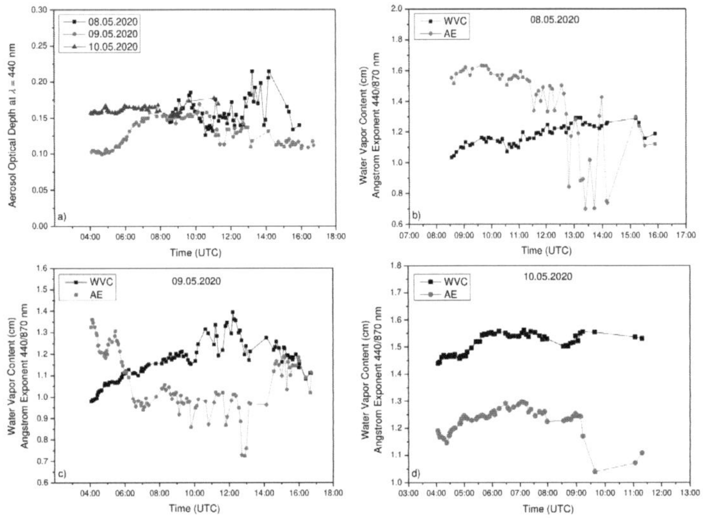

Figure 2. AERONET daily variations in the aerosol optical depth (AOD) at $\lambda = 440$ nm on 8–10 May 2020 (a), columnar water vapor content (WVC) and Ångström exponent (AE) obtained on 08 May (b), 09 May (c) and 10 May 2020 (d).

On May 9, AE passes through a minimum (figure 2). At the same time, the relative contribution of the small particles to the scattering process also passes through a minimum (figure 3). On May 10, AE is slowly oscillating around a constant level in the period between 05:00 and 09:00 UTC (figure 2), and AODf and AODc (AODf >AODc) are approximately constant.

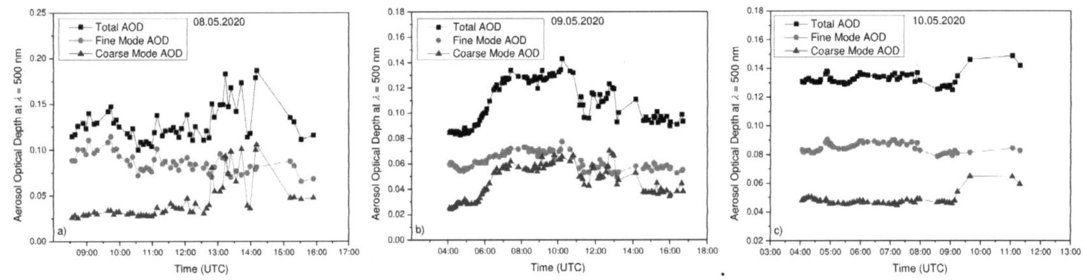

Figure 3. AERONET daily variations of the aerosol optical depths of the fine aerosol fraction (AODf), and the coarse aerosol fraction (AODc), determined by CE318-TS9 on 08 May (a), 09 May (b) and 10 May (c) 2020.

3.2. Determining the origins and trajectories of the air masses reaching Sofia region on 8, 9 and 10 May 2020 and the corresponding weather conditions and ABLD behavior

To avoid ambiguities in identifying the aerosol matter observed, we used the HYSPLIT model to determine the 72-hour backward trajectories of the aerosol-carrying air masses reaching Sofia on May 8, 9 and 10 at 09:00 UTC. The air masses arriving at heights between 500 m and 3000 m AGL contained mostly continental aerosols that in some cases might have been enriched with marine ones. The initial points and trajectories of the air masses arriving at the heights of 4000 m, 5000 m and 6000 m AGL are traced in figure 4. It is seen that those arriving on May 8 at all heights originated 72 hours earlier in the Atlantic Ocean and then passed over Great Britain, the Scandinavian Peninsula (at

4000 m and 5000 m AGL), North Sea (at 6000 m), Baltic Sea, and Central Europe. During this long travel, most of the drifting large-particle aerosol fraction, and especially that of marine origin, seemingly sedimented so that the small-particle fraction became prevailing. The 72-hour travel of the air mass arriving at the heights of about 4000 m and 5000 m on May 9 began in the Atlantic Ocean and passed over Western and Central Europe, while the one at 6000 m originated in Algeria and passed over Morocco, Western and Central Europe, which is consistent with the assumption that a mixture was observed of continental aerosol and Saharan dust components. The air mass arriving at the heights of about 5000 m and 6000 m AGL on May 10 originated in the Atlantic Ocean and passed mainly over South-Western Europe; at the other height, the air mass traveled mostly over Morocco and South-Western Europe. The influence of the Saharan dust is not clearly seen in the AOD and AE data on 10 May. As a whole, the information provided by the HYSPLIT model is in accordance with the values measured of the AOD and AE. It is also consistent with the predictions based on the aerosol type classification developed in [5]. The weather parameters during the experimental days (table 1) have close values and point to a relatively calm atmosphere with wind speed of 0 – 2 m/s and atmospheric pressure of 1010 – 1021 hPa. The temperatures measured are near the dew point in the morning and well above it afterwards – one may thus assume that condensation and aggregation occurred in the early morning during the experiments.

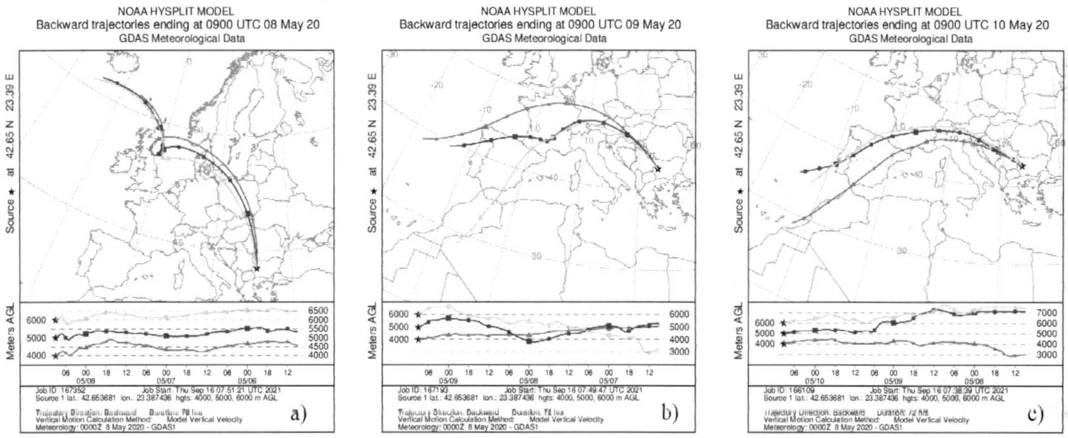

Figure 4. 72-hour back-trajectories calculated by the HYSPLIT model of air masses ending at 09:00 UTC over Sofia on (a) 08 May, (b) 09 May and (c) 10 May 2020.

The BLD behavior outlined on the basis of the WRF-GDAS data (figure 5) on May 8 shows a rapid and continuous increase reaching a maximum at about 1320 m AGL at 09:00 UTC, while on May 9 and 10 it is characterized by an initial smoother run followed by a secondary sharper rise. The similarity in the BLD behavior seems natural, taking into account the similar weather conditions and aerosol content in the atmosphere on these days. The BLD maximum of about 1300 m AGL (on May 9) and 1450 m AGL (on May 10) is reached at 12:00 UTC on both days. Higher atmospheric instability and vertical mixing coefficient are seen on May 8. The decreasing trend seen in the $PM_{2.5}$ and PM_{10} number and mass concentrations on May 10 after the morning peak (figure 1c, f) could be related to the increasing BLD [6] in the case of relatively more stable atmospheric conditions.

In general, the studies conducted and the results obtained show that on ordinary springtime days the aerosol mass over Sofia consists mainly of small-size particles of continental origin. This is to be expected considering that Sofia is located relatively far from large sea basins, deserts, etc. (see also [7,8]), while in countries situated near deserts and surrounded by seas the aerosol mass should contain mainly marine type particulate matter and, relatively often, desert dust [5, 9, 10]. Certainly, extraneous aerosol events such as Saharan dust intrusions have also been observed over Sofia [11, 12].

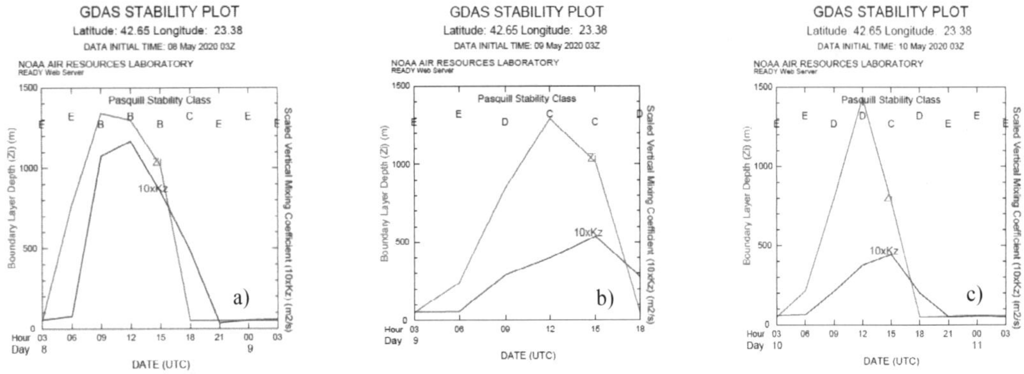

Figure 5. WRF-GDAS-model-based forecast of the BLD diurnal cycle (m AGL) on 08 (a), 09 (b), and 10 (c) May 2020. The vertical mixing coefficient behavior and the Pasquill stability class are also presented.

4. Conclusions

The experimental and modeling results obtained demonstrate the potential of the versatile approach used by us in studying the aerosol characteristics and dynamics based on modern experimental facilities, data from worldwide networks of sensors, and sophisticated and accurate models of the atmospheric processes. The characteristics determined are valuable as concrete case quantities, but they can also be useful in confirming some theoretical or modeling results. It is shown that the aerosol mass over Sofia usually consists of mostly small-size particles of continental origin because of the relatively long distance away from large sea basins, deserts, etc. The existence is also proved of a time correlation of the AOD at all wavelengths used with the number and mass aerosol concentration, and, to a weaker extent, with the WVC. Further, the WVC and AE at the 440/870 nm wavelength pair are anti-correlated probably due to a passing moist air mass containing coarse marine aerosol particles, accompanied in some cases by tiny amounts of Saharan dust, and/or to water vapor condensation and/or particle aggregation leading to a rise in the aerosol particle size.

Acknowledgments

The work of the team from IE-BAS was supported by the Ministry of Education and Science of Bulgaria (support for ACTRIS BG, part of the Bulgarian National Roadmap for Research Infrastructure) and by the European Commission under the Horizon 2020 – Research and Innovation Framework Program, Grant Agreement number 871115 (ACTRIS IMP). The authors acknowledge AERONET-Europe, part of ACTRIS IMP project, for providing calibration service. The NOAA Air Resources Laboratory provision is acknowledged for the HYSPLIT transport and dispersion model and WRF-GDAS model, as well as the READY website used in this publication.

References

[1] Holben B N et al 1998 *Remote Sensing of Environment*, **66** 1–16
[2] Archived weather data for Sofia, Bulgaria (https://www.stringmeteo.com/synop/bg_stday.php), provided by the Bulgarian National Institute of Meteorology and Hydrology
[3] Stein A F et al 2015 *Bull. Am. Meteorol. Soc.* **96** 2059–77
[4] Rolph G, Stein A and Stunder B 2017 *Environ. Modell. Softwares* **95** 210–28
[5] Toledano C, Cachorro V E, Sorribas M, Berjón A, de la Morena B A, de Frutos A and Gouloub P 2007 *Q. J. R. Meteorol. Soc.* **133** 795-807
[6] Savov P, Kolev N, Evgenieva Ts, Vatzkitcheva M and Danchovski V 2016 *C. R. Acad. Bulg. Sci.* **69** (1) 19–24
[7] Borbely-Kiss I, Bozo L, Koltay E, Meszaros E, Molnar A and Szabo Gy 1991 *Atmos. Environ. Part A. General Topics* **25** (3-4) 661–8

[8] Crippa M et al 2013 *J. Geophys. Res.: Atmospheres* **118** 1950–63
[9] Guerrero-Rascado J L et al 2011 *J. Geophys. Res.* **116** D02209
[10] Pio C A, Santos I M, Anacleto T D, Nunes T V and Real R M 1991 *Atmos. Environ. Part A. General Topics* **25** (3-4) 669 – 680
[11] Peshev Z, Dreischuh T, Evgenieva Ts, Deleva A, Tonev D and Stoyanov 2016 *J. Appl. Remote Sens.* **10** 036009
[12] Peshev Z, Deleva A, Vulkova L and Dreischuh T 2021 *J. Phys.: Conf. Ser.* **1859** 012031

Journal of Physics: Conference Series **2240** (2022) 012029 doi:10.1088/1742-6596/2240/1/012029

Conversion of CO_2 in stabilized low-current arc discharge at atmospheric pressure

V Ivanov[1], S Lazarova, S Iordanova, Ts Paunska, N Georgiev and St Kolev[2]

Faculty of Physics, St. Kl. Ohridski University of Sofia, 5 J. Bourchier Blvd., 1164 Sofia, Bulgaria

E-mail: vladislavvi@phys.uni-sofia.bg, skolev@phys.uni-sofia.bg

Abstract. One of the direct ways to achieve CO_2 conversion is through the use of a non-isothermal plasma environment. In this work, we present an experimental study of the dissociation of CO_2 in a magnetically-stabilized arc discharge in a cross-flow configuration, where the CO_2 gas flow is perpendicular to both the arc and the external magnetic field. The system works at atmospheric pressure. The study examines the effect of the gas flow and discharge current on the quantities of most interest, namely, the CO_2 conversion percentage and the energy efficiency of the process. The experimental results show that the conversion tends to increase with the current, while it drops as the gas flow rate is increased. The efficiency seems to decrease with the conversion percentage, as it increases with the flow rate.

1. Introduction

One of the promising paths of converting CO_2 is by using non-isothermal plasma discharges. The most effective way to perform this is through discharges at low pressure [1], where one can achieve high values of both the conversion and the efficiency. An alternative to these low-pressure gas reactors are those working at moderate pressures and at atmospheric pressure [2, 3, 4]. Conversion devices working at atmospheric conditions, for example, do not require a vacuum system and can achieve high efficiency, but the possible CO_2 conversion values are typically not very high.

The system used in our experiments also works at atmospheric pressure. It is expected to have similar qualities to other atmospheric arc devices, but we argue that it has the advantages of low cost and scalability. Our configuration is that of a magnetically stabilized arc discharge at low currents ($\sim 10^2$ mA) working at low power (< 300 W). This study focuses on the measurement of two quantities, the CO_2 conversion X_{CO_2} (%), and the energy efficiency for the conversion η (%). The goal of our research is to find the conditions under which we can achieve the highest values of both these parameters using our experimental device. In the results section, we present measurements for both the conversion and the efficiency using our discharge device at different arc currents and different gas flow rates.

The configuration of a magnetically-stabilized arc was studied theoretically in [5] as a means of revealing the general trends in the arc behavior and the conditions for arc stabilization. These results

[1] To whom any correspondence should be addressed.
[2] To whom any correspondence should be addressed.

Content from this work may be used under the terms of the Creative Commons Attribution 3.0 licence. Any further distribution of this work must maintain attribution to the author(s) and the title of the work, journal citation and DOI.
Published under licence by IOP Publishing Ltd

were obtained in argon, but it is expected that the gas flow-arc interaction should remain qualitatively similar so that the results in [5] can be used to provide a general understanding of the experimental results obtained here.

2. Experimental setup and measurement method

The experimental setup consists of several devices shown in figure 1 (a). The CO_2 gas flow is controlled and monitored using a mass-flow controller (MFC Bronkhorst EL-FLOW F-201CM).

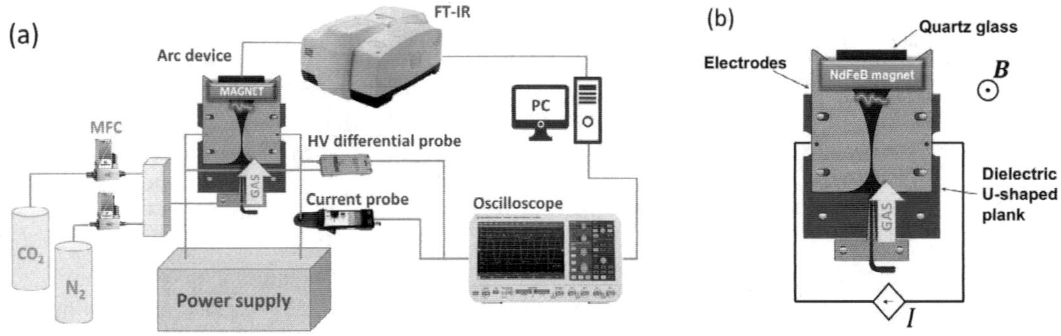

Figure 1. A schematic of the experimental setup (a), an illustration of the discharge device (b).

The gas coming from the flow controllers is directly fed into the arc discharge device, which performs the actual CO_2 conversion. The arc device is placed in a sealed glass chamber. The discharge can be monitored through the walls of the chamber.

The voltage drop across the discharge is measured using a high-voltage (HV) differential probe – Pintek DP-30K, and the discharge current is measured using a DC clamp probe – Pintek PA-699. All electrical measurements are recorded using a digital oscilloscope. The values of the discharge current and voltage drop are used for computing the total power introduced to the discharge.

The entire gas system remains at atmospheric pressure during the operation. After a certain period of time from the start, we know for certain that the input gas has propagated throughout the entire system. These times are dependent on the input flow rate and were measured separately. After the system is "full" of the treated gas, we extract a sample with a sampling pump, which is used to determine the composition of the gas by performing Fourier transform infrared spectroscopy (FTIR). The sample is pumped into a gas cell with an optical length of 10 cm, a diameter of 6 cm, and walls of CaF_2. The CO_2 conversion X_{CO_2} is derived from the intensity of a single line at 2209 cm^{-1} in the infrared part of the absorption spectrum of the CO molecule.

2.1. Discharge device

Our gas conversion reactor is a modification of the common diverging electrode gliding arc design, where a magnetic field is used to create a stationary arc. It is similar to the configurations in [6] and [7]. The device comprises two knife-shaped electrodes sandwiched between two quartz glass plates, as shown in figure 1 (b). The electrodes and the glasses are fastened together between two dielectric U-shaped pieces. A pair of NdFeB magnets are placed on the outside of the two quartz plates to create a strong permanent magnetic field (0.5 – 0.6 T) in the gap between the electrodes.

The gas is introduced from a nozzle at the bottom of the device. The diameter of the nozzle opening is specially chosen to ensure a laminar gas flow. Initially, the arc is ignited at the closest distance between the electrodes (approx. 5 mm), but because of the drag created by the gas flow it gets "pushed" upward (downstream) and reaches the region of the strong magnetic field. At that stage, the arc begins to experience a magnetic force with a magnitude of IB in a direction perpendicular to the arc current I and the magnetic flux density B. The directions of the magnetic field and the current are so chosen that the magnetic force is directed downward (upstream). In this setup, the arc becomes stationary at a certain position, where the drag force and the magnetic force cancel out. We refer to this process as magnetic stabilization. The currents used in our experiment are low ($\sim 10^2$ mA), but because of the strong magnetic field ($\sim 10^{-1}$ T), the magnetic force is sufficient to stabilize the arc even for flow rates as high as 7 l/min. In our experimental setup, the arc remains stationary just under the magnets and has a length of 1.5 cm to 2 cm and a diameter ranging from 1 cm to 2 mm. One of the advantages of this type of configuration is that most of the input gas (up to 70-80%) passes through the arc, which is arguably beneficial to conversion. Another obvious benefit is that the device works at low currents, therefore, low power.

The disadvantages of this type of setup arise from the fact that the arc introduces a large amount of heat to the region near the magnets, where it remains localized throughout the entire operation. This leads to wear of the quartz glasses and also to a significant heating of the magnets, which may reduce their magnetization. For the latter reason, we added an additional water-cooling system for the magnets, so that their temperature remains under the so-called maximal rated operating temperature ($\approx$ 150°C).

2.2. Power supply

A schematic of the power supply is shown in figure 2. To initiate and sustain the arc discharge, we use several $5-15$ kV high voltage step-up transformers. Each of these has a fixed value of the maximal current in the secondary winding, so it acts as an AC current source. We switch between the transformers to select different maximum discharge currents in the range: $25-100$ mA.

To rectify the current in the secondary winding, a Graetz circuit is used and the voltage is filtered by a large inductance of 40 H.

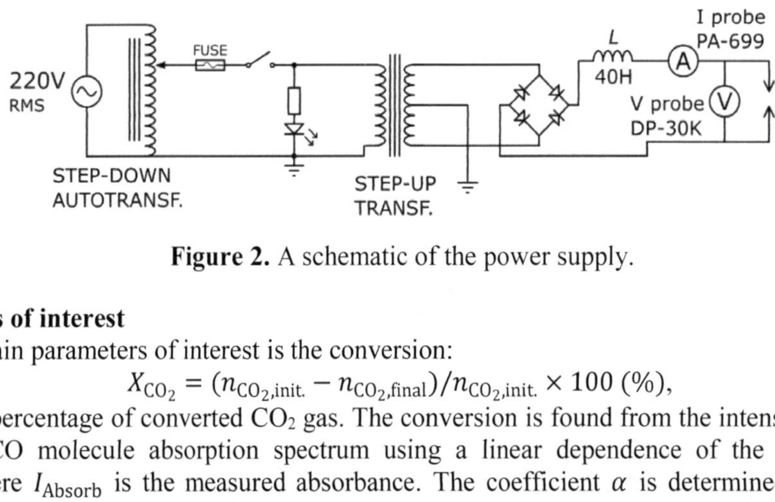

Figure 2. A schematic of the power supply.

3. Quantities of interest

One of the main parameters of interest is the conversion:

$$X_{CO_2} = (n_{CO_2,\text{init.}} - n_{CO_2,\text{final}})/n_{CO_2,\text{init.}} \times 100\ (\%),$$

which is the percentage of converted CO_2 gas. The conversion is found from the intensity of a specific line in the CO molecule absorption spectrum using a linear dependence of the form: $I_{\text{Absorb}} = \alpha X_{CO_2,0}$, where I_{Absorb} is the measured absorbance. The coefficient α is determined from separate measurements using a calibration CO gas mixture. When we calculate the value of the conversion, it is essential to account for the effect of gas expansion. For the dissociation process $CO_2 \rightarrow CO + (1/2)O_2$, the expansion coefficient is 1.5. Because our measurement is done in a gas cell with a fixed volume, we need to correct the measured value of the converted gas, following the relation ([8]): $X_{CO_2,0} = 1-(1-X_{CO_2})/(1+1/2\,X_{CO_2})$, and thus $X_{CO_2} = 2X_{CO_2,0}/(3-X_{CO_2,0})$, where $X_{CO_2,0}$ is the conversion calculated directly from the absorbance, and X_{CO_2} is the actual conversion value accounting for the gas expansion. The energy efficiency η is defined as the ratio of the energy spent for the conversion of X_{CO_2}, to the input energy $\eta = X_{CO_2}\Delta H_R/(SEI \times (1/22.4)\ (\text{mol/Ln})) \times 100\ (\%)$. Here $\Delta H_R = 279.8$ kJ/mol is the reaction enthalpy for the CO_2 splitting reaction, and SEI (J/Ln) is the so-called specific energy input, defined as the energy delivered per liter of gas $SEI = P\ (\text{J/s})/(Q\ (\text{Ln/s}))\ (\text{J/Ln})$, where Q (Ln/s) is the measured input flow rate. The unit Ln (normal liter) is equal to one liter at normal conditions (0 °C, 1 atm). The average power P delivered to the discharge is calculated as $P = (\mathrm{N}^{-1}) \sum_i^N U_i I_i$, with the sample data collected from the oscilloscope.

3.1. Uncertainty analysis

The uncertainties are calculated following the guidelines in [9]. The relative uncertainty in the X_{CO_2} values is around 4%, while the uncertainty in η is approximately 6%. The conversion uncertainty is a function of the standard deviations for both the absorbance value and the absorbance-to-conversion coefficient. The uncertainties in the electrical quantities are computed using the uncertainties of the measuring instruments.

4. Experimental results

The results from our experimental measurements are shown in figure 3. In figure 3 (a), one can see that the conversion X_{CO_2} decreases with the gas-flow rate. The efficiency has the reverse tendency, as it increases with the drop in conversion. This relation between the efficiency and conversion is common for these types of discharges.

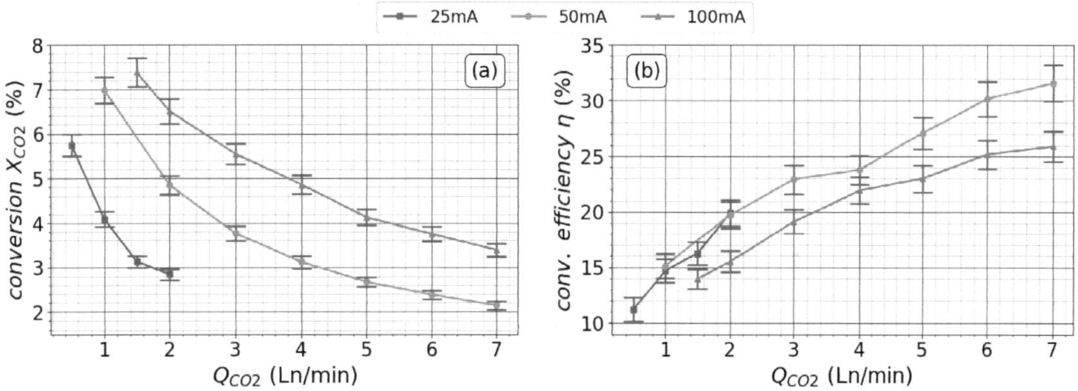

Figure 3. The conversion $\mathbf{X_{CO_2}}$(a), energy efficiency $\boldsymbol{\eta}$ (b), as a function of the input flow rates, at different currents.

The decrease in the conversion rate in figure 3 (a) can be explained if we refer to the results in [5]. The conditions here correspond to the case of a 4-mm distance between the dielectric glasses in [5]; qualitatively, the discharge in the experiments shows the same behavior – the arc is stabilized near one of the dielectric walls and not in the middle. In [5], it is shown that the increase in the gas velocity, i.e., the gas flow, leads to a decrease in the gas temperature, a thinner arc column (reduced radius) and to the effect mentioned whereby the arc gets pushed to one of the side walls. These effects can lead to the conversion reduction seen in figure 3 (a) in several ways: i) the arc column has a smaller cross-section in the flow direction, so that a lesser amount of gas passes through it and therefore less of the gas is treated, ii) the gas that enters the arc column remains in the arc region for a shorter time, iii) the near arc region has a lower average temperature. Using the experimental data, we made some rough estimates for the amount of gas bypassing the arc. Assuming a plug flow (constant transverse velocity distribution), we estimated that up to 30% of the inlet flow can bypass the arc at 100 mA currents, and up to 70%, at lower currents (25 ÷ 50 mA). This is related to the smaller arc radii at lower currents and contributes to the reduction in conversion seen in figure 3 (a). We should stress that our estimations are made assuming the arc radius to be the radius of the glowing arc channel. For a fixed flow rate, increasing the current leads to a higher gas temperature in the arc and a larger arc radius, which results in an increase in conversion.

In figure 3 (b), we see that for a given current the efficiency increases with the gas-flow rate. This could be explained by reasoning that higher gas velocities reduce the gas temperature and increase the deviation from thermal equilibrium. In [10], it is claimed that non-thermal plasma is generally more efficient than plasma in thermal equilibrium. Further, from figure 3 (b), we see that the efficiency seems to slightly decrease with the current. The discharge current has a greater effect on the conversion than on the efficiency, as seen from both figure 3 (a) and figure 3 (b).

5. Conclusions

The results of this study lead to the conclusion that the optimal conditions, where both the conversion and efficiency are high, are at $50 \div 100\,\mathrm{mA}$ for $2 \div 5\,\mathrm{Ln/min}$. The values obtained for the conversion and efficiency are close to those presented in [11, 12], although in those studies the configurations examined are not exactly of a magnetically-stabilized arc, but are still variations of the gliding arc concept and are at similar power values. We observed that the bypass flow (the flow that does not pass through the arc) is not negligible and is in the range from 30% to 70%. This is slightly better compared to a gliding arc reactor without dielectric side walls, where the bypass can be around 80% [12], but still leaves a significant part of the flow untreated.

We were also able to show that the results from this study match the predictions in [5]. The simulation results from [5] give us an estimate for the arc parameters in the case of a laminar flow under similar conditions. The predicted behavior of the arc-to-wall attachment matches our experimental observations. In a separate computer modeling, also related to [5], we estimated that at atmospheric pressure and for current of 100 mA the gas temperature in and around the arc is close to 0.1 eV and is significantly lower than the electron temperature, which is around 3 eV. Although that study considers argon, we assume that indeed the real arc is non-thermal and that its temperature does not exceed by an order of magnitude the computed one, because of the identical current values, geometries and gas velocities in both the experiments and the simulations.

Also, the experiments performed enabled us to determine some of the issues related to our discharge device. These are mainly to do with heat loss and structural wear, and will require some modifications of the discharge apparatus in the future.

Acknowledgments

This work was supported by the Bulgarian National Science Fund, Ministry of Education and Science, research grant KP-06-OPR 04/4 from 14.12.2018 and by the European Regional Development Fund within the Operational Program "Science and Education for Smart Growth 2014 - 2020" under the Project CoE "National center of mechatronics and clean technologies " BG05M2OP001-1.001-0008-C01.

References

[1] Snoeckx R and Bogaerts A 2017 *Chem. Soc. Rev.* **46** 5805–63
[2] Gangoli S P, Gutsol A F and Fridman A A 2010 *Plasma Sources Sci. Technol.* **19** 065003
[3] Ramakers M, Trenchev G, Heijkers S, Wang W and Bogaerts A 2017 *ChemSusChem* **10** 2642–52
[4] Kalra C, Gutsol A and Fridman A A 2005 *IEEE Trans. Plasma Sci.* **33** 32–41
[5] Ivanov V, Paunska Ts, Tarnev Kh and Kolev St 2021 *Plasma Sources Sci. Technol.* **30** 085007
[6] Hui L and Mingfeng X 2012 *Plasma Sci. Technol* **14** 712–15
[7] Feng Z, Saeki N, Kuroki T, Tahara M and Okubo M 2011 *IEEE Trans. Plasma Sci.* **39** 2846–47
[8] Snoeckx R, Heijkers S, Van Wesenbeeck K, Lenaerts S and Bogaerts A 2016 *Energy Environ. Sci.* **9** 999–11
[9] BIPM, IEC, IFCC, ILAC, ISO, IUPAC, IUPAP and OIML 2008 *Evaluation of measurement data – Guide to the expression of uncertainty in measurement* (JCGM 100)
[10] Fridman A 2008 *Plasma Chemistry* (Cambridge: Cambridge University Press)
[11] Li L, Zhang H, Li X, Huang J, Kong X, Xu R and Tu X 2020 *J. CO2 Util* **35** 28–37
[12] Sun S, Wang H, Mei D, Tu X and Bogaerts A 2017 *J. CO2 Util* **17** 220–234

Organic bulk heterojunction solar cells spin-coated in ambient environment on flexible and glass substrates

R Gergova[1,4], M Sendova-Vassileva[1], G Popkirov[1], Hr Dikov[1], G Grancharov[2] and P Terziyska[3]

[1]Central Laboratory of Solar Energy and New Energy Sources, Bulgarian Academy of Sciences, 72 Tsarigradsko Chaussee, 1784 Sofia, Bulgaria
[2]Department of Functional and Nanostructured Polymers, Institute of Polymers, Bulgarian Academy of Sciences, Acad. G. Bonchev St., Bl. 103-A, 1113 Sofia, Bulgaria
[3]Institute of Solid State Physics, Bulgarian Academy of Sciences, 72 Tsarigradsko Chaussee, 1784 Sofia, Bulgaria

E-mail: rositsa.gergova@gmail.com

Abstract. Organic solar cells based on Poly (3-hexylthiophene) [6,6]-phenyl C61-butyric acid methyl ester (P3HT-PCBM) were fabricated in standard architecture by spin-coating on glass and on flexible polyethylene terephthalate (PET) substrates in ambient laboratory environment. In addition to the conventional DC characterization, impedance spectroscopy was used in order to study the capacitive behavior of the devices and obtain their electrical parameters. In combination with the electrical characterization methods, laser-beam-induced current (LBIC) measurements were used as a non-destructive method of mapping the photocurrent distribution and analyzing local device failures and macroscopic defects (shunts, dark spots, defects). The results of the study can be used to develop guidelines for the development and design of P3HT:PCBM solar cells in ambient environment.

1. Introduction

The organic photovoltaic devices could expand the potential application of solar energy as they can be readily deposited by low-cost deposition methods on various rigid and flexible substrates. They may be used to power applications where the maximum power conversion efficiency is not as important as their functionality over the product lifetime [1]. Organic solar cells in general experience stability problems when exposed to outdoor conditions, but their lifetime in a mild ambient environment can be fairly long [2] if they are properly encapsulated. Therefore, these devices present a good opportunity for indoor applications; e.g., implementation in flexible or low-power smart electronic devices [3, 4].

We present studies on the performance of P3HT:PCBM organic bulk heterojunction solar cells, as they remain the polymer-fullerene solar cells that are the most stable and reproducible in ambient environment. Usually, the organic solar cells are fabricated in an inert atmosphere because of their sensitivity to oxygen and humidity; this increases the cost of the fabrication process. Recently, a 4-% efficiency was reported for P3HT:PCBM-based solar cells deposited on glass substrates in ambient laboratory environment [5]. These results indicate a potential for cost reduction in the fabrication

[4] To whom any correspondence should be addressed.

Content from this work may be used under the terms of the Creative Commons Attribution 3.0 licence. Any further distribution of this work must maintain attribution to the author(s) and the title of the work, journal citation and DOI.
Published under licence by IOP Publishing Ltd

process of this type of devices. In our study, we aim to optimize the process of development of standard architecture devices deposited on flexible PET substrates in air as compared to similar devices deposited on rigid glass substrates.

2. Experimental

All of the investigated devices were fabricated on 8-pixel patterned glass and PET substrates in the standard device configuration, ITO/PEDOT:PSS/P3HT:PCBM/ZnO/Ag. The substrates were cleaned by ultra-bath sonication with deionized water followed by acetone and isopropanol cleaning, each procedure lasting 10 minutes. After drying the substrates in a nitrogen flow, UV ozone treatment was applied for 15 minutes to improve the wettability. A poly(3,4-ethylenedioxythiophene) polystyrene sulfonate (PEDOT:PSS) layer was spin-coated in air environment on top of the indium tin oxide (ITO) electrode and was then annealed at 120 °C for 15 minutes under a nitrogen flow. The active layer blend was prepared from a 1:1 weight ratio of P3HT and PCBM in a 12-mg/ml chlorobenzene solution and was thereafter spin-coated in ambient atmosphere. A solution containing ZnO nanoparticles was spin-coated on top of the dry active layer. After drying by nitrogen, the devices were transferred to a RF magnetron sputtering system for deposition of the back-silver electrodes. The solar cells were encapsulated before testing with cover glass or laminating foil by using UV-curable epoxy resin and were subjected to post-deposition annealing for 10 minutes at 120 °C.

DC electrical characterization was performed by a Keithley 2400 source meter; 100 mW/cm^2 of illumination intensity was provided by a halogen lamp and a Si photodiode was used as reference. The AC characterization was performed in the time domain by a fast Fourier transform (FFT) impedance spectrometer. The impedance spectra were taken by applying a multisine perturbation signal with an amplitude of 25 mV over a frequency range of 1 mHz – 100 kHz and were subsequently analyzed using EIS spectrum analyzer software [6]. LBIC measurements were performed by a homemade instrument which consists of a 1-mW red laser diode (650 nm wavelength) attached to a computer-controlled moving X-Y stage. The LBIC measurements were carried out at short-circuit conditions without correction for reflections. The thickness of the P3HT:PCBM active layer was measured by spectroscopic ellipsometry on separate samples deposited on glass within the same batch of the studied devices. A Woollam M2000D ellipsometer was used with a wavelength range from 193 nm to 1000 nm.

3. Results and Discussion

3.1. Solar cell optimization and performance

To determine the dependence of the device efficiency on the substrates used, we fabricated devices with identical standard architectures on PET and glass substrates. Figure 1 presents the *J-V* characteristics obtained under illumination and in the dark of devices deposited on PET and glass substrates. It is evident that the device deposited on the flexible substrate has a lower fill factor (FF) than the device with a rigid substrate, but otherwise the short-circuit current (I_{sc}) and the open-circuit voltage (V_{oc}) of both solar cells are comparable. The FF of the device deposited on a glass substrate was 58.7 %, while that of the flexible device was 34.7 %. A very high apparent series resistance of 82 Ω cm^2 of the flexible device was estimated from the slope of the *J-V* curves, which is also evident from the dark *J-V* characteristics (figure 2). In our experiment, the devices deposited on glass substrates had an approximately 1 % conversion efficiency, which is related to the low current density because the thickness of our active layers was ~50 nm, as estimated by spectroscopic ellipsometry. The power conversion efficiency of the devices deposited on PET was approximately 0.6 %. Although the optimum active layer thickness requirement for indoor applications is about 200 nm, thick active layers in organic solar cells may lead to a poor charge extraction and thus hinder the optimization of the fabrication process for flexible devices. Typically, organic solar cells deposited on flexible substrates underperform as compared to those deposited on glass [7]. There are many factors contributing to this relatively poor performance. Some are related to the properties of the selected transparent electrodes and charge transport materials, while others have to do with the type of the deposition process, device processing temperatures and encapsulation.

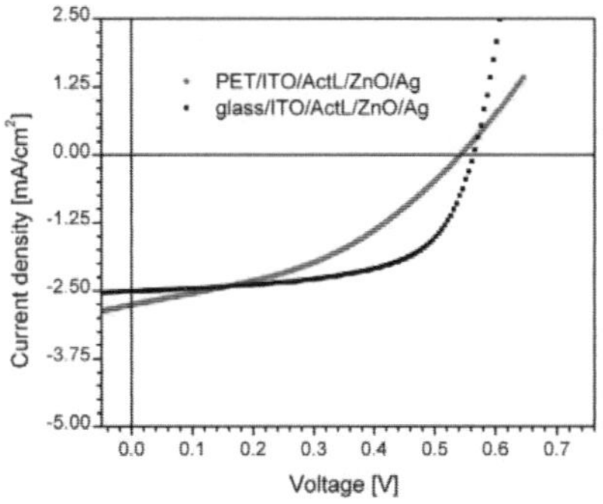

Figure 1. *J-V* characteristics of devices deposited on PET and glass substrates under illumination.

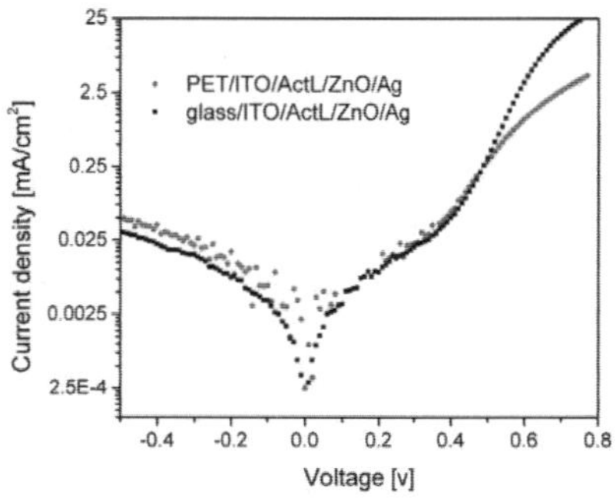

Figure 2. Dark *J-V* curves of the devices presented in figure 1.

LBIC was used for mapping the current extracted from the solar cells and for characterizing the local defects caused by inhomogeneities of some of the layers. In figure 3, an LBIC map is presented of a typical solar cell with eight distinct pixels fabricated on a glass substrate. It can be seen that the pattern of the current corresponds to the pattern of the inhomogeneities observed in the photograph on the right. In this particular case, the active layer of the device was homogeneous, but the ZnO electron transport layer did not cover evenly the active area of the pixels, so that the active layer may have been partially damaged by the magnetron sputtering of the back electrode. The dark parts of the bottom four pixels correspond to parts where the carriers were not efficiently extracted and the *J-V* characteristics resulted in kinks forming double S-shaped curves in the solar-cell working area. In this solar cell, the first top pixel on the left exhibited similar *J-V* characteristics to the device presented in figure 1, which was deposited on a glass substrate. The devices presented in figure 1 were optimized for a better performance and had reasonable homogeneity of the layers; however, in order to be able to fabricate a flexible solar cell with performance comparable to the rigid solar cell, the origin of the relatively low fill factor remains to be addressed.

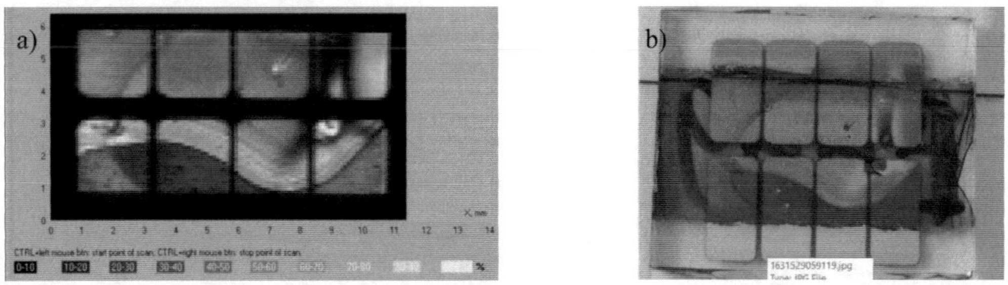

Figure 3. a) LBIC map of the current distribution and homogeneity-related defects of a typical solar cell deposited on glass substrates; b) photograph of the measured solar cell.

3.2. Impedance spectroscopy measurements

Impedance spectroscopy measurements were carried out for different DC bias applied from 0 V to 600 mV under illumination in order to resolve the resistive and capacitive contributions related to the interfacial charge transport properties. In figure 4, the Nyquist diagrams of both types of devices are shown for a) 0 V DC bias equivalent to short-circuit current, and b) 550 mV applied voltage, close to the open-circuit voltage of the cells. As we have reported previously [8], in the EIS spectra of the solar cells with glass substrates, only one process is visible on the measurement time scale as a single semicircle in the Nyquist plot over the entire voltage range from short-circuit to open-circuit conditions. These spectra were modelled by a simple *R*||CPE circuit responsible for a single time-constant, connected in series with a resistance R_s; which accounts for the sheet resistance of the ITO, the connecting wires and the contact resistances of the different layers [8]. However, the spectra of the solar cells deposited on PET substrates could not be fitted with a single time-constant – two processes were observed, although they were not fully resolved in the phase-shift Bode plot (cf. figure 4 d). These spectra were best fitted with the equivalent circuit model presented in figure 4 c), which consists of two *R*||CPE circuits connected in series. We suggest that the additional *R*||CPE feature found at high frequencies is related to the high series resistance observed in the *J-V* curves (figure 1 a) and is responsible for the low fill factor seen in the case of PET device.

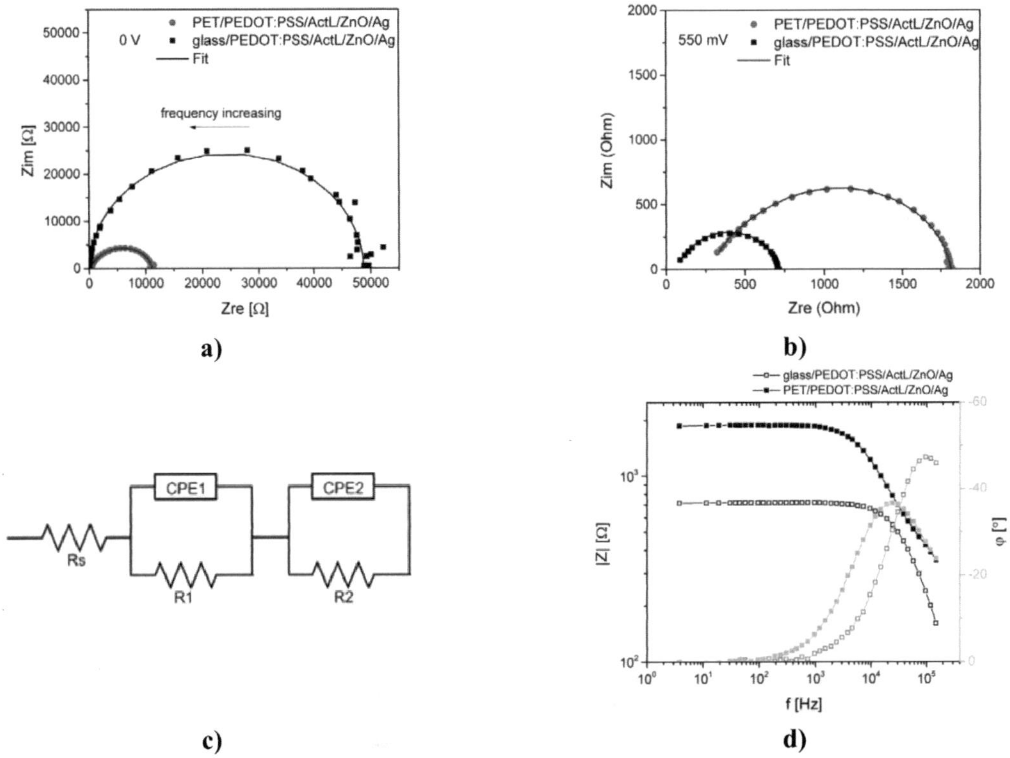

Figure 4. Impedance spectra of solar cells deposited on glass and PET substrates: a) Nyquist diagram of the devices under illumination at 0 V; b) at 550 mV; c) equivalent circuit used for device modelling and d) Bode plot of the impedance spectra taken at 550 mV.

Under conditions close to short circuit (figure 4 a), the parallel resistance from the impedance spectra represents the shunt resistance of the solar cell. It is evident from the plot that the shunt resistance of the rigid device is about five times larger than that of the flexible solar cell. The values of the parallel resistance obtained from the impedance spectra at 0 V for both devices were consistent with the values of the shunt resistance determined by the slope of the *J-V* curves. In figure 4 b), the Nyquist diagrams are shown of the devices under conditions close to the open-circuit voltage. It can be seen from the plot that the series resistance of the flexible device is much higher than R_s of the device with a glass substrate. This points to a higher sheet resistance of the ITO electrode and higher resistive losses at the contact interfaces, which are almost constant as the forward bias applied varies from short circuit to open circuit [9].

Table 1 shows the electrical parameters obtained from the fit of the impedance spectra from 0 to 600 mV of the device deposited on a PET substrate. The R_1||CPE_1 represents the additional feature appearing at high frequencies, which we tentatively attribute to be the cause of the increased series resistance of the cell. As we can see, the parallel resistance R_1 decreases from 0 to 400 mV, while it remains nearly constant with the further increase of the voltage and the parameter n_1 of the constant phase element reaches the value of a pure capacitor. We speculate that this behavior may be caused by a defective area which dominates in one of the contact interfaces when the current flowing through the cell is large. However, the impedance data does not allow for a direct determination which interface is responsible for the observed effect. Since we do not observe this feature in the samples deposited on glass substrates, we believe that it corresponds to the ITO/PEDOT or PEDOT:PSS/P3HT:PCBM interface of the flexible solar cells. Future work will be focused on optimizing these interfaces related to the flexible substrate in order to achieve improved device performance. This may include replacing the ITO with a PEDOT:PSS-based conductive polymer electrode in order to improve the electrode brittleness and adhesion to the PET substrate.

Table 1. Parameters obtained from fitting the impedance spectra of the device deposited on PET substrate with the equivalent circuit presented in figure 4 c).

Vapp mV	Rs Ω	R1 Ω	R2 Ω	CPE1 $F \cdot s^{n-1}$	n1	CPE2 $F \cdot s^{n-1}$	n2
0	349.05	803.74	10025	3.4896E-8	0.93372	2.2105E-8	0.88882
100	340.1	615.66	11464	3.0588E-8	0.95023	2.1362E-8	0.89587
200	337.1	593.59	9763.6	5.217E-8	0.90388	2.0845E-8	0.90997
300	332.74	467.29	5239.5	8.1918E-8	0.87277	2.2281E-8	0.92243
400	332.35	136.15	2762.5	2.7784E-8	1	2.7389E-8	0.91347
500	303.82	135.55	1691.6	2.2809E-8	1	3.308E-8	0.91577
550	302.37	137.68	1373.1	2.1661E-8	1	3.5244E-8	0.91976
600	302.37	137.67	1089.3	2.0456E-8	1	3.6823E-8	0.92432

4. Conclusions

We studied the fabrication of standard architecture P3HT:PCBM-based solar cells by spin-coating in ambient environment as a possible method for reducing the production costs of flexible organic solar cells.

By using impedance spectroscopy, we were able to distinguish different resistive and capacitive behavior of the devices and determine the possible origin of the lower fill factor typical for the solar cells deposited on PET substrates, which points to an important area of further device optimization.

Acknowledgments
The authors acknowledge the support of National Research Program "Low Carbon Energy for Transport and Domestic Use (E+)" under grant DO1-214/2018, Bulgarian Academy of Sciences, Bulgaria.

References

[1] Subramaniam S V, Kutsarov D, Sauermann T and Meier S B 2020 *Energy Technol.* **8** 2000234
[2] Peters C H, Sachs-Quintana I T, Kastrop J P, Beaupre S, Leclerc M and McGehee M D 2011 *Adv. Energy Mater.* **1** 491
[3] Park S et al 2018 *Nature* **561** 516
[4] Yang S-S, Hsieh Z-C, Keshtov M L, Sharma G D and Chen F-C 2017 *Sol. RLL* **1** 1700174
[5] Anjusree S, Arya K R and Bikas C Das 2020 *RSC Adv.* **10** 24882
[6] Bondarenko A S and Ragoisha G A, EIS Spectrum Analyzer, http://www.abc.chemistry.bsu.by/vi/analyzer
[7] Sun Y, Liu T, Kan Y, Gao K, Tang B and Li Y 2021 *Small Sci.* **1** 2100001
[8] Gergova R, Sendova-Vassileva M, Popkirov G, Dikov Hr, Atanasova M-D and Grancharov G 2021 *J. Phys. Conf. Series* **1762** 012039
[9] Leever B J, Bailey C A, Marks T J, Hersam M C and Durstock M F 2012 *Adv. Energy Mater.* **2** 120

Conversion of CO_2 in a pulsed arc discharge

V Vasilev[1], S Iordanova, E Vladkov and St Kolev[2]

Faculty of Physics, St. Kliment Ohridski University of Sofia, 5 J. Bourchier Blvd., 1164 Sofia, Bulgaria

E-mail: vesko.vass@abv.bg, skolev@phys.uni-sofia.bg

Abstract. We present a study of CO_2 gas conversion into a mixture of CO_2, CO and O_2 by means of a pulsed arc discharge at atmospheric pressure. The CO_2 molecule is dissociated as a result of the interaction with the highly energetic electrons and excited species produced in the plasma domain. The plasma conversion of CO_2 provides a relatively simple approach to CO_2 conversion without the need of special materials required by chemical reactors. In this study, the discharge operates at atmospheric pressure, which leads to a reduced efficiency compared to the low pressure operation due to the additional electron losses via elastic collisions. The operation at atmospheric pressure, however, provides simplicity, robustness and low cost of the discharge system, which makes it very attractive from an industrial point of view. Further, we use a pulsed discharge, which, in comparison to the DC discharges, provides additional possibilities for optimizing the dissociation process by controlling precisely the energy input into the plasma. This is achieved by controlling the pulse repetition rate and duty ratio in addition to the discharge current and gas flow characteristics. We present results for the CO_2 dissociation rate and energy efficiency for a wide range of different discharge parameters.

1. Introduction

In this work, we analyze the conversion efficiency of CO_2 in a pulsed arc discharge at atmospheric pressure. The CO_2 conversion has been studied intensively in the last 20 years thanks to the increasingly urgent need of lowering the CO_2 concentrations in the atmosphere. Plasma conversion at atmospheric pressure is a technology that can provide considerable practical advantages for industrial applications, like compact size, low cost, no need of rare materials, operation insensitive to contaminations of the working gas, etc. A key milestone in the possible application of such technology is achieving a high dissociation rate of CO_2 conversion and a high energy efficiency.

We use a pulsed arc discharge as an alternative option, which may provide certain advantages compared to other arc discharges. It is known that the highly non-isothermal plasmas can be more efficient for direct dissociation of CO_2 compared to plasmas close to thermal equilibrium [1]. One way of achieving a lower gas temperature and, thus, a higher deviation from thermal equilibrium, is to control the heating time of the gas by controlling the discharge operation time. Therefore, we have developed a system for pulsed operation, whereby we can control the pulse length, the duty cycle and the arc current during the pulse. Similar discharges with pulse lengths in the nanosecond range have already been used in similar applications [2-7] and have shown promising results of a relatively high efficiency. In this study, we analyze the performance of the system producing pulses in the microsecond range.

[1] To whom any correspondence should be addressed.
[2] To whom any correspondence should be addressed.

Content from this work may be used under the terms of the Creative Commons Attribution 3.0 licence. Any further distribution of this work must maintain attribution to the author(s) and the title of the work, journal citation and DOI.
Published under licence by IOP Publishing Ltd

2. Experimental setup and data processing

A general view of the experimental setup, including the diagnostic devices, is shown in figure 1. The major components are: a high-voltage pulse source, electrical measurement circuit, gas supply and outlet gas analysis. The CO_2 gas is supplied through a Bronkhorst EL-FLOW F-201CM mass flow controller and the system operates at atmospheric pressure. The gas is fed through an alumina ceramic (Al_2O_3) tube with an internal diameter of 2 mm, which also contains the anode. The power supply includes two current-limited high-voltage sources (V_1 up to 14 kV/10 mA and V_2 up to 2.6 kV/5 A) and a Behlke HTS181-01-HB-C high-voltage solid-state switch controlled by a Tektronix AFG1062 pulse generator. All electrodes are tungsten wires with a diameter of 0.9 mm. The ignition electrode is positioned closer to the anode (around 0.5 mm), while the cathode is above the alumina tube outlet at a distance of 1.5 mm from the anode and the tube, i.e., it is positioned exactly at the gas outlet. The role of the ignition electrode is to provide an initial short spark around the anode thus allowing the ignition of the wider gap between the cathode and the anode. When the ignition electrode produces a spark, the current through it stops due to the presence of the capacitor. At this stage, the cathode-anode gap is ignited due to the applied 12-kV high voltage. Within several hundred of nanoseconds, the discharge current increases and the discharge voltage (cathode-anode) drops below 1 kV. At this moment, the low-voltage (high-current) source V_1 starts supplying current to the discharge, so that the discharge voltage drops further to below 400 V. The current in V_1 is limited by a 660 Ω resistor. During the initial ignition, when the voltage is up to 12 kV, the lower voltage power supply is protected by a fast high-voltage diode. The complete electrical circuit has additional components related to additional protection, not shown here.

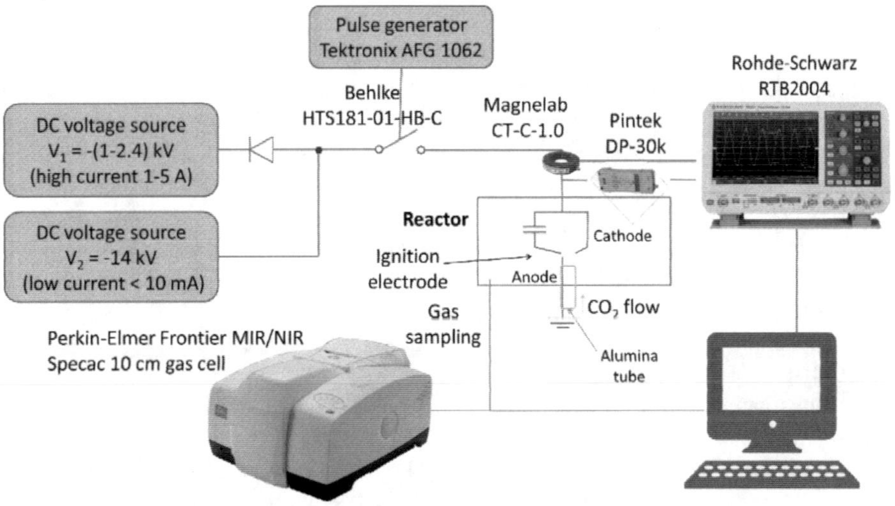

Figure 1. Schematic presentation of the experimental set-up.

The electrical measurements are performed by a Pintek DP-30k high-voltage probe and a Magnelab CT-C-1.0-BNC Rogowski current probe. The signals are acquired by a Rohde-Schwarz RTB2004 digital oscilloscope. The power deposited in the discharge is calculated from the current and the voltage signals by numerical integration of the power.

The measurements of the dissociation rate are carried out by infrared gas analysis based on a general purpose Perkin-Elmer Fourier spectrometer Frontier Mid-IR and Near-IR System equipped with a Specac 10-cm gas cell with CaF_2 windows. The spectrometer provides determination of the CO_2 gas dissociation by actually measuring the absorbance of the CO resulting from the process $CO_2 \to CO + (1/2)O_2$ and ignoring the minor quantities of carbon produced in the discharge. In general, the CO_2 conversion is defined as $X_{CO_2,0} = |(n_{CO_2,\text{final}} - n_{CO_2,\text{init.}})/n_{CO_2,\text{init.}}| \times 100\%$, which is the percentage of converted CO_2 gas. The conversion is found from the absorbance values of a specific CO line in the spectrum using a linear dependence of the form: $A = \alpha X_{CO_2,0}$, where A is the measured absorbance.

The coefficient α is determined by calibrating the gas analyzer by a reference CO/CO_2 mixture. In addition, the CO_2 dissociation leads to an increase in the number of molecules and to gas expansion, since the whole system is kept under atmospheric pressure. Since the gas sample in the gas cell is extracted from this expanded gas, to derive the correct dissociation rate the above expression should be modified [8]. The final expression used is therefore:

$$\mathrm{X}_{\mathrm{CO}_2} = \frac{2X_{\mathrm{CO}_2,0}}{3 - X_{\mathrm{CO}_2,0}}, \tag{1}$$

where $X_{\mathrm{CO}_2,0}$ is the conversion calculated directly from the absorbance, and $\mathrm{X}_{\mathrm{CO}_2}$ is the actual conversion value, accounting for the gas expansion.

The other important quantity of interest is the energy efficiency η, which is defined as $\eta = (\mathrm{X}_{\mathrm{CO}_2}\Delta H_R)SEI \times 100\ (\%)$. Here $\Delta H_R = 279.8\ \mathrm{kJ/mol}$ is the reaction enthalpy for the CO_2 splitting reaction, and SEI (J/mol) is the so-called specific energy input, defined as the energy delivered per mole of gas:

$$SEI = \frac{P\ (\mathrm{J/s})}{Q\ (\mathrm{Ln/s}) \times (1/22.4)\ (\mathrm{mol/Ln})} \left(\frac{\mathrm{J}}{\mathrm{mol}}\right), \tag{2}$$

where Q (Ln/s) is the measured input flow rate and P is the applied average power.

The data processing includes also uncertainty analysis. The relative uncertainty in the dissociation rate $\delta\mathrm{X}_{\mathrm{CO}_2}$ is relatively small – $\delta\mathrm{X}_{\mathrm{CO}_2}$~8% for confidence level (CL) of 95%. It is based on the statistical uncertainty derived from multiple measurements of a calibration gas. The efficiency measurements presented here are subject to significant uncertainty (around 20% for CL = 95%), as shown in figures 2 – 5, mainly due to the significant uncertainty in the averaged power. This is due to i) the significant electrical noise produced by the discharge and ii) the lack of exact reproducibility of the consecutive pulsed discharges, as well as inconsistent discharge ignition timing (slightly delayed ignition). All this leads to a spread of the measurement data and, thus, to a significant uncertainty in the calculated averaged power. The uncertainty value for the power is based on the statistical error of multiple measurements under the same conditions; a global estimate of about 12% is obtained. The uncertainty due to the measurement instruments (probes and oscilloscope) remains much smaller compared to the statistical error (discharge reproducibility).

3. Results and discussion

The results presented here show the dependencies of the major quantities of interest – $\mathrm{X}_{\mathrm{CO}_2}$ and η – for various discharge parameters. This work is a first systematic study of the set-up and includes a relatively limited number of different discharge conditions. The discharge parameters varied are the pulse repetition rate, the signal duty cycle, the gas flow and the discharge current; the results are shown in figures 2 to 4. The uncertainty in the dissociation rate $\delta\mathrm{X}_{\mathrm{CO}_2}$ is not plotted for better visualization of the graphs; the uncertainty in the η values is only plotted.

Figure 2 shows the dependencies on the repetition rate of the high voltage pulses. Generally, the conversion increases with the repetition rate in the range considered here. The dependence is very strong at lower repetition rates below 3 kHz. This is attributed mainly to the fact that the gas velocity is sufficiently high so that the gas is displaced at a longer distance compared to the arc size, i.e., between two consecutive pulses there is a part of the gas which is untreated. This can be argued by simple estimations. If we consider a gas flow of 1 L/min and an approximate cross section of the alumina tube outlet of 2.5 mm^2, the average gas velocity will be roughly 7 m/s. For a repetition rate of 1 kHz and a duty cycle of 3%, the pulse duration is 6 μs, while the time between pulses is 194 μs. The distance traveled by the gas for 194 μs is 7 mm. If we assume an arc size of roughly 1 mm, which is the order of magnitude of the visible radiating part of the arc, one can easily conclude that at this repetition rate a considerable part of the gas remains untreated. For 5 kHz, the distance is in the order of 1.4 mm; i.e., comparable to the arc size. Therefore, this leads to the weaker increase of the conversion between 3 kHz and 5 kHz. The strong dependence of the conversion on the repetition rate is also observed below 1 kHz (not shown here).

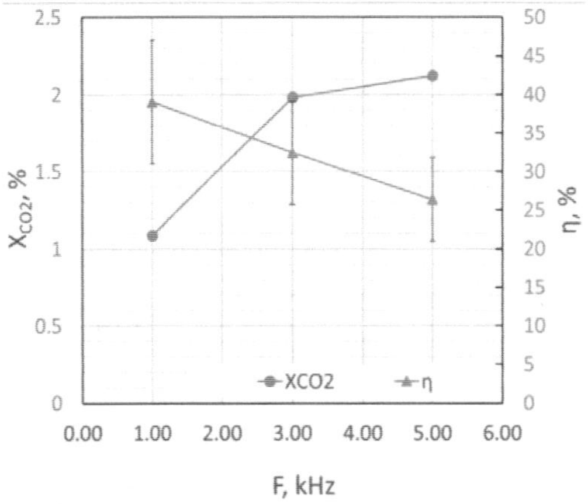

Figure 2. CO_2 conversion X_{CO_2} and energy efficiency η dependence on the pulse repetition rate *F*. Gas flow 1 Ln/min, duty cycle 3%, V_1 1.1 kV (discharge current ~1-1.2 A), δX_{CO_2}~8% (CL 95%).

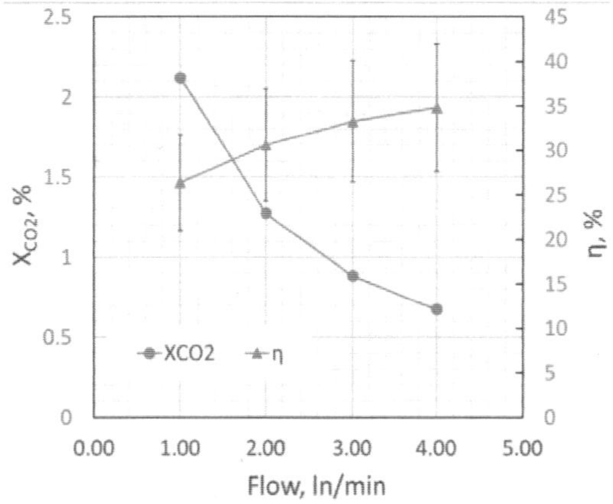

Figure 3. CO_2 conversion X_{CO_2} and energy efficiency η dependence on the gas flow. Repetition rate 5 kHz, duty cycle 3%, V_1 1.1 kV (discharge current ~1-1.2 A), δX_{CO_2}~8% (CL 95%).

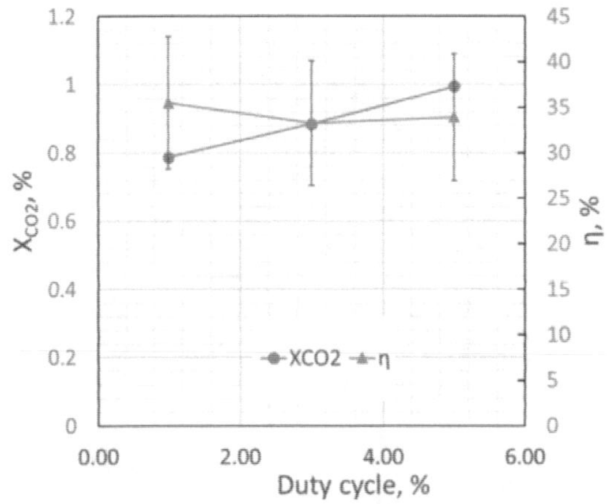

Figure 4. CO_2 conversion X_{CO_2} and energy efficiency η dependence on the duty cycle. Repetition rate 5 kHz, gas flow 3 Ln/min, V_1 1.1 kV (discharge current ~1-1.2 A), δX_{CO_2}~ 8% (CL 95%).

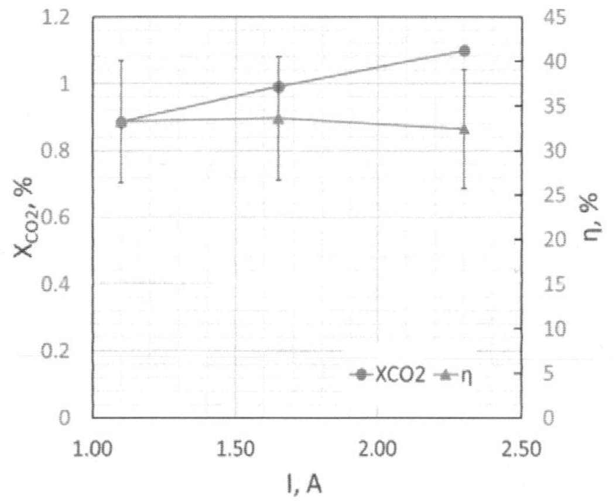

Figure 5. CO_2 conversion X_{CO_2} and energy efficiency η dependence on the discharge current (set by V_1 voltage). Repetition rate 5 kHz, gas flow 3 Ln/min, duty cycle 3%, δX_{CO_2}~8% (CL 95%).

The same reasoning can readily explain the trend in X_{CO_2} observed in figure 3 – the decrease as the gas flow is increased is again expected to be due to the longer time between pulses, which allows a certain part of the gas to pass during the off time of the discharge; therefore, the dissociation decreases with the gas flow rate.

In addition to the above effect, we should stress that the CO_2 dissociation depends also on the electron density, the gas temperature and the gas residence time in the active plasma. For example, in figure 4 we see an increase of X_{CO_2} with the duty cycle and the pulse length at a constant repetition rate.

Obviously, the increase of the pulse duration is increasing the dissociation rate, but the dependence is not proportional – increasing the duty cycle five times leads to only 20% (1.2 times) increase in X_{CO_2}. The increase of the discharge current (figure 5) also shows a similar trend – increasing the current twice results in a 25% increase in X_{CO_2}.

In what concerns the energy efficiency η, it falls in the range between 25% and 40%. A higher gas flow leads to a higher efficiency (figure 3), while the higher repetition rate reduces the efficiency. Overall, the efficiency remains high and provides an attractive starting point for further improvements of the setup mainly towards achieving a higher dissociation rate.

4. Conclusions

We present an experimental study of a direct CO_2 plasma dissociation in a pulsed arc discharge. The dependence is examined on several discharge parameters – pulse repetition rate, signal duty cycle, gas flow and arc current.

Overall, the dissociation rate shows clear trends with the above discharge parameters: increasing the repetition rate, the duty cycle and the arc current leads to an increase in the dissociation rate, while it decreases as the gas flow is increased. These trends can be explained reasonably, but the weak increase with the duty cycle and the arc current increase poses additional questions and requires additional analysis and measurements. The dissociation rate remains relatively low, but we expect to improve it by proper modifications in the discharge geometry ensuring less untreated gas.

The energy efficiencies obtained for the CO_2 dissociation are high, in the order of 30-40%, which is among the best results reported in the literature for CO_2 dissociation by atmospheric pressure discharges. These results present an excellent starting point for further improvements of the developed experimental set-up mainly towards achieving a higher dissociation rate.

Acknowledgments

This work was supported by the Bulgarian Ministry of Education and Science under the National Research Programme E+: Low Carbon Energy for the Transport and Households, grant agreement D01-214/2018 and by the European Regional Development Fund within the Operational Program "Science and Education for Smart Growth 2014 - 2020" under Project CoE "National center of mechatronics and clean technologies" BG05M2OP001-1.001-0008-C01.

References

[1] Fridman A 2008 *Plasma Chemistry* (Cambridge university press)
[2] Bak M, Im S-K and Cappelli M 2015 *IEEE Trans. Plasma Sci.* **43** 1002
[3] Martini L M, Lovascio S, Dilecce G and Tosi P 2018 *Plasma Chem. Plasma Process.* **38** 707
[4] Moss M S, Yanallah K, Allen R W K and Pontiga F 2017 *Plasma Sources Sci. Technol.* **26** 035009
[5] Pachuilo M, Levko, Raja L and Varghese P L 2017 AIAA 2017-1968 *55th AIAA Aerospace Sciences Meeting*
[6] Pietanza L, Colonna G, D'Ammando G, Laricchiuta A and Capitelli M 2015 *Plasma Sources Sci. Technol.* **24** 042002
[7] Heijkers S, Martini L M, Dilecce G, Tosi P and Bogaerts A 2019 *J. Phys. Chem.* C **123** 12104
[8] Snoeckx R, Heijkers S, Van Wesenbeeck K, Lenaerts S and Bogaerts A 2016 *Energy Environ. Sci.* **9** 999–1011

Vertical and near-horizontal scanning lidar observations of aerosol loads over Sofia, Bulgaria, in the presence of Saharan dust

Z Peshev[1,3], A Deleva[1], A Chaikovsky[2], V Pescherenkov[2], L Vulkova[1] and T Dreischuh[1]

[1]Acad. E. Djakov Institute of Electronics, Bulgarian Academy of Sciences, 72 Tsarigradsko Chaussee, 1784 Sofia, Bulgaria
[2]B.I. Stepanov Institute of Physics, National Academy of Sciences of Belarus, 68-2 Nezavisimosti av., 220072 Minsk, Belarus

E-mail: zypeshev@ie.bas.bg

Abstract. Intrusions of extraneous aerosols and their interactions with locally produced aerosols result in serious climatic, ecological and health effects and consequences. In this work, we analyze results of lidar observations during Saharan dust loads in the atmosphere above the city of Sofia. Two types of measurements are described, namely, vertical lidar sounding and near-surface aerosol mapping by quasi-horizontal lidar scanning. The temporal dynamics of the aerosol density vertical distribution is displayed by color-mapped time series of range-corrected lidar signals. The resulting time-averaged profiles of the backscattering coefficients at two lidar wavelengths (532 nm and 1064 nm) were retrieved and the topological parameters of the detected aerosol fields were determined. The aerosols' microphysical characteristics and their height distribution were estimated by calculating the backscatter-related Ångström exponent. Presented and analyzed are results of vertical profiling of the total and mode volume concentration for the aerosol/dust fractions obtained by using the inversion code LIRIC-2 based on combined synergistic use of data from the CALIPSO satellite on-board lidar and sun-photometer data. The quasi-horizontal mapping of near-ground aerosols was carried out over a large urban area of the city of Sofia of about 25 km^2 within a 50° angular sector in a south-western direction. The results presented in the work demonstrate the capabilities of the combined approach including lidar, radiometer, satellite measurements and theoretical modeling in conducting effective ecological monitoring of the atmospheric aerosol pollution.

1. Introduction

Aerosols are major atmospheric pollutants and a key factor in the alarming climate change and the deteriorating environmental conditions on a global scale [1]. The aerosol impacts depend on their optical and microphysical properties, mass stratification, and altitude of occurrence. The lidar technology allows one to evaluate the aerosol characteristics listed above over broad areas with a high range and time resolution [2]. A variety of ground-, air-, and space-based lidars are involved in the detection and characterization of atmospheric aerosols of natural and/or anthropogenic origin, such as

[3] To whom any correspondence should be addressed.

Content from this work may be used under the terms of the Creative Commons Attribution 3.0 licence. Any further distribution of this work must maintain attribution to the author(s) and the title of the work, journal citation and DOI.
Published under licence by IOP Publishing Ltd

general and desert dust, volcanic ash, sea salts, fire smoke, water and mixed aerosols [3-6]. For more effective and comprehensive monitoring and characterization of the large-scale distribution and dynamics of atmospheric aerosols on a continental scale, lidar stations have been united into cooperative structures, such as the European Aerosol Research Lidar Network (EARLINET) [7] and the Aerosol, Clouds and Trace Gases Research Infrastructure (ACTRIS) [8]. Airborne Saharan dust, due to its huge amounts and areas of distribution in the atmosphere and its associated climatological and environmental impacts, is one of the main types of aerosols subject of lidar research, including within the aforementioned research structures [9, 10].

The sun-radiometer stations organized in the global Aerosol Robotic Network AERONET [11] provide data on a significant set of column-averaged physical, optical and micro-physical parameters of atmospheric aerosols. Inversion codes and algorithms, such as LIRIC, GARRLiC, POLIPHON [12-14], have been developed, synergistically combining the sun-photometer and lidar data, for successful vertical profiling of the size/shape-mode and/or total volume concentrations of the aerosols.

The atmospheric aerosols, particularly those of anthropogenic origin, are predominantly concentrated in the near-surface atmospheric layer thus deteriorating the air quality and causing serious environmental and health problems in densely populated areas and such with high industrial activity. The aerosol content of the air directly on the surface is monitored by local *in situ* measuring stations of aerosol concentration (PM_{10} and $PM_{2.5}$). However, lidars with (quasi)horizontal scanning of the probing laser beam have become indispensable tools when one needs fast (including real-time) and accurate measurement, initial characterization, visualization and mapping of the aerosol density distribution over large areas (up to tens of square kilometers) [15-17].

To identify the aerosol type and origin, online available modeling resources are widely and actively used. For general air transport tracking, the most applicable one is the HYbrid Single-Particle Lagrangian Integrated Trajectory (HYSPLIT) model, developed by the Air Resource Laboratory of the National Oceanic and Atmospheric Administration (NOAA), USA [18, 19]. The specialized models of the Barcelona Supercomputing Center BSC-DREAM8b and the new NMMB/MONARCH (formerly NMMB/BSC-Dust) are commonly used for the distribution and mass concentration of Saharan dust [20-22].

In this work, we describe and analyze the results of two types of lidar observations during Saharan dust loads in the atmosphere above the city of Sofia, namely, vertical lidar sounding carried out on 31.08.2020, and aerosol mapping by near-horizontal lidar scanning within an angular sector performed on 20.02.2019. An essential part of the studies reported is the vertical profiling of the total and mode volume concentration of the aerosol/dust fractions achieved by using the LIRIC-2 inversion code based on the synergistic use of CALIPSO satellite on-board lidar and sun-photometer data.

2. Vertical lidar profiling of aerosol properties

The remote sensing vertical observation was performed at 1064 nm and 532 nm using an elastic backscatter lidar based on a Nd:YAG laser (see, e.g., [6]). Hereinafter, the elevation (height, altitude) values are quoted in AGL (above ground level), unless explicitly quoted in ASL (above sea level).

2.1. Aerosol/dust layering and air transport forecasting

Figure 1 (a) displays the height mass distribution and temporal dynamics of the aerosol field detected in terms of color-mapped time series of range-corrected lidar signals (RCS) at 1064 nm as measured on 31.08.2020 in the time interval 15:40-17:20 UTC (universal time coordinated). On the right-hand side of the map, a color bar of the aerosol density in arbitrary units is shown. The color map reveals the presence of aerosols from the earth's surface up to a height of about 4 km. This height exceeds the top of the atmospheric boundary layer (ABL) over Sofia, which in summer is at about 1.8 km. Most of the aerosol mass is concentrated at heights between 1 km and 3 km. Above 4 km, the atmosphere is free of significant amounts of aerosols. Figure 1 (a) also shows a distinct layer of higher density in the height range 2-3 km with a nearly homogeneous structure and well-defined boundaries, indicating a calmer atmosphere with virtually no convective processes. The particles above ABL are very likely of non-local origin because practically most of the aerosols originating from local sources and human activities are normally confined within the ABL.

Figure 1 (b) presents the BSC-DREAM8b model vertical profile of dust mass concentration above Sofia and the forecast color map (inset) for Saharan dust spreading in the atmosphere over the Mediterranean Basin and Europe. In accordance with the model predictions, the atmosphere above the city was loaded with Saharan dust (on the map, Sofia region is marked by a red circle). The profile of the dust concentration (on the left-hand side of the figure) shows that on 31.08.2020 the desert dust was located in a layer with boundaries at approximately 2 km and 4.5 km ASL (1.4 km and 3.9 km AGL, respectively), which is in agreement with the experimental data in figure 1 (a). The HYSPLIT backward trajectories in figure 1 (c) reveal that the air masses reaching Sofia at the heights of 1.7 km and 2.5 km during the lidar observations have been moving just above Sahara surface and, partly, close to the Mediterranean Sea surface. Those at an altitude of 3.7 km have passed for a long time through the desert ABL, whose top reaches 5 km. Based on the data analysis from the two models, we assume that the atmosphere above Sofia in the altitude range 1-4 km was loaded with Saharan dust possibly mixed with marine aerosols.

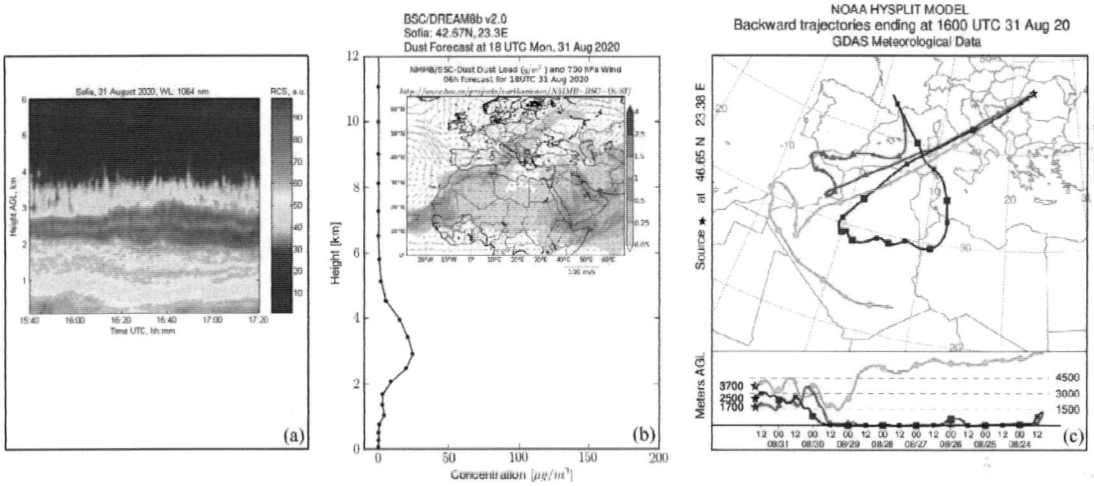

Figure 1. (a) Color-mapped RCS time series, as measured at 1064 nm on 31.08.2020; (b) BSC-DREAM8b model forecast of the Saharan dust concentration profile above Sofia with regional dust load map (inset); (c) HYSPLIT-backward trajectories ending above Sofia during the measurement.

2.2. Vertical lidar profiling of aerosol/dust optical and microphysical properties

The Klett-Fernald's inversion algorithm was applied to retrieve the atmospheric aerosol backscatter coefficient (BSC), which is directly proportional to the aerosol density [23,24]. Figure 2 (a) displays the integral BSC vertical profiles obtained at 532 nm and 1064 nm, which depict the time-averaged aerosol stratification in the atmosphere, along with the time-resolved one presented as RCS map in figure 1 (a). They clearly show that on 31.08.2020 aerosols were registered from the surface up to a height of 4.5 km, predominantly in the altitude range of 1.8 – 4 km with a center of mass at about 2.5 km. The calculated BSC profiles are in good agreement with the RCS map shown above and with the BSC-DREAM8b dust concentration profile in figure 1 (b). The maximum BSC values obtained of 2.76/2.26 M m^{-1} sr^{-1} at 532/1064 nm indicate a weak to moderate dust load.

Lidar measurements performed simultaneously at two laser wavelengths make it possible to calculate the backscatter-related Ångström exponent (BAE), whose value is inversely proportional to the aerosol particle size and is thus regarded as an indicator of the dominant particle size fractions in an aerosol layer. Figure 2 (b) shows the time-averaged BAE vertical profile calculated by using the BSC profiles in figure 2 (a). The profile is shown up to 4.5 km, since due to the low aerosol content above this height, i.e., the low signal-to-noise ratio, the inaccuracy in determining the BAE rises significantly. In general, the shape of the BAE profile is close to mirror symmetrical to those of the BSC, with the lowest values at the maximum of the layer, rising significantly in its periphery. The average BAE value for the 0.3-4.5 km height interval is 0.51/±0.25. The mean BAE values for some characteristic areas of the aerosol layer are also determined. For the near-ground interval up to 1.5 km, largely overlapping with the local seasonal ABL, the BAE has a mean value of 0.55/±0.18; in the height range of maximum aerosol density 1.5-3.5 km, it is 0.34/±0.09; while in the upper part of the layer at 3.5-4.5 km it is 0.78/±0.26. The values after the slashes are the corresponding standard deviations. According to the Sofia AERONET station data for the day (not shown here), the value of the columnar 440-870 nm Ångström exponent decreases steadily from 1.3 in the morning to about 0.5 at the time of the lidar measurement. The average BAE value of 0.51 obtained for the whole layer is in good accordance with the latter. In the local ABL, values of BAE near and above 1 are normally observed, so that the value of 0.55 calculated for the lowermost part of the layer indicates a downward penetration of larger dust size modes and their dominance over the local aerosols. In the central and densest part of the detected aerosol/dust layer, the average BAE value obtained of 0.34 and the minimum one of about 0.2 are markedly low, evidencing the presence of quite large fractions of dust particles, likely exceeding considerably 1 μm. This is in good agreement with the columnar profile of the particle size distribution according to the sun-photometer data of the time of lidar measurements that exhibit a pronounced maximum at particle radius values of about 5-6 μm. The presence of such large particle size modes is in very good agreement with the HYSPLIT diagrams of the backward trajectories ending above Sofia in the 2-3.5 km height range (figure 1 (c)). These trajectories pass close to Sahara surface for about six days; thus, the respective air masses have had the possibility to capture and transport very large dust particles. In contrast, the trajectories corresponding to heights above 3.5 km pass through the Saharan ABL, which explains the higher BAE mean of 0.78 consistent with less coarse dust particles.

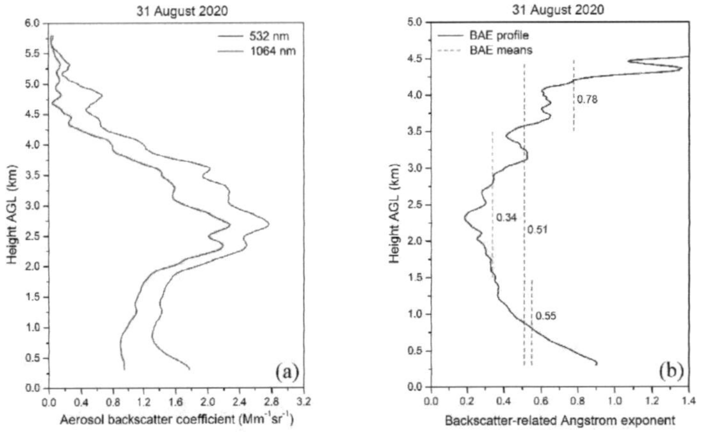

Figure 2. Vertical profiles of the aerosol backscatter (a) and Ångström exponent (b).

2.3. Vertical profiling of the aerosol concentration by means of the LIRIC-2 inversion code, based on combined use of satellite-lidar and sun-radiometer data

This section presents results of calculating the vertical profiles of the mode volume concentrations of aerosols present in the atmosphere over Sofia region on 31.08.2020. Calculations were performed with the newly developed version LIRIC-2 of the LIdar-Radiometer Inversion Code (LIRIC) [14], based on synergistic use of data from satellite-lidar and ground-based sun-radiometer measurements. It is worth noting that LIRIC is an inversion algorithm providing simultaneous separate height profiling of the volume concentration for fine and coarse aerosol fractions, distinguishing between the spherical and non-spherical modes of the latter [12]. The lidar measurements data used in the calculations were obtained by the on-board Cloud-Aerosol Lidar with Orthogonal Polarization (CALIOP) of the Cloud-

Aerosol Lidar and Infrared Pathfinder Satellite Observation (CALIPSO) satellite during its passage over the region of Sofia (time: 12:03:18-12:04:14 UTC; distance 62 km); these were combined with sun-photometer data from the Sofia AERONET station at time close to the above one (13:57:37 UTC). The CALIOP lidar has three operating registration channels: one at 1064 nm and two at 532 nm, with parallel (532P) and cross (532C) polarization.

The first three (a-c) panels of figure 3 show the calculated RCS height profiles normalized by reference values for 1064 nm, 532 nm (P) and 532 nm (C). To verify and validate the calculation results, normalized RCS profiles are also shown as obtained from the actual measurements by the corresponding three CALIOP lidar channels. The aerosol optical thickness (AOT) values according to AERONET data are given inside the panels, while those for the integrated backscattering B_{int} are according to CALIOP data. The values of their ratios $P = AOT/B_{int}$ are also shown, being an integral version of the lidar (i.e. extinction to backscatter) ratios. As can be seen, the coincidence of the calculated and the measured normalized RCS profiles is fairly good, with some small deviations mainly in the altitude range 3.5-5 km. We ascribe these to differences between the columnar aerosol content in the CALIOP measurement zone and that above the AERONET station in Sofia, which are at a distance of 62 km, as well as to the time difference of about two hours between the two measurements. In addition, the one-minute averaging interval of the signal with which CALIOP operates corresponds to a scanned band with a length of about 4 km, which can also be a source of a small error. Given these circumstances, it can be argued that LIRIC-2 retrieves aerosol parameters with high accuracy. Figure 3 (d) presents the height profiles of the aerosol mode volume concentrations c_k for both the fine and the coarse aerosol/dust fractions calculated by means of LIRIC-2; for the latter, the profiles corresponding to spherical and non-spherical (spheroid) aerosol particle modes are distinguished and shown. The height profile of the total aerosol volume concentration is also displayed, representing the sum of the three partial aerosol modes.

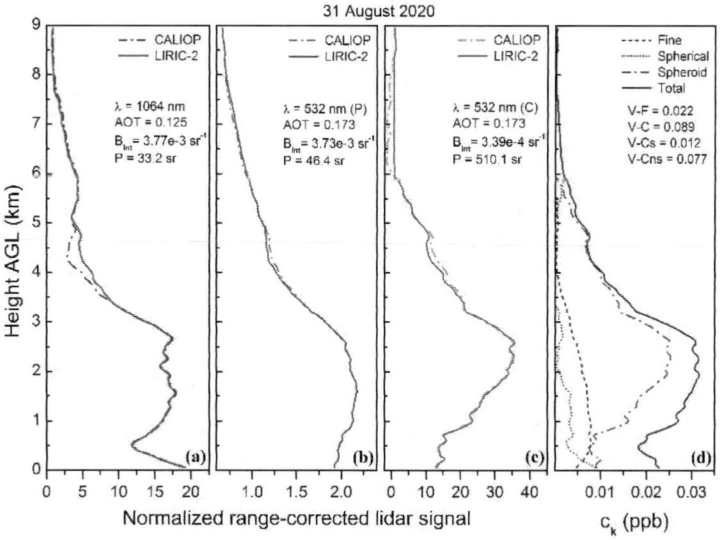

Figure 3. Normalized vertical RCS profiles from CALIOP lidar measurements and calculated by using the LIRIC-2 inversion code ones, at 1064 nm (a) and 532 nm for parallel (b) and cross (c) polarization; LIRIC-2 calculated vertical profiles of aerosol volume concentration c_k for fine (F) and coarse (C) [spherical (Cs) and spheroid (Cns)] particle fractions, as well as for the resulting total aerosol concentration (d); V-F, V-C, V-Cs, and V-Cns are the corresponding columnar volume concentrations [$\mu m^3/\mu m^2$].

The aerosol concentration values are presented in dimensionless units, parts-per-billion (ppb, 10^{-9}). The distinction shown between the spherical and the non-spherical fractions is based on data from (de)polarization measurements by the CALIOP lidar with cross-polarized channels 532P and 532C. As can be seen in figure 3 (d), the aerosol concentration reaches maximum values of around 0.03 ppb in the altitude range 1-3 km, then gradually decreases to zero at about 8 km. A distinct feature of the retrieved vertical aerosol distribution is the definite dominance of the non-spherical (spheroid) aerosol fractions over both the fine and the spherical coarse modes, with maximum concentration values of about 0.025 ppb in the 2-3 km range and an integral one of 0.089 ppb, representing about 85% and 80% of the maximum and the total integral concentration, respectively. Such aerosol mode distributions are typical of desert aerosols, in which large irregularly shaped particles predominate. The profile of the spheroid coarse fractions in figure 3 (d) and the depolarized profile of the normalized RCS in figure 3 (c) are consistent with the aerosol distribution in figures 1 (a) and 2 (a) measured by the ground-based lidar, as well as with the BSC-DREAM forecast profile (figure 1 (b)). It is also in good agreement with the backward trajectories shown in figure 1 (c); these point to the possible presence of very large dust particles in the dust plume detected above Sofia that have been captured off the Sahara desert surface during a period of several days. We attribute the small differences in the shape of the discussed profiles to the above-noted spatial and temporal differences between the measurements of the ground-based lidar and sun-radiometer and the satellite lidar. They are related to the fact that the measurements were performed during the period of initial active dust load typically characterized by highly variable vertical distribution and dynamics of the dust concentration in the different loading zones. We note that the LIRIC inversion code, and its version LIRIC-2 in particular, allow retrieving of altitude profiles and integral values of virtually all radiative-forcing significant aerosol parameters, thus being a unique and powerful tool for complete characterization and analysis of atmospheric aerosols, as well as for assessing their impacts on climatic and environmental processes on a regional and on a larger scale.

3. Near-horizontal lidar scanning observations over the city of Sofia

Near-surface atmospheric measurements over urban areas aimed at assisting the air-quality monitoring attain increasing societal significance due to the strong and direct impact on the human health of aerosol pollutions (natural or anthropogenic) in the low troposphere. The horizontal or near-horizontal lidar scanning/mapping is an effective approach to detecting aerosol loads over broad areas and characterizing them in terms of local density and temporal dynamics [15-17].

The measurement described below demonstrates the capabilities of the elastic-scattering lidar to follow the temporal changes of the aerosol field density. It was performed on 20.02.2019 at the laser wavelength of 1064 nm in south-southwest directions from the lidar station toward Vitosha Mountain. A sector of 50° was studied at a step of 2°; a surface area of about 25 km^2 was scanned in the time interval 18:28-18:41 local time (16:28-16:41 UTC). A low angle (about 6°) with respect to the horizon was used because of the adjacent high buildings present in the specific angular sector. The results are shown in figure 4 (a) as a color map (superimposed on the topological map of Sofia region) obtained from the BSC values calculated along 23 sounding directions. On 20.02.2019, an aerosol field with an inhomogeneous structure lacking a strong contrast was registered low above the scanned sector. The highest aerosol load (dark-red) area extending to about 2 km from the lidar station arises from the scanned section covering densely populated residential areas with heavy street traffic. Further, the density of the aerosols gradually decreases (green-yellow); it is the lowest (blue) beyond the ring road (the red stripe at about 4 km from the lidar) approaching Vitosha Mountain.

Figures 4 (b) and 4 (c) present data on the air transport of dust/aerosol related to the time of lidar measurements, as well as data from in-situ measurements of the PM_{10} concentration in that period by automated measuring stations (AMSs) in Sofia. The BSC-DREAM8b map in figure 4 (b) shows that on 20.02.2019 a massive Saharan dust plume spread in the atmosphere over the Eastern Mediterranean and parts of Southeast Europe, including Bulgaria. It is important to note that wintertime dust intrusions from North Africa to Europe, and particularly to the Balkans, are extraordinary seasonal events as they are common in spring and summer. Recently, such unusual wintertime penetrations have become more frequent and intensive, pointing to a climate change and deviations in the seasonal regime of the intercontinental air circulation systems [6, 25]. The forecast Saharan dust concentration profile displays the presence of desert dust (with concentration below 10 mg/m^3) evenly distributed from ground level (600 m ASL) to a height of about 2.5 km ASL (about 2 km AGL) above the city. The HYSPLIT trajectories calculated for a period of 150 hours (figure 4 (c)) reveal that the air masses over Sofia in the altitude range 0.7-2 km AGL prior to the lidar scanning passed a few days earlier low over North Africa. Therefore, it is reasonable to assume that they contained desert dust. This assumption is supported by the PM_{10} mass concentrations data registered by the Kopitoto AMS in Sofia (figure 4 (b)), available at the European Environment Agency database [26] (the lidar scanning time is approximately marked by a green vertical line). The Kopitoto AMS is located in Vitosha Mountain at about 800 m AGL above the city (1321 m ASL, 42°38′13.2″N, 23°14′38.4″E) and is classified as an elevated rural station providing reference aerosol background measurements. It is necessary to specify that the height of the winter ABL above Sofia is estimated at 900 – 1200 m, so that the anthropogenic aerosols contribution to the PM_{10} data provided by this station in wintertime can be neglected. The annual mean values for PM_{10} registered by Kopitoto AMS in 2019 is 14 μg/m^3 [6]. As shown in figure 4 (b), the hourly mean value of the PM_{10} mass concentrations at Kopitoto AMS (about 30 μg/m^3) for the period 19-21 February 2019 is significantly higher than those normally recorded at this station in the winter. Here it is appropriate to note that these days fall in the period 3-21 February 2019, when a long wintertime Saharan dust transport over Bulgaria took place [6]. Comparing the data from Kopitoto AMS with the forecasts of the BSC-DREAM8b and HYSPLIT models presented above, one can conclude that the results of the *in situ* measurements by the Kopitoto reference AMS confirm the presence of Saharan dust in the lower troposphere above Sofia during the lidar observation described here.

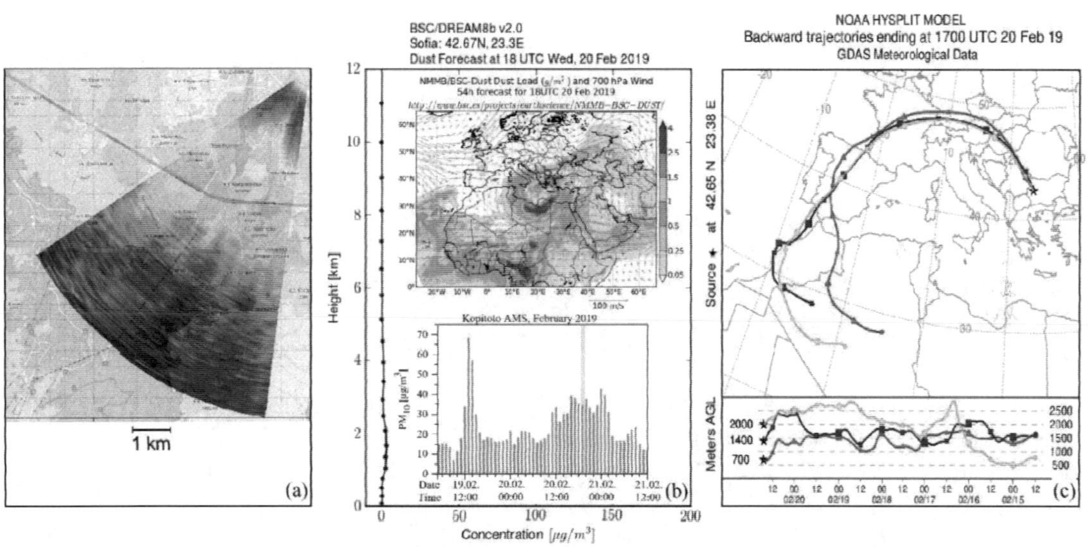

Figure 4. (a) Color map of the near-surface aerosol density distribution as measured on 20 February 2019; (b) BSC/DREAM8b model forecast of Saharan dust concentration profile above Sofia with regional dust load map (inset, top) and hourly mean values of the PM_{10} mass concentration measured by the Kopitoto AMS (inset, bottom), the vertical green bars indicate the time interval of lidar scanning; (c) HYSPLIT-backward trajectories ending above Sofia during the measurement.

4. Conclusions

The airborne Saharan dust registered during the presented measurements had a significant perturbative effect on the aerosol composition and structure/layering in the lower troposphere (particularly, in the ABL) over the measurement site region, strongly influencing the optical, microphysical, topological and dynamic characteristics of the local atmospheric aerosols.

Combining data obtained in parallel from ground-based and satellite lidar measurements, sun-photometer data, *in situ* measurement data, forecast/modeling ones, and aerosol concentration retrievals by using the LIRIC inversion code, systematically compiled and analyzed, represented a powerful and efficient tool for complex study and characterization of the atmospheric aerosols and their possible climatological, ecological and health effects.

The lidar mapping of the aerosol density distribution in the air layer adjacent to the ground performed by means of quasi-horizontal lidar scanning clearly showed differentiation in the aerosol pollutions over the studied sector, correlating well with the presence or absence of polluting sources.

Acknowledgements

This research was supported partly by the Bulgarian National Science Fund (Grant No. KP-06-N28/10/2018), by the Bulgarian Ministry of Education and Science in the framework of the National Science Program "Environmental Protection and Reduction of Risks of Adverse Events and Natural Disasters" (Agreement No. D01-279/2021), and by the research agreement between the Bulgarian Academy of Sciences and the National Academy of Sciences of Belarus. The equipment used is of ACTRIS BG, part of the Bulgarian National Roadmap for Research Infrastructure. The authors acknowledge Atmospheric Science Data Center for providing the CALIPSO data, AERONET-Europe for providing calibration service in the framework of the ACTRIS IMP project funded by the European Union under Grant Agreement No 871115, as well as data and images from the NMMB/BSC-Dust and BSC-DREAM8b models operated by the Barcelona Supercomputing Center and the NOAA Air Resources Laboratory for the provision of the HYSPLIT transport and dispersion model and READY website used in this publication.

References

[1] Tomasi C, Fuzzi S and Kokhanovsky A, ed 2017 *Atmospheric Aerosols: Life Cycles and Effects on Air Quality and Climate* (Berlin: Wiley-VCH)
[2] Comerón A, Muñoz-Porcar C, Rocadenbosch F, Rodríguez-Gómez A and Sicard M 2017 *Sensors* **17** 1450Z4
[3] Groß S et al 2011 *Tellus B: Chemical and Physical Meteorology* **63** 706–724
[4] Mona L et al 2012 *Advances in Meteorology* **2012** 356265
[5] Groß S et al 2012 *Atmos. Environ.* **48** 85–96
[6] Deleva A, Peshev Z, Vulkova L and Dreischuh T 2021 *J. of Applied Remote Sensing* **15** 024517
[7] Pappalardo G et al 2014 *Atmos. Meas. Tech.* **7** 2389–2409
[8] Aerosol, Clouds and Trace Gases (ACTRIS) Research Infrastructure. Available online: https://www.actris.eu (accessed on 30.08.2021).
[9] The EARLINET Publishing Group 2000–2015 2018 EARLINET SaharanDust 2000–2015 (Hamburg: World Data Center for Climate at DKRZ)
[10] Peshev Z, Deleva A, Vulkova L and Dreischuh T 2021 *J. Phys.: Conf. Ser.* **1859** 012031
[11] Holben B N et al 1998 *Remote Sensing of Environment* **66** 1–16
[12] Chaikovsky A et al 2016 *Atmos. Meas. Tech.* **9** 1181–1205
[13] Tsekeri A et al 2017 *Atmos. Meas. Tech.* **10** 4995–5016
[14] Chaikovsky A, Chaikovskaya L, Denishchik-Nelubina N, Fedarenka A and Oshchepkov S 2018 *EDJ Web of Conferences* **176** 08007
[15] Dreischuh T et al 2016 *Aerosols – Science and Case Studies* ed. K Volkov (Rijeka: InTech) pp 85–107
[16] Ong P M, Lagrosas N, Shiina T and Kuze H 2019 *Atmosphere* **11** 36
[17] Dreischuh T et al 2021 *J. Phys. Conf. Ser.* **1859** 012028
[18] Stein A F et al 2015 *Bull. Am. Meteorol. Soc.* **96** 2059–77
[19] Rolph G, Stein A and Stunder B 2017 *Environ. Modell. Softwares* **95** 210–228
[20] Basart S, Pérez C, Nickovic S, Cuevas E and Baldasano J M 2012 *Tellus* B **64** 18539
[21] Pérez C et al 2011 *Atmos. Chem. Phys.* **11** 13001–27

[22] Haustein K et al 2012 *Atmos. Chem. Phys.* **12** 2933–2958
[23] Klett J 1981 Appl. Opt. **20** 211-220
[24] Fernald F 1984 *Appl. Opt.* **23** 652
[25] Varga G 2020 *Environ. Intern.* **139** 105712
[26] European Environment Agency (EEA), Download of air quality data, https://discomap.eea.europa.eu/map/fme/AirQualityExport.htm (accessed on 02.08.2021)

Development of lidar for remote methane sensing using an optimal configuration of high-power laser diodes

V Pencheva[1], S Penchev and T Dreischuh

Emil Djakov Institute of Electronics, Bulgarian Academy of Sciences
72 Tsarigradsko Chaussee, 1784 Sofia, Bulgaria

E-mail: vasilka@ie.bas.bg

Abstract. A broadband differential absorption lidar (DIAL) is developed making us of the properties of high-power pulsed laser diodes for the purpose of monitoring methane, a major atmospheric greenhouse biogas. The detection range is optimized for the CH_4 Q-branch $2\nu_3$ overtone spectrum centered at 1.667 μm and mixed with water-vapor absorption bands of. The resonance absorption evaluated of the DIAL signal is an integral sum of the multiple CH_4 resonance absorption lines modulating the characteristic broad laser line and simultaneously compensating the interfering water-vapor absorption. A compact DIAL scheme is developed using an optimal configuration of high-power laser diodes matching the selected spectral parameters.

1. Introduction

Methane plays a key role in the enhancement of the greenhouse effect, amounting to 20 % of the increased atmospheric infrared radiation trapping during the past 200 years. It also affects the oxidizing capacity of the atmosphere and the lifetimes of other strong greenhouse gases [1]. Methane is contained in the continental shelf in the form of methane hydrates and markedly large amounts of it are trapped in the permafrost regions of Siberia, America, and Europe [2]. With global temperature rise, the need for monitoring methane will likewise grow over facilities of sewage treatment, hazardous zones of natural gas and petrol production, rice fields, stock- breeding and particularly over the hard-to-access permafrost regions. Conversely, the advances of ecological remote sensing can be implemented for prospecting for energy resources.

Atmospheric lidar (light detection and ranging) remote sensing provides highly selective real- time data of high spatial resolution unattainable by in-situ analyzers. Concretely, the differential absorption lidar (DIAL) traces atmospheric gases by their resonance molecular absorption properties. Such lidar emits two wavelengths tuned on and off the spectrum of the monitored gas. The measurement is unaffected by systematic external influence, such as, disturbing electric fields, temperature, pressure etc. This study advances DIAL remote sensors of atmospheric methane by way of extending the spectral applications of high-power pulsed laser diodes of microjoule range.

2. Utilization of high-power broadband laser diodes for DIAL gas analysis

The recent advances of spectral applications of broadband lasers in atmospheric optics have been facilitated by the accumulation of accurate spectroscopic data, in particular HITRAN database of

[1] To whom any correspondence should be addressed.

Content from this work may be used under the terms of the Creative Commons Attribution 3.0 licence. Any further distribution of this work must maintain attribution to the author(s) and the title of the work, journal citation and DOI.
Published under licence by IOP Publishing Ltd

molecular absorption. We contributed to the pioneering studies [3-6] utilizing high-power laser diodes for remote sensing of atmospheric major greenhouse gases. As a main goal, the integral absorption of laser radiation modulated by multiple resonance lines within a broader spectral band is less critical than measurement of the laser intensity within an individual absorption line.

2.1. CH_4 DIAL on paired laser diodes

The proposed dual-wavelength setup is based on high-power laser diodes matching the strong Q-branch $2\nu_3$ overtone CH_4 spectrum around 1.667 µm. The CH_4 and H_2O absorption profiles vs. gas abundance, lidar path and wavelength within the referenced spectral band are simulated using prior spectral data, namely the HITRAN'2012 update. The gas concentration is retrieved from the integral laser radiation attenuation by the multiple resonance absorption lines in the DIAL signal [7], given by the ratio of lidar "On/Off" spectral intensities (figure 1).

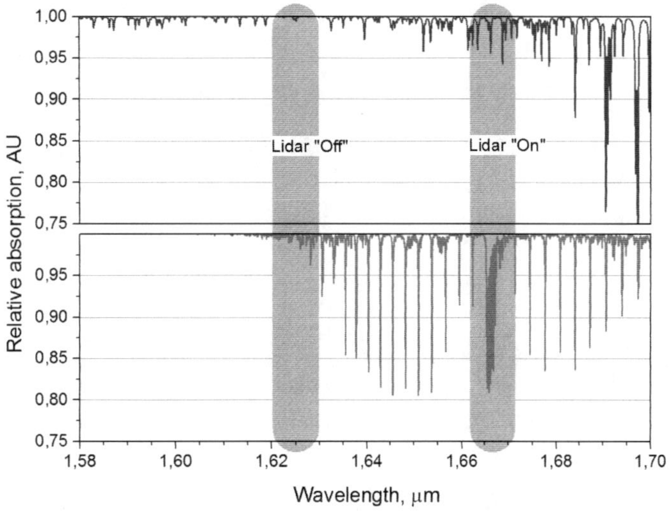

Figure 1. Absorption spectra of CH_4 10 ppm concentration and H_2O 3 g m^{-3} concentration on 1 km lidar path, where "lidar-on" 1.667 µm and "lidar-off" 1.625 µm wavelengths are selected.

Integral absorption ~4% of the lidar returns "on-spectrum" by CH_4 10 ppm concentration is shown in figure 1. It is noted that the spectral band lacks significant absorption by carbon dioxide. A single absorption line varies strongly with the pressure gradient resulting in a considerable measurement error. Detecting the integral absorption spectrum avoids this effect; as well, selective tuning between lines of different intensities is not necessary, which is a particular advantage for transition to unattended operation. The parameters of HITRAN molecular database are known with better than 1% uncertainty [4] and the dependence on the ambient temperature and the non-resonant atmospheric absorption properties, such as Rayleigh and Mie scattering, introduce a negligible error. However, the interfering H_2O absorption spectra are responsible for a systematic error of nearly 20%. At this stage of the study, taking into account the atmospheric humidity would allow for a higher accuracy of the DIAL measurement. Consequently, except for the product of CH_4 concentration and lidar path (equation A.1), the calibration curves depend also on the atmospheric humidity (to be published). An attractive opportunity of DIAL scheme implementing commercial pigtailed, paired laser diodes (https://www.laserdiode.com/) would still require one of the pair to be replaced to match the lidar "On" wavelength of 1.667 µm.

2.2. *Advanced CH_4 DIAL on single broadband laser diode*

The proposed technique optimizes the optical setup by integrating two spectral channels of selected wavelengths within the broad line of a single pulsed laser diode, providing sufficient power for each channel. Simultaneously, the interfering absorption by water vapor in the lidar returns is compensated, so that the DIAL signal depends on the properties of the referenced CH_4 $2\nu_3$ overtone spectrum (see Appendix). Such a system provides the best design relevant to "simplicity (price) vs. performance" as a customer recommendation to the laser diode industry.

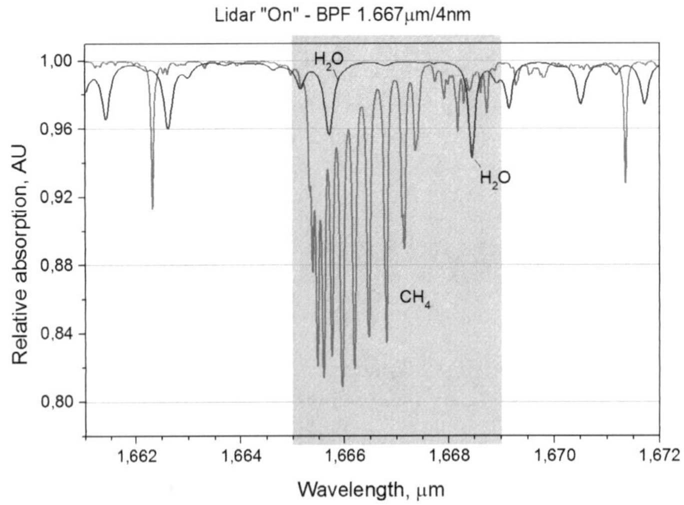

Figure 2. CH_4 molecular absorption bands of 10 ppm concentration mixed with H_2O of 10 gm^{-3} concentration (58% relative humidity at 20°C) centered at 1.667 μm wavelength on a 1 km lidar path.

Plotted in figure 2, the lidar "On" returns are centered at the 1667 nm wavelength confined by a band pass filter of 4 nm. The resultant CH_4 resonance absorption band (red line) is free of CO_2 absorption and coincides with separate absorption lines of water vapor (blue line). The integral absorption function (see Appendix) in the range of the laser line provides the molecular absorption magnitude, amounting to few percent of the lidar "On" signal (figure 3). Thus, the absorption limit of 1% on 1 km lidar track is equivalent to the background CH_4 1.7 ppm concentration level [8, 9]. Here, the residual H_2O absorption (figure 4) need not be monitored for accurate calibration of the absorption function reducing the systematic error. The DIAL measurement is immune to atmospheric humidity variations and depends solely on the product of CH_4 concentration and lidar path (equation A.3).

It is noted as a basic advantage of the method that the lidar parameters evaluated are unaffected by the varying atmospheric conditions, such as pressure and temperature. A patent pending optical design customizes the available technology of the high-power pulsed laser diodes, e.g., on the basis of CVLL 350-CL90 series: https://www.laserdiode.com/. The laser beam is coupled to the corresponding dual-wavelength lidar channels by a beam splitter combined with band pass and notch filters. Prospective examples of Alluxa ultra-narrow spectral filters (https://www.alluxa.com/optical-filter-catalog/) are close to the required parameters of DIAL application.

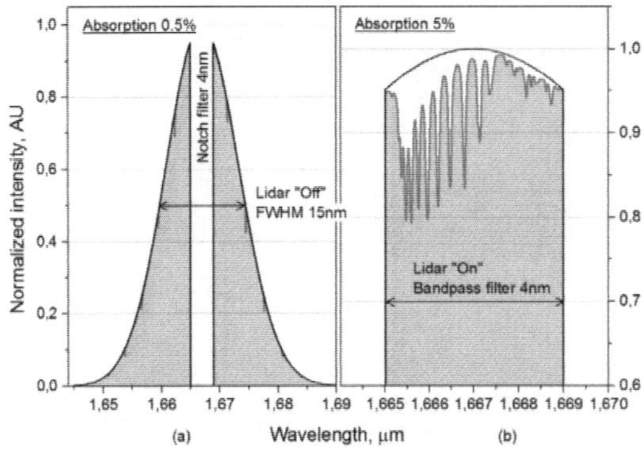

Figure 3. Resonance absorption by CH_4 of 10 ppm concentration modulating the laser intensity in the lidar "On" spectrum of 4 nm, vs. the remaining lidar "Off" range of the laser line on 1 km lidar path.

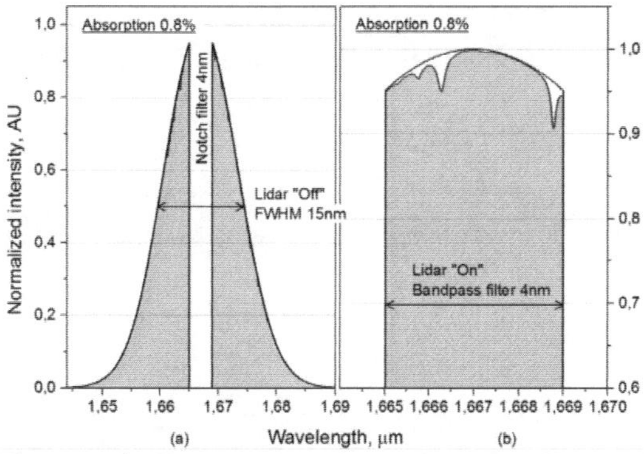

Figure 4. Resonance absorption by H_2O of 10 g m^{-3} concentration modulating the laser intensity in the lidar "On" spectrum of 4 nm, vs. the remaining lidar "Off" range of the laser line on 1 km lidar path.

3. Conclusions

The paper discusses the DIAL method as designed for remote sensing of the atmospheric content of the methane greenhouse gas utilizing the spectral and energy properties of high-power pulsed laser diodes. The lidar is optimized for the strong Q-branch $2\nu_3$ overtone CH_4 NIR absorption spectral band around 1.667 μm bearing in mind that the commercially available, sensitive, uncooled photodetectors are favorable for the selected spectrum. The spectral result specifies parameters of resonance absorption for measurement of the gas abundance on the background level and above. The integral molecular spectrum revealed within the characteristic broad laser line centered at the 1 667 μm wavelength is analyzed explicitly. The interfering H_2O absorption bands are evaluated to minimize the introduced systematic error. An optimal design of the DIAL scheme integrates a single laser diode with its beam confined in dual-wavelength channels of the optical system by a setup of narrowband spectral filters. The projected next-stage of the study targets realization of commercial devices matching the selected spectral parameters. Such a compact lidar would be capable of unattended operation, particularly for airborne nadir and path-integrated monitoring of the diurnal variations, peak emissions and long-term trends of atmospheric methane affecting the global climate.

Appendix: Absorption function

The integral absorption of the laser radiation over the measured lidar path is given by the convolution integral, C, integrated numerically in the frequency range of the laser line. Gaussian laser mode is assumed, modulated by multiple molecular absorption lines of Lorentzian profile:

$$C=\int_{\nu} f_{\mathrm{G}} \exp\left(-K\sum_{n} f_{\mathrm{L}}\right)\mathrm{d}\nu=\int_{\nu}\exp\left[-4\ln 2\left(\frac{\nu-\nu_{\mathrm{l}}}{\Delta\nu_{\mathrm{l}}}\right)^{2}-K\sum_{n} S_{\mathrm{n}}\frac{\Delta\nu_{\mathrm{a}}^{2}/4}{\left(\nu-\nu_{\mathrm{n}}\right)^{2}+\Delta\nu_{\mathrm{a}}^{2}/4}\right]\mathrm{d}\nu \tag{A.1}$$

with, $\Delta\nu_{\mathrm{l}}$ is the laser linewidth; ν_{n}, the absorption line frequencies; $\Delta\nu_{\mathrm{a}}$, the absorption linewidth at pressure 1 atm; S_n, the absorption line strengths; and K, the product of the gas concentration and the lidar path. H_2O is measured as absolute humidity (g m^{-3}) and CH_4 is given as parts per million.
The exponent in the DIAL signal is replaced by difference for small amplitudes of integral absorption:

$$I_{\mathrm{DIAL}}=C_{\mathrm{on}}/C_{\mathrm{off}}=\frac{\int_{\nu}\beta f_{\mathrm{G}}\left(1-K\sum_{n} f_{\mathrm{L}}\right)\mathrm{d}\nu}{\int_{\nu}(1-\beta) f_{\mathrm{G}}\mathrm{d}\nu}, \tag{A.2}$$

where $\beta(\nu)$ is a step function assuming values of 0 and 1.
In case of equal residual H_2O absorption in both channels, we obtain:

$$I_{\mathrm{DIAL}} = \frac{\int_{\nu} \beta f_{\mathrm{G}} \left[1 - \left(K \sum_{n} f_{\mathrm{L}} \right)_{\mathrm{H_2O}} - \left(K \sum_{n} f_{\mathrm{L}} \right)_{\mathrm{CH_4}} \right] \mathrm{d}\nu}{\int_{\nu} (1-\beta) f_{\mathrm{G}} \left[1 - \left(K \sum_{n} f_{\mathrm{L}} \right)_{\mathrm{H_2O}} \right] \mathrm{d}\nu} \cong 1 - \frac{\int_{\nu} \beta f_{\mathrm{G}} \left(K \sum_{n} f_{\mathrm{L}} \right)_{\mathrm{CH_4}} \mathrm{d}\nu}{\int_{\nu} \beta f_{\mathrm{G}} \mathrm{d}\nu} \tag{A.3}$$

where, $\int_{\nu} (1-\beta) f_{\mathrm{G}} \left[1 - \left(K \sum_{n} f_{\mathrm{L}} \right)_{\mathrm{H_2O}} \right] \mathrm{d}\nu = \int_{\nu} \beta f_{\mathrm{G}} \left[1 - \left(K \sum_{n} f_{\mathrm{L}} \right)_{\mathrm{H_2O}} \right] \mathrm{d}\nu$; $\int_{\nu} \beta f_{\mathrm{G}} \left(K \sum_{n} f_{\mathrm{L}} \right)_{\mathrm{H_2O}} \mathrm{d}\nu << \int_{\nu} \beta f_{\mathrm{G}} \mathrm{d}\nu$

Acknowledgements
This research was supported by the Bulgarian Ministry of Education and Science (support for ACTRIS BG, part of the Bulgarian National Roadmap for Research Infrastructure).

References

[1] IPCC 2007 Summary for Policymakers *Climate Change 2007: The Physical Science Basis. Contribution of Working Group I to the Fourth Assessment Report of the Intergovernmental Panel on Climate Change* eds. Solomon S, Qin D, Manning M, Chen Z, Marquis M, Averyt K B, Tignor M and Miller H L (United Kingdom and New York NY USA: Cambridge University Press, Cambridge) pp 1–18
[2] Miller S M, Taylor M A and Watts J D 2018 *Earth & Space Science News* **99** (7) 15–19
[3] Penchev S, Pencheva V and Naboko S 2012 *Comptes rendus de l'Académie Bulgare des Sciences* **65** (5) 669–74
[4] Thomas B, David G, Anselmo C, Coillet E, Rieth K, Miffre A, Cariou J P and Rairoux P 2013 *Journal of Molecular Spectroscopy* **291** 3–8
[5] Thomas B, David G, Anselmo C, Cariou J P, Miffre A and Rairoux P 2013 *Applied Physics* B **113** 265–75
[6] Riris H, Numata K, Li S, Wu S, Ramanathan A, Dawsey M, Mao J, Kawa R and Abshire J B 2012 *Applied Optics* **51** (34) 8296–305
[7] Platt U 1994 Differential optical absorption spectroscopy (DOAS) *Air monitoring by spectroscopic techniques* ed Zigrist M W (New York NY: John Wiley & Sons) p 31
[8] Smith F A, Elliott S, Blake D R and Rowland F S 2002 *Environmental Science & Policy* **5** 449–61
[9] Bamberger I, Stieger J, Buchmann N and Eugster W 2014 *Environmental Pollution* **190** 65e74

Thermodynamic analysis of the processes during electron beam melting and refining of copper

V Stefanova[1,3], K Vutova[2,4] and V Vassileva[2]

[1]University of Chemical Technology and Metallurgy, 1756 Sofia, Bulgaria
[2]Institute of Electronics, Bulgarian Academy of Sciences, 72, Tsarigradsko Chaussee, Sofia 1784, Bulgaria

E-mail: vps@uctm.edu, katia@van-computers.com

Abstract. In the paper, results are presented and discussed obtained by thermodynamic analysis of the refining processes of copper aimed at removing metal impurities and oxygen via electron beam melting (EBM) in vacuum. Thermodynamic evaluations of the possible chemical interactions between the base metal (Cu) and the metal impurities present in Cu that can proceed during the refining of copper are made based on the calculated values of the Gibbs free energy and equilibrium constants taking into account the physical state of copper and of the metal impurities during EBM.

1. Introduction

Copper is a basic metal in energy, energy systems and networks, architecture, electronics and telecommunications, transport and other branches of the industry. Copper consumption continues to grow every year; it is estimated that by 2035 copper consumption will increase by more than 40 % compared to 2020 [1]. According to the International Copper Study Group, copper ore reserves are steadily decreasing [2]. This requires the development and application of innovative methods for recycling copper scrap. It is known that copper can be recycled multiple times without losing its characteristics. At present, approximately 50 % of the copper in Europe is produced by recycling [2]. Recycling requires up to 85 % less energy than the conventional copper producing techniques.

With the development of new branches of electronics, the requirements for the purity of copper are constantly increasing. The processes of refining copper from impurities, both non-metals and metals, are labor-intensive and expensive and require the use of modern metallurgical methods of melting metals, such as laser, plasma and electron beam melting (EBM) [3, 4].

Of these methods, the EBM method is particularly appropriate as it combines the advantages of vacuum environment and high-energy special electrometallurgy [3, 5]. In [6] the influence was studied of the EBM parameters on the process of refining copper from impurities. On the basis of the calculated values of the partial pressures of the metals and their oxides and the conducted experiments, a possible mechanism for the removal of impurities at a certain technological regime was proposed.

[3] To whom any correspondence should be addressed.
[4] To whom any correspondence should be addressed.

Content from this work may be used under the terms of the Creative Commons Attribution 3.0 licence. Any further distribution of this work must maintain attribution to the author(s) and the title of the work, journal citation and DOI.
Published under licence by IOP Publishing Ltd

This article is a continuation of the research conducted so far by the team. It reports on a thermodynamic analysis of the process of refining liquid copper from metal impurities and oxygen taking into account the physical state during melting in different technological modes.

2. Thermodynamic analysis of the process of refining at EBM

Under EBM conditions, the refining process takes place mainly on the reaction surfaces of the liquid metal (its interface with the vacuum) in three reaction zones [3]. Depending on the thermodynamic conditions of the EBM and the type of the metal impurities, the refining process can take place by: (i) degassing (removal of components with a partial pressure higher than the partial pressure of the base metal); (ii) distillation (evaporation of the more volatile compounds from the metal components).

Usually, the refined metal contains a significant amount of O_2 which is found either in the form of a solid solution or in the form of oxides of the base metal and the metal impurities. Their oxidative refining is based on the differences in the affinity of different metals to oxygen. From a thermodynamic point of view, the refining process is possible when the dissociation pressure of the metal oxide of the base metal (*MeO*) is much higher than that of the metal impurity (RiO_2) [7].

The equilibrium constant (K_{eq}) of the metal impurity oxide dissociation reaction: $RiO_2 \leftrightarrow Ri + O_2(g)$ in the liquid pool is:

$$K_{eq} = \frac{P_{O2} C_{Ri}}{C_{RiO2}} \quad (1)$$

The pressures of oxygen (P_{O2}) and the concentration of the metal-impurity (C_{Ri}) at equilibrium in the base metal oxide and metal-impurity oxide system can be expressed by equation (1).

The influence of temperature (*T*) on the efficiency of the refining process is not straightforward. According to Le Chatelier's principle, the increase in temperature in endothermic reactions pulls the reaction to the right-hand side. On the other hand, increasing the temperature increases the solubility of the metal oxides in both the base metal and the slag.

3. Results and discussion

The thermodynamic analysis of the possible chemical interactions occurring during the refining of copper from impurities, such as Se, Te, Bi, As, Sb, Pb, Sn, Ni, Zn, Ag, under EBM conditions is performed on the basis of the Gibbs free energy (ΔF) and the equilibrium constants (K_{eq}). The analysis was carried out using the professional thermochemical calculation program HSC Chemistry ver.7.1, Reaction Equation module, taking into account the physical state of the metal impurities and their oxides during e-beam melting.

The calculations were performed under EBM conditions: at temperatures of 1460 K, 1500 K, 1600 K, 1670 K and 1800 K and operating pressure in the vacuum chamber of 10^{-3} Pa. These parameters correspond to the actual conditions of melting and refining of copper scrap in the electron beam furnace.

The copper content in the investigated material is 99.45 % Cu. The impurities with higher concentrations are: O_2 (0.2251 %), Ni (0.1 %), As (0.07 %), Ag (0.014 %), Se (0.0382 %), Pb (0.021 %) and Bi (0.015 %). The other impurities are: Sb (0.0095 %), Te (0.0068 %), Sn (0.0034 %), Zn (0.0023 %); the Co and Cd content is less than 2 ppm, therefore these elements are not taken into consideration in the further analysis.

Under EBM conditions, the liquid metal is a complex system of Cu, Cu_2O, metal impurities and their oxides, and they are in liquid or solid state depending on the temperature of refining and their individual properties.

Figures 1 and 2 show the melting temperatures and densities of liquid Cu and the impurities present in it.

It can be seen that in the temperature range 1460 – 1800 K most of the oxides have a melting point lower than that of Cu and Cu_2O. Exceptions are the oxides of Ni and Zn, which have a very high melting point. Nickel forms a solid solution with Cu, which at temperatures above 1357.87 K transforms completely into a liquid phase. In addition, ZnO, unlike NiO, has a significantly lower density than that of the base metal so that during refining ZnO will float to the reaction surface in a solid state.

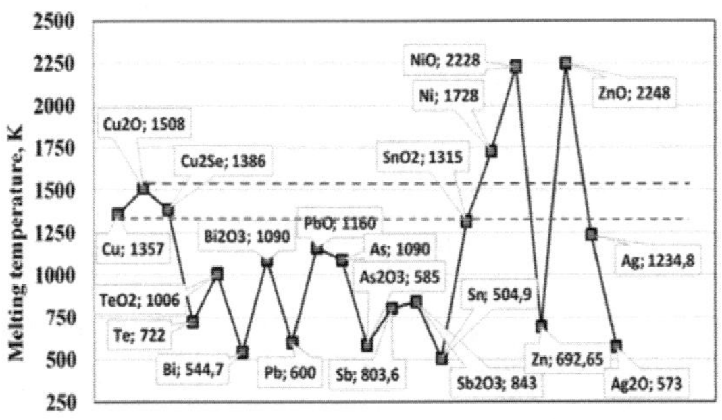

Figure 1. Melting temperatures of Cu, metallic inclusions and their oxides.

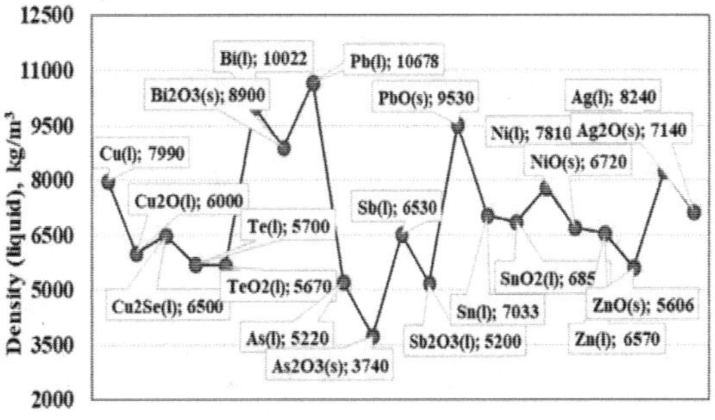

Figure 2. Density of liquid Cu, metallic inclusions and their oxides.

Therefore, during the EBM, the metal impurities and their oxides, with the exception of ZnO, are in a liquid state. They will move to the molten highly superheated reaction surface of Cu, where, depending on the thermodynamic conditions of refining, the oxidation processes, the dissociation of the formed oxides and the interaction between the molten metals and the copper oxide will take place. At operating temperatures above 1500 K, Cu_2O is in a liquid state, while at 1460 K it is in a solid state.

The density of Pb, Bi and their oxides is significantly higher than that of Cu (figure 2). Despite their low melting point, they will be difficult to move to the reaction surface.

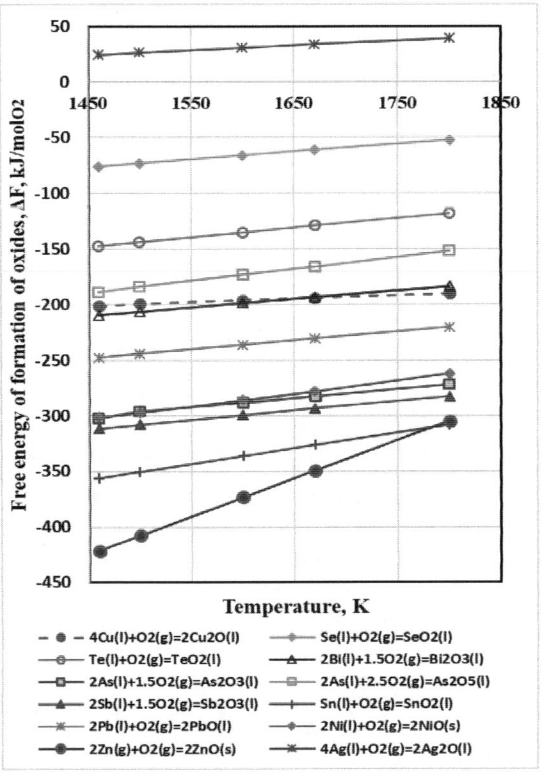

Figure 3. Free energy diagrams of formation of Cu_2O and metallic impurities oxides in vacuum.

Thermodynamic evaluation of the possible chemical interactions between the base metal (Cu) and the metal impurities present in it is made on the basis of the following equations:

$$2Cu + 0.5O_2 = Cu_2O + \Delta F_{T,\ Cu/Cu2O} \quad (2)$$

$$Ri + 0.5O_2 = RiO + \Delta F_{T,\ Ri/RiO} \quad (3)$$

$$Cu_2O + Ri = RiO + 2Cu + \Delta F_T \ , \quad (4)$$

where $\Delta F_{T\ Cu/Cu2O}$, $\Delta F_{T\ Ri/RiO}$ and ΔF_T are the free energies of the respective processes.

The possible oxidation reactions of the base metal (Cu) and metal-impurity (R_i) were investigated. Figure 3 shows the temperature dependences of $\Delta F_{T\ Cu/Cu2O}$ and $\Delta F_{T\ Ri/RiO}$. The indices (*s*), (*l*) and (*g*) refer to the substance being in a solid, liquid or gaseous state, respectively. According to the thermodynamic laws, the course of a given reaction is only possible when the calculated value of the free energy is negative.

The diagram allows estimation of the energy of formation of Cu_2O and oxides of metal impurities under the conditions of vacuum (10^{-3} Pa) within the temperature range 1460 K – 1800 K. It is seen that the probability of oxidation of Zn(*g*) to ZnO(*s*) is the highest. As the temperature increases, the affinity of zinc to oxygen decreases significantly. In the whole temperature range, ΔF of formation of SnO_2(*l*), Sb_2O_3(*l*), As_2O_3(*l*), NiO(*s*) and PbO(*l*) are higher than that of the oxidation of Cu(*l*) to Cu_2O(*l*). The thermodynamic probability of oxidation of bismuth to Bi_2O_3 is almost the same as that of copper, while the probability of oxidation of arsenic to As_2O_5(*l*), tellurium to TeO_2(*l*) and selenium to SeO_2 is significantly lower. The positive values of ΔF in the case of silver indicate that it does not oxidize.

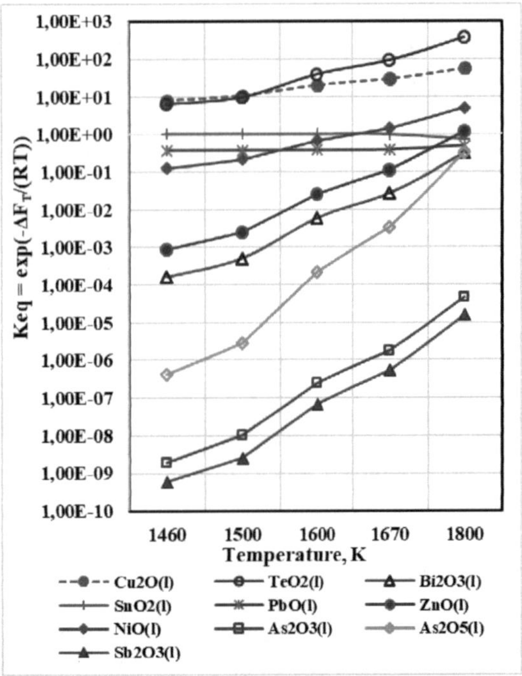

Figure 4. Equilibrium constants of oxygen for Cu_2O and metallic impurities (*RiO_2*) in the temperature range 1460 – 1800 K in vacuum.

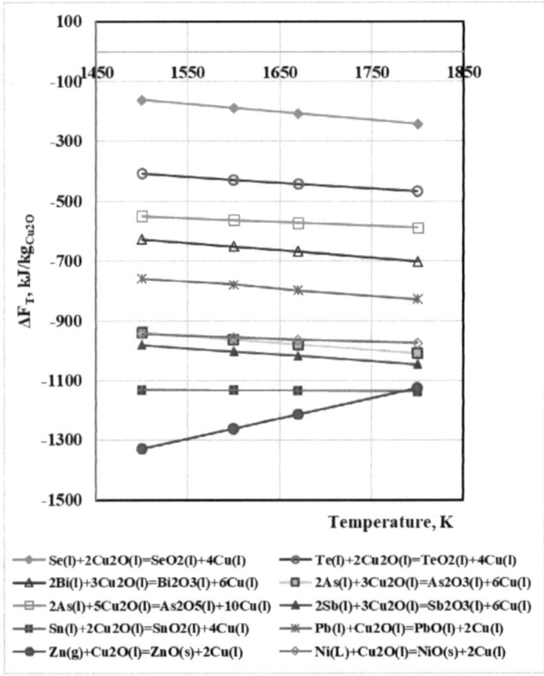

Figure 5. Free energy variations for interactions between Cu_2O and metal impurities in vacuum.

The oxidation reactions are reversible. The free energy of the dissociation ΔF_T^0 is calculated following the equation:

$$\Delta F_T^0 = -RTlnK_{eq} = -RTlnP_{O2} \quad (5)$$

where *R* is the gas constant. Then the equilibrium constant (K_{eq}) is calculated from:

$$K_{eq} = \exp((-\Delta F_T^0/(RT)) \tag{6}$$

The calculated values K_{eq} of the dissociation reactions are shown in figure 4. The analysis of the obtained dependences shows that, with the exception of TeO_2, the values of the equilibrium constant (K_{eq}) are lower than that of copper oxide and decrease in the following sequence: $Cu_2O(l)$ > $SnO_2(l)$ > $PbO(l)$ > $NiO(l)$ > $ZnO(l)$ > $Bi_2O_3(l)$ > $As_2O_5(l)$ > $As_2O_3(l)$ > $Sb_2O_3(l)$. It means that the stability of these oxides increases and when the pressure of the O_2 in the system is higher than the equilibrium pressure of the oxide, a reverse process will take place (oxidation of the metal).

The possibility of chemical interactions between Cu_2O and metal impurities (equation (4)) is also evaluated. The calculations at T = 1460 K were performed under the condition that the copper oxide is in a solid state ($Cu_2O(s)$) and above this temperature the copper oxide is in a liquid state ($Cu_2O(l)$) (figure 1). The calculated free energy values for the investigated temperatures are shown in figure 5. The analysis of the results obtained shows that in the studied temperature range all impurities, with the exception of Ag, will interact with the copper oxide regardless of its phase state. The interaction with Zn, which is in a gaseous state under melting conditions, is the most probable. As the temperature increases, the possibility of interaction decreases significantly. This is due to the high partial pressure of zinc oxide, which will be removed by distillation.

Unlike Zn, the free energy of interaction of copper oxide with the other impurities increases slightly as the temperature increases. The free energy values (ΔF_T) for interactions between Cu_2O and metal impurities decrease in the order ZnO > SnO_2 > Sb_2O_3 > As_2O_3 > NiO > PbO > Bi_2O_3 > As_2O_5 > TeO_2 > SeO_2. With the exception of Se, the calculated values of Gibbs energy are very high. Therefore, in the conditions of EBM, oxidation reactions of the impurities with O_2 dissolved in copper in the form of copper oxide will predominantly take place (figures 3 and 5).

4. Conclusions

Based on the calculated values of the Gibbs energy and equilibrium constants, the thermodynamic probability of the following processes is estimated: oxidation of the base metal and the impurities (Se, Te, Bi, As, Sb, Pb, Sn, Ni, Zn, Ag) present in it; dissociation of the produced oxides and chemical interactions between the metal impurity and Cu_2O taking into account the physical state of the metals and their oxides.

It is found that under the conditions of vacuum in the temperature range 1460 – 1800 K, the probability of $Zn(g)$ oxidation to $ZnO(s)$ is the highest. Oxidation of $As(l)$ to $As_2O_5(l)$, $Te(l)$ to $TeO_2(l)$ and $Se(l)$ to $SeO_2(l)$ is less likely than that of the base metal. Only Ag is not oxidized out of all the impurities present in copper. The high values of the Gibbs energy of the chemical interactions between the liquid melts of $Cu_2O(l)$ and the metal impurities indicate that from a thermodynamic point of view the probability of these processes is much higher than the direct oxidation of these impurities. The further removal of the impurities is controlled by the partial pressure of their oxides.

The thermodynamic analysis shows that in order to determine the thermodynamic conditions of the refining process in vacuum and at high temperatures it is necessary to take into account the physical state of the individual impurity and its physical properties, such as melting temperature and density.

Acknowledgements

The work was supported by the Bulgarian National Science Fund under contract DN17/9.

References

[1] Biwas A K and Davenport W G 2013 *Extractive Metallurgy of Copper,* Third edition (Elsevier)

[2] International Copper Study Group 2020 *The World Copper Factbook* 2020, http://www.icsg.org/index.php/component/jdownloads/finish/170/3046

[3] Mladenov G, Koleva E, Vutova K and Vassileva V *in book Practical Aspects and Application of Electron Beam Irradiation* ed M Memtanu and M Brasoveanu (Transworld Research Network India) chapter 3 pp 43–93

[4] Kalugin A 1980 *Electron beam melting of metals* (Moscow: Metall. Publ. House, in Russian)

[5] Vutova K, Vassileva V, Stefanova V, Amelnerkar D and Tanaka T 2019 *Metals* **9** 683
[6] Vassileva V and Vutova K 2020 *J. Phys.: Conf. Ser.* **1492** 012014
[7] Vaniukov A V and Zaitzev V J 1993 *Theory of pyrometallurgical processes* (Moscow: Metall. Publ. House, in Russian)

Turbulent flow influence on the discharge parameters of a magnetically-stabilized gliding arc discharge

P Tsonev[1,3], V Ivanov[1], St Kolev[1], Kh Tarnev[2] and Ts Paunska[1,4]

[1]Faculty of Physics, St. Kl. Ohridski University of Sofia, 5 James Bourchier Blvd., 1164 Sofia, Bulgaria
[2]Department of Applied Physics, FAMI, Technical University of Sofia, 8 St. Kl. Ohridski Blvd., 1000 Sofia, Bulgaria

E-mail: tzonev.plamen@gmail.com, cwalchew@phys.uni-sofia.bg

Abstract. A DC gliding arc discharge stabilized by a magnetic field in a gas flow is studied. The discharge is bounded by dielectric walls channeling the gas flow. The arc, the magnetic field and the gas flow are mutually perpendicular so that the Lorentz force arising due to the arc current and magnetic field interaction is directed in a direction opposite to the gas flow. The effects of the gas flow turbulence on the plasma parameters and the discharge behavior are investigated by means of a numerical simulation. The results pertain to different intensities of the turbulence and two distances between the dielectric plates where the arc is stabilized, namely, 2 mm and 4 mm. The results show the significant effect of the turbulent transport on the arc stabilization and, in general, the need of a stronger magnetic field for stabilizing the arc for the same gas flow.

1. Introduction

The arc discharges at low currents (< 1 A) are not a representative example for all types of arcs, but the recent increase in their usage in applications have made them a popular subject for researchers. Because of their simple inexpensive design and the ability to operate at atmospheric pressure, they are becoming an attractive option as a source for non-isothermal plasma, especially in the field of plasma chemistry [1]. In recent years, they have been used for gas treatment, including for dissociation of CO_2.

In this work, we investigate a gliding arc discharge (GAD) in an external magnetic field perpendicular to the arc current and to the gas velocity. As a result, the arising Lorentz force acts in a direction opposite to the gas drag force. When these two forces become equal, the arc is stabilized. This modified arc construction has the potential of treating much larger amounts of gas compared to a classical GAD.

In our previous work [2], we presented a 2D numerical model of a magnetically-stabilized GAD in argon. The modelling domain lies in a plane perpendicular to the arc current, i.e., the approximation is adopted of a very long arc column and, thus, zero gradients along the arc current; i.e., we neglect the electrode regions as they have a minor contribution to the interaction with the flowing gas. The model is based on the fluid plasma theory with the Navier-Stokes equations solved to describe the plasma

[3] To whom any correspondence should be addressed.
[4] To whom any correspondence should be addressed.

Content from this work may be used under the terms of the Creative Commons Attribution 3.0 licence. Any further distribution of this work must maintain attribution to the author(s) and the title of the work, journal citation and DOI.
Published under licence by IOP Publishing Ltd

Journal of Physics: Conference Series **2240** (2022) 012035 doi:10.1088/1742-6596/2240/1/012035

behavior. The results in [2] have shown different arc behavior for different distances between the dielectric flat side-walls that bound the domain. While at a distance of 2 mm between the walls, the arc is symmetrical and well stabilized, for distances of 20 mm (condition corresponding to arc without walls) the arc has an unstable behavior and vortices are formed in the arc's wake. The arc can be stabilized if the gas velocity is low enough and the magnitude of the magnetic field is large enough so that the Lorentz force and the gas drag are in equilibrium.

In [2], the results and the corresponding conclusions are based on numerical simulations assuming a laminar gas flow. Here, we analyze the effect of turbulent gas motion by using a Reynolds averaged Navier-Stokes (RANS) model [3, 4]. After a detailed analysis of the conditions, the SST RANS model was chosen. The influence of the turbulence on the plasma parameters and the conditions for arc stabilization are thus investigated.

2. Numerical model

In the current study, the model in [2] is further developed by modifying the gas flow description in order to account for the turbulent effects. The model is 2D with a magnetic field in argon and the geometry, the magnetic field, and the gas flow configuration are the same as in [2]. Here, we will only make a short summary of the model; more details can be found in [2]. The numerical model is implemented in Comsol Multiphysics® (version 5.3).

2.1. Plasma model

The plasma parameters are obtained by using the particle balance equations of charged and neutral particles, the electron energy balance equation, the condition for quasi-neutrality and the condition for an ideal gas. The particle fluxes are expressed from the equations of motion by using the drift-diffusion approximation. In the presence of a magnetic field, there is an additional flux related to the **E**×**B** drift. It is important to mention that the model approximates the central part of a DC discharge with a classical design with diverging electrodes. The external circuit considered in the model is a simple voltage source connected in series to a large ballast resistor limiting the discharge current. In addition, the current density conservation equation is solved for the derivation of the electric potential in the plasma domain.

In the simulation, the argon gas chemistry is included, with the species being electrons *e*, argon atomic ions Ar^+, molecular argon ions Ar_2^+, excited neutral atoms Ar(4s) and neutral argon atoms in ground level Ar.

More details and the descriptions of the approximations in the derived equations in the model are given in [2].

2.2. Gas flow model

For the proper description of the stabilized arc, it was shown in [2] that the plasma model should be coupled to a gas flow model. Here we use the Comsol Multiphysics® for the whole system of equations. The SST RANS model [3, 4] considers a weakly compressible flow due to the low variation of the gas pressure in the discharge domain. In addition, the gas heat-transfer equation is solved. As a result, the dynamic viscosity η, gas density ρ and heat capacity C_p are function of the gas temperature T_g during the discharge evolution.

As it is known, the RANS models use an approximation in which the motion of the fluid is described by a time-averaged component of the viscosity and a nonlinear term accounting for the turbulent fast fluctuations in the flows. After estimating the conditions in our experiment, it was decided that the SST model is appropriate for our purposes. The turbulence parameters are the turbulent kinetic energy k and the specific turbulence dissipation rate ω, which determines the rate at which the turbulent kinetic energy is converted to thermal internal energy. The SST model combines the advantages of the two widely used models k-ω and k-ε. The former precisely describes the boundary layer and the latter accurately approximates the turbulent layer in the second one. The SST model is widely used since it is more stable and with good convergence and accuracy.

Generally, the turbulent gas flow is characterized by a flatter gas velocity profile due to the flow separation from the boundary. Within the simulations at different settings suitable for our conditions, the SST model showed proper behavior and, as demonstrated in figure 1, the resulting turbulent gas profile velocity is flatter than the parabolic profile for a laminar flow. Figure 1 shows normalized profiles, but the simulation is for the same gas flow.

The turbulence quantities of interest in the model are the turbulence intensity I_t and the turbulence intensity scale L_t. The values for L_t are taken from the relation for mixing flow: $L_t = 0.07 * L$, where L is the layer width [3].

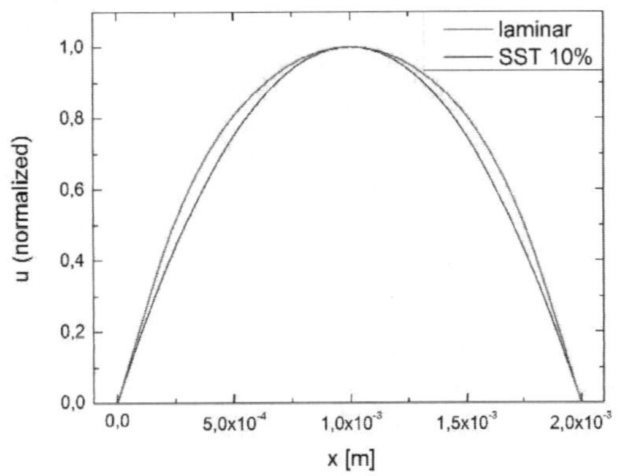

Figure 1. Normalized profiles of the gas velocities in laminar and turbulent flows (SST model) with $I_t = 10$ %.

In this work we consider turbulent intensities up to 50% at the gas inlet. Since this value might seem too high, we must provide a more detailed explanation. From the literature [3, 5], it is known that for still fluids this value is less than 1 %. For a simple geometry, like pipes, and a fluid with low Reynolds numbers the values are about 1 – 5 %. The typical turbulence intensity values for conditions corresponding to a highly turbulent flow are between 5 and 20%. Here we show results for I_t = 10, 30 and 50 %, with the higher values (50 %) corresponding to a turbulence intentionally intensified by using multiple nozzles at the gas inlet. Additional CFD simulations of various nozzle geometries and configurations were performed showing that values of I_t = 50% are achievable with simple nozzle designs.

3. Results and discussion

The calculated set of results assumes a single value of the discharge current I = 400 mA, a maximum magnetic flux density of B = 0.6 T, two distances between the dielectric walls W_d = 2 and 4 mm, a mass flow rate of 0.0145 kg/s corresponding to an averaged velocity u_{av} = 2.18 m/s and a turbulence intensity I_t = 10, 30 and 50 %. The simulations show that at W_d = 2 mm the arc is not stabilized and goes out of the domain, while at W_d = 4 mm it is stable and stationary. Figure 2 shows 2D plots (arc cross-section) of the electron density n_e distribution at various I_t. It is apparent that the arc is stabilized at different positions and has different behavior for the different values of the turbulence intensity. While at low turbulent intensity the solution resembles the one obtained in [2] with a laminar gas flow, namely, an arc stabilized near one of the walls, at higher I_t the behavior is rather different. At 30 % and 50 %, the arc is stabilized in the middle of the domain between the dielectric walls. We should stress here that we assume cooled walls so that the boundary condition for the gas temperature there is 293 K. Taking that into account that one of the major consequences of a highly turbulent flow is an enhanced thermal transport, we can conclude that the arc is not stabilized near the walls since the intensive cooling there will lead to lower gas temperature and plasma conductivity. As a result, the arc is stabilized in the middle between the dielectric walls, where significant gas temperature can be achieved and, consequently, arc contraction and operation at lower overall power. The stabilization in the middle, however, leads to a higher drag due to the arc-gas friction from both sides of the arc and to a higher pressure drop due to the arc presence. This results in stabilization at a higher "y" value, i.e. at higher magnetic field and Lorentz force. The latter is related to the imposed **B** profile – Gaussian, centered at $y = 0$. The particular values of the n_e maxima ($(x, y)_{ne,max}$) are as follows: for $I_t = 10$ % $(x, y)_{ne,max} = (0.67, -7.82)$ mm; for $I_t = 30$ % $(x, y)_{ne,max} = (0, -5.84)$ mm and for $I_t = 50$ % $(x, y)_{ne,max} = (0, -5.92)$ mm.

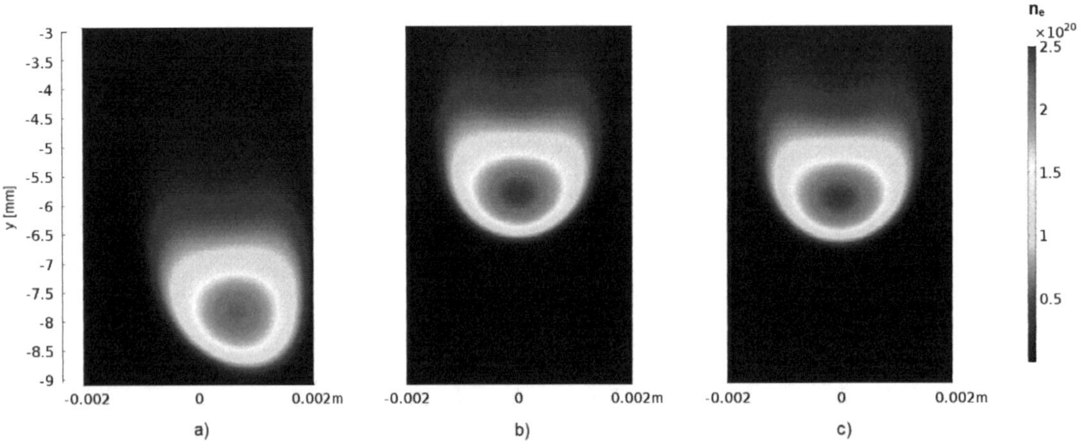

Figure 2. 2D plots of the electron density at $W_d = 4$ mm, for $I_t = 10\%$ in (a), $I_t = 30\%$ in (b) and $I_t = 50\%$ in (c). This is only a part of the modelling domain, which is 60-mm long in total.

The maximum values of the gas temperature are $T_{gmax} = 3684$ K for $I_t = 10\%$, but for $I_t = 30\%$ and $I_t = 50\%$, respectively, $T_{gmax} = 3548$ K and 3552 K. Their positions coincide with the maximum of the electron densities. Note that the behavior simulated here and in [2] for a low turbulent or a laminar flow, i.e., stabilization near a wall, was observed experimentally in our set-up of a stabilized GAD with $W_d = 4$ mm [6]. There is no physical process determining to which wall the arc will be attached – the displacement towards one of the walls is random. There are two reasons why the arc is stabilized at the wall and not in the center – the almost parabolic profile of the gas velocity and the considerable distance between the arc and the dielectric walls. With this semi-laminar velocity profile, the arc attachment to the wall provides a lower overall gas drag compared to an arc in the center. The mechanism of the arc attachment to the wall is also discussed in [2]. In case of high turbulence (figures 2 b) and 2 c), the flatter profiles of the gas velocity and the increased thermal conductivity do not allow a stabilization near the wall.

In figures 3 a) and b), we show the gas velocities of the cases of $I_t = 10\%$ and $I_t = 30\%$. In the case of high turbulence, the gas flow around the arc is symmetrical and the velocity is higher. This, combined with the increased turbulent thermal transport, is probably the reason for the slightly better cooling and the lower gas temperatures. For more clarity, figure 3 (c) presents the normalized profiles of the velocity along the x axis taken at the y position of the velocity maxima for the three cases.

In the numerical results of this investigation, we found that under the same condition but for $W_d = 2$ mm, the Lorentz force magnitude is not sufficient to stabilize the arc. The drag force is then dominant and it carries the discharge downstream.

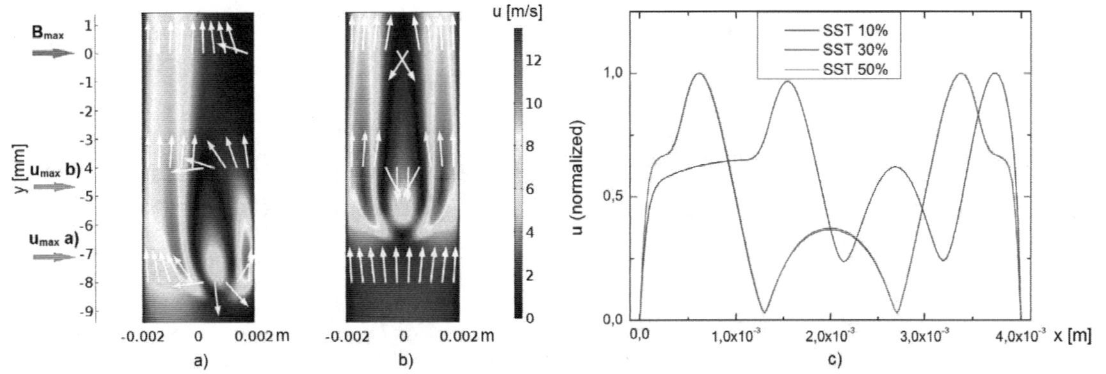

Figure 3. A 2D plots of the gas velocity magnitude at $W_d = 4$ mm for $I_t = 10\%$ in (a) and for $I_t = 30\%$ in (b). The arrows visualize the gas flow direction; (c) normalized profiles of the gas velocity along the x axis at the y positions corresponding to the maximum gas velocity as denoted by the green arrows.

4. Conclusions

In this paper, we present a study on the effect of the turbulent transport on the stabilization of a low-current arc (400 mA) in a magnetic field. The arc, the gas flow and the magnetic field are mutually perpendicular so that the Lorentz force is directed in a direction opposite to the gas flow drag. This discharge configuration aims to improve the treatment rate of a gas and, particularly, of CO_2 conversion.

The modelling results show that the arc behavior is qualitatively affected by the turbulent transport. At a high turbulent intensity, the arc is stabilized at the middle between the side walls at an axial position farther from the stabilization position under low turbulence or laminar flow conditions. However, a higher magnetic field is needed to stabilize the arc.

Acknowledgments

This work was supported by the Bulgarian National Science Fund, Ministry of Education and Science, research grant KP-06-OPR 04/4 of 14.12.2018 and by the European Regional Development Fund within the Operational Program "Science and Education for Smart Growth 2014 - 2020" under Project CoE "National center of mechatronics and clean technologies" BG05M2OP001-1.001-0008-CO1.

References

[1] Fridman A 2008 *Plasma Chemistry* (Cambridge: Cambridge University Press)
[2] Ivanov V, Paunska Ts, Tarnev Kh and Kolev St 2021 *Plasma Sources Sci. Technol.* **30** 085007 17pp.
[3] CFD Module User's Guide, COMSOL:Version 5.3 (2017)
[4] Versteeg H K and Malalasekera M 2007 *An Introduction to Computational Fluid Dynamics*, (Harlow, Pearson Education Limited)
[5] https://www.cfd-online.com/
[6] Ivanov V, Lazarova S, Iordanova S, Paunska Ts, Georgiev N and Kolev St 2021 *J. Conf. Series* /submitted/

Simulations of a capacitor with graphene electrodes

V Vachkov and Zh Kiss'ovski[1]

Faculty of Physics, St. Kl. Ohridski University of Sofia
5, J Bourchier Blvd., Sofia 1164, Bulgaria

E-mail: kissov@phys.uni-sofia.bg

Abstract. In this work, the behavior of a graphene-based capacitor was investigated by using an electromagnetic (EM) simulator. The graphene electrodes were simulated as a porous structure with a varying pore diameter as derived from experimental data. The main issue in modeling such a structure is the large number of objects with nano- or micro-dimensions (~millions), which imposes a significant computation challenge for the simulator. A solution was developed by introducing an externally created matrix of the electrode with a fixed number of elements, which allowed the successful and accurate simulation of the complex porous-based structure. The pores of the electrodes are modeled as cylindrical holes in a solid graphene parallelepiped, while the capacitance is determined for different dimensions of pores of electrodes with a corresponding shape of the dielectric medium. The results obtained show a significant increase in the capacitance compared to a capacitor with flat metal electrodes.

1. Introduction

Graphene structures, such as nanotubes, vertical graphene sheets, and graphene foams formed on the electrodes of batteries and super-capacitors, are widely applied in the automotive industry for power supplies of electric vehicles and for hybrid energy storage systems [1-3]. Their unique applications are due to the distinctive qualities of the graphene structures, which provide chemically stable electrodes for batteries and capacitors with high conductivity and enormous surface area (up to 2630 m^2/g [4]).

Novel capacitor electrodes from materials that have a large specific surface are presently the subject of intensive research. These materials allow the production of electrochemical capacitors (EDLC) with a large capacitance, which accumulate an enormous amount of charge in their double layer [5]. Other processes that significantly increase the capacitance of electrochemical capacitors [6] are intercalation of ions in the electrode material [7] and pseudo-capacitance redox processes [8]. Many different designs, electrode structures and types of electrolytes of capacitors with graphene structures on the electrodes have been studied [9-12]; however, to the best of our knowledge, no studies have so far been published involving a complete model of the behavior of such porous graphene-based capacitors due to their intricate morphology and the complex processes taking place on the electrodes. Carbon nanowires deposited on nickel foam showed a significant increase in the specific surface area of electrodes and, consequently, a greater capacitance of the modelled capacitor [13]. Vertical graphene sheets [14] have been deposited by microwave-plasma-enhanced CVD on nickel plate electrodes, which resulted in a further increase in the surface area of the capacitor electrodes. The performance of capacitors with vertical graphene structures on the electrodes has been investigated by representing the space between

[1] To whom any correspondence should be addressed.

Content from this work may be used under the terms of the Creative Commons Attribution 3.0 licence. Any further distribution of this work must maintain attribution to the author(s) and the title of the work, journal citation and DOI.
Published under licence by IOP Publishing Ltd

the graphene sheets as pores of different diameter and height. The pores can be classified depending on their diameter as micro- (< 2 nm), meso- (2 – 50 nm) and macro-pores (> 50 nm) [15]. In the present study, the effect of macro-pores on the surface area of graphene electrodes was investigated and the consequent increase in capacitance was modelled by an EM simulator. Modeling a large number of objects with nanometer dimensions imposes a significant challenge for the simulation of the graphene electrodes. Therefore, in our model, we study the increase in the electrode surface area caused by the largest pores with dimensions ~1 μm observed in SEM images (figure 1 a) of deposited vertical graphene structures [14]. These structures contain also micro- and meso-pores (figure 1 b), to be considered in future modeling. The process of parameterization of the pore sizes is implemented by introducing an externally created matrix of the electrode as a parallelepiped with a fixed number of identical elements.

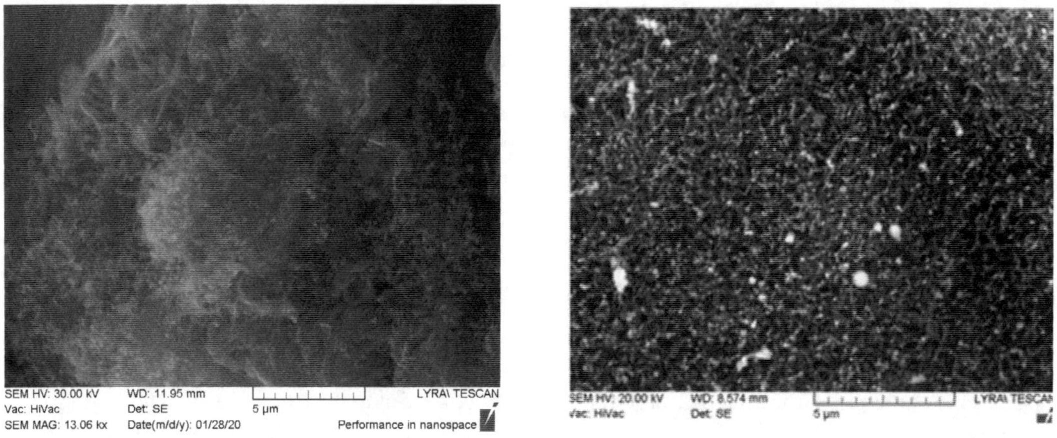

a) **b)**

Figure 1. Vertical graphene structures deposited on a Ni substrate with macro **a)** and meso and micro pores **b)**.

2. Simulation model and results

An electromagnetic simulator was used to investigate the capacitance of a capacitor with graphene structures on the electrodes. The finite element method (FEM) was used to analyze the electric field distribution and the capacitance at three different pore diameters in the graphene electrodes. The pores are assumed to be of cylindrical shape, allowing for an increase in the electrode area compared to the flat metal electrode capacitors. The capacitor model includes a solid parallelepiped with square surface as a graphene electrode with up to 64 holes (figure 2 a). The diameter ($d = 2r$, r being the radius) of the pores are 0.4 μm, 0.85 μm, and 1.7 μm representing the largest holes in the graphene structures (figure 1a) with a depth (h) of 0.5 μm.

Every pore increases the electrode surface area by a factor k, which is the ratio of the surrounding wall and the base area of the cylinder to the area of its base:

$$k = \frac{\pi r(r+2h)}{\pi r^2} = \frac{r+2h}{r} = 1 + 2h/r \qquad (1)$$

The increased surface area S' due to all holes (N) in the parallelepiped is given by the formula:

$$S' = S + N\,2\pi rh, \qquad (2)$$

where S is the surface of the electrode without holes. The number of holes N depends on the dimensions of the square with area $S = L \times L = L^2$ m^2; its maximum for a cylindrical structure is $N_{max} = (L/2r) \times (L/2r)$, which leads to the following result for the increased area S':

$$S' = S + (L/2r)^2\,2\pi rh = S + L^2 \pi h/2r = S(1 + \pi h/2r) \qquad (3)$$

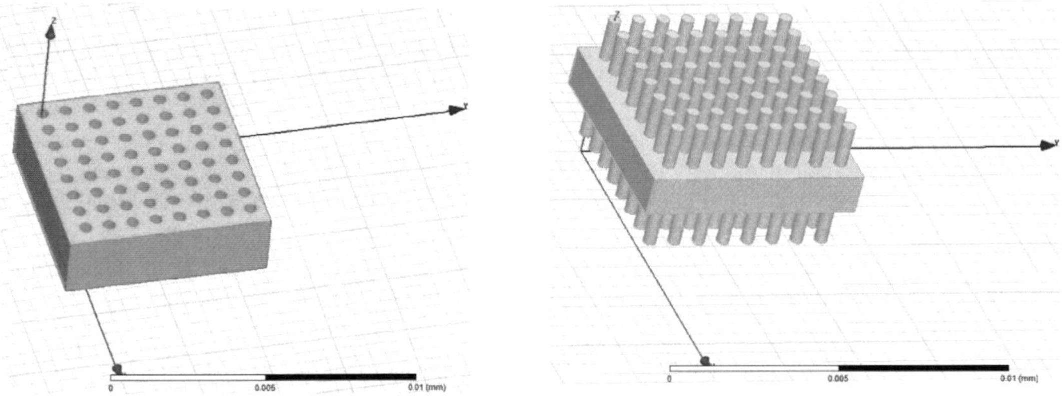

a) **b)**

Figure 2. In the capacitor model, the graphene electrode of the capacitor is a solid parallelepiped with up to 64 holes **a)**, and a dielectric medium **b)** between electrodes with permittivity $\varepsilon_r = 10.2$.

The DC voltage applied between the electrodes is 4 V in all simulations. The physical characteristics of the solid graphene parallelepiped in the model are chosen from experimentally measured data [16]. The geometry (figure 2 b) of the dielectric filling ($\varepsilon_r = 10.2$) between the two electrodes is in accordance with the pores diameter. Using electrodes with a large number of pores with a small diameter and a large depth will result in a high increase in the electrode area and, hence, in the capacitance of the capacitor. The distance between the electrodes is $d = 0.00032$ mm in the three simulations performed.

The model is applied for a capacitor with 16 pores with a diameter of 0.4 μm embedded in both electrodes, which are parallelepipeds with dimensions 0.00171 mm×0.00171 mm×0.001 mm. The simulation model of the capacitor taking into account the location of the pores and the complicated field distribution between the electrodes yields a capacitance of 3.81×10^{-15} F (figure 3). The theoretical estimation of the capacity taking into account only the increase of the electrode area due to pores is close to that obtained by the model:

$$C = \frac{\varepsilon_0 \varepsilon_r S'}{d} = \frac{\varepsilon_0 \varepsilon_r (S + N\, 2\pi r h)}{d} = 3.67\times10^{-15}\text{F} \tag{4}$$

A capacitor with holes diameter of 0.85 μm and the same electrode dimensions has a capacity of 2.15×10^{-15} F, while for pore diameters of 1.7 μm, the capacity is 1.10×10^{-15} F (figure 3).

The result obtained for the capacitance (figure 3) of an electrode with 16, 4 and 1 pores shows that it decreases with increase of the pore diameter. The large number of meso and micro-pores of graphene electrodes (figure 1 b) also will be taken into account in our next capacitor model, as they lead to the largest increase in the capacitance due to their high depth/radius ratio (4).

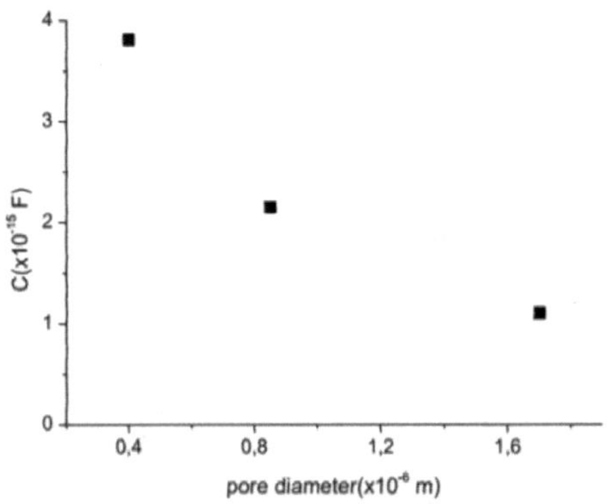

Figure 3. Capacitance dependence on the diameter of pores in the electrodes.

3. Conclusions

An EM simulator was used to investigate the capacitance of a graphene-based capacitor with porous electrodes and dielectric filling. The application of an EM simulator to model a capacitor with electrodes with vertical graphene structures is inevitable due to the necessity of considering a large number of pores with different sizes and the complicated distribution of the electric field between the electrodes, although such models are still to be well developed. In this paper, we present a new solution for obtaining the capacitance using a special externally situated matrix of the electrode with a fixed number of elements. An innovative solution was designed to overcome the computational challenges of simulating large amounts of micro-structures. The electrodes were modeled as a solid graphene parallelepiped with cylindrical holes in its body. The effect of the parameters of holes (diameter, depth, distance) on the capacitance was investigated by the simulator, which takes into account the complicated distribution of the electric field arising from the shape of the electrodes.

Acknowledgements
This work is supported by the EPLUS National Science Program (NIS N2848) and the PEGASUS H2020 project (NIS N3257).

References

[1] Velasco A, Ryu Y K, Bosca A, Ladron-de-Guevara A, Hunt E, Zuo J, Pedros J, Calleab F and Martinez J 2021 *Sustainable Energy Fuels* **5** 1235–1254
[2] Zuo W, Li R, Zhou C, Li Y, Xia J and Liu J 2017 *Adv. Sci.* **4** 1600539
[3] El-Kady M F, Shao Y and Kaner R B 2016 *Nature reviews: Materials* **1** 1
[4] Gadipelli S and Gu Z X 2015 *Progress in Materials Science* **69** 1–60
[5] Wu T-H, Chang C-T, Wang C-C, Parwaiz S, Lai C-C, Chen Y-Z, Lu S-Y and Chueh Y-L 2018 *Nanoscale Research Letters* **13** 242
[6] Conway B E 1999 *Electrochemical supercapacitors: scientific fundamentals and technological applications* (New York: Plenum Press)
[7] Liu Y, Jiang S P and Shao Z 2020 *Materials Today Advances* **7** 100072
[8] Jiang Y and Liu J 2019 *Energy Environ. Mater.* **2** 30-37
[9] Yang C 2021 *Smart Grid and Renewable Energy* **12** 1-15
[10] Li Z, Xu K and Pan Y 2019 *Nanotechnol Rev.* **8** 35–49
[11] Dias A et al 2021 *Chemical Engineering Journal* (2021) (doi.org/10.1016/j.cej.2021.133153) 133153
[12] Zhao C and Zheng W 2015 *Frontiers in Energy Research* **3** 23
[13] Marinov S, Vachkov V, Djermanova N and Kiss'ovski Zh 2020 *J. Phys. Conf. Series* **1492** 012033
[14] Marinov S, Vachkov V and Kiss'ovski Zh 2020 *J. Phys. Conf. Series* **1492** 012032
[15] Ferrer P R, Mace A, Thomas S N and Jeon J-W 2017 *Nano Convergence* **4** 29
[16] Tatarova E et al 2017 *Sci. Rep.* **7** 10175

Model-based multicriterial optimization of electron beam melting and refining of copper

E Koleva[1,2,3], L Koleva[1], V Vassileva[2], K Vutova[2] and Ts Tsonevska[2]

[1]University of Chemical Technology and Metallurgy – Sofia, 8, Kliment Ohridski Blvd., Sofia 1756, Bulgaria
[2]Institute of Electronics, Bulgarian Academy of Sciences, 72, Tsarigradsko Chaussee, Sofia 1784, Bulgaria

e-mail: eligeorg@uctm.edu

Abstract. Experimental results on the quality of electron beam melting and refining of copper samples by implementing the “method of washers” are analyzed by using an empirical model estimation approach. The influence is studied of the process parameters – melting power and time of refining, on the residual content of impurities, the refining rate, the overall removal efficiency and the material losses. A graphical user interface is designed and developed for calculation, investigation and multicriterial optimization of the quality of the refined copper samples.

1. Introduction

The aim of a metallurgical refining process is removing the metallic impurities and reducing the non-metallic components present in the raw material. The actual surface-to-volume ratio of the liquid pool controls the limitations of the removal by: (i) evaporation from the liquid metal surface of high vapor pressure elements and (ii) mass-transport process within the molten metal. Other important factors influencing the refining process are the stirring rate of the melt, the temperature distribution in the treated samples, the form of the crystallization front, the time during which the refined metal is in a liquid state, the hard inclusion floatation, the removal of dissolute gases, the evaporation conditions and the evaporated material condensation.

Thanks to its taking place in vacuum and its excellent refining capacity and heat source flexibility, the electron beam melting (EBM) is efficiently used for production of ultra-pure refractory and reactive materials [1]. In [2], the thermodynamic conditions and the kinetic limitations on the refining processes flow and the possible refining mechanisms in electron beam refining of copper were analyzed.

The removal of oxygen is one of the main issues connected with the refining of copper; it is accomplished during the electron beam melting process by: (i) degassing of oxygen that is in the form of a solid solution; (ii) removing the bound oxygen via evaporation of metal oxides and volatile or unstable at melting temperatures oxides of metallic impurities with higher affinity to the oxygen; (iii) direct electron beam irradiation.

[3] To whom any correspondence should be addressed.

Content from this work may be used under the terms of the Creative Commons Attribution 3.0 licence. Any further distribution of this work must maintain attribution to the author(s) and the title of the work, journal citation and DOI.
Published under licence by IOP Publishing Ltd

In this paper, we report on an empirical model estimation studies of the influence of the process parameters – melting power and refining time – on the refined ingot quality characteristics, the content of impurities after the electron beam (EB) refining process, the refining rate, the overall removal efficiency and the material losses. The optimal technological conditions are found through implementing multicriterial parameter optimization approaches aiming at fulfilling the simultaneous requirements for minimal residual contents of the investigated impurities, maximal refining rates, overall removal efficiency and minimal material losses. The refining process mechanisms are discussed in the cases of different technological conditions during EBM of copper and oxygen content reduction. A graphical user interface was designed and developed in view of supporting further investigations, training young specialists and assisting the EBM operators in choosing appropriate processing parameter settings.

2. Experimental results

The experiments on e-beam melting and refining (EBMR) of cylindrical copper samples [2] (height of 50 mm, diameter of 60 mm) were conducted on the ELIT-60 60-kW e-beam melting furnace operating in the Physical Problems of the EB Technologies Laboratory of the Institute of Electronics, Bulgarian Academy of Sciences. The installation consists of a melting vacuum chamber, one electron gun, a vacuum system, and feeding and extraction systems [3]. The operation pressure in the chamber is 3 – 6×10^{-3} Pa. The copper material investigated contained 99.83 % Cu and inclusions (such as Sn, Sb, Pb, Bi, Ni, Ag, As, Zn and O_2). The initial material was processed at e-beam melting powers of 7.5 kW (T = 1400 K), 10 kW (T = 1500 K) and 15 kW (T = 1700 K) for different refining times: 3 min, 10 min, 15 min, 20 min and 30 min. The experiments were carried out using the so-called "method of washers" [4] and the chemical compositions of the metal samples before and after EBM were determined using emission spectral analysis [2]. Table 1 summarizes the experimental data for the concentrations of the impurities considered before EBM (C_0), and the minimal and maximal concentrations after EBM (C_{min} and C_{max}). As seen, the impurities controlled are refined to a different extent, or their content even increases due to the evaporation of the material.

Table 1. Impurities' concentrations.

Impurities	Sn	Sb	Pb	Bi	Ni	Ag	As	Zn	O_2
Before EBMR, C_0, ppm	27	42	63	1.6	55	8	16	7	1500
After EBMR, C_{min}, ppm	<1	4.4	4.7	1.2	40.5	6.9	7.2	5	10
After EBMR, C_{max}, ppm	10.7	26.6	36.1	2.3	49.0	8.2	11.7	95	300

The minimal concentration values of the investigated impurities were obtained for different EBM process parameters, namely, different refining times and different EB powers (temperatures). Consequently, a compromise should be found in terms of minimizing all residual impurities' concentrations simultaneously. Based on the experimental data, the percentage of the overall removal efficiency η and the material losses weight W_{loss} were also evaluated; they can be considered as additional criteria in the EBMR process optimization. The material losses were estimated by weighing the initial copper ingots and those obtained after EBMR. These losses are mainly due to evaporation, but also to splashes during the refining processes. The values of the evaluated overall removal efficiency [2] vary in the range of 74.2 % – 94.7 %, while those of the material losses weight, within the interval 0.14 g – 30.21 g.

3. Results and discussion

3.1. *Regression models*

Based on the experiments performed and the characterization data obtained, we applied a statistical approach to estimating the dependencies of the inclusion concentrations C_i in the copper samples after

EBM, the overall removal efficiency η and the weight material losses W_{loss}. The influence was investigated of the EB process parameters listed in table 2.

The estimated regression models are given in table 3, together with the model determination coefficients R^2 and the adjusted determination coefficients R^2_{adj}, which are measures for the model accuracy. They are estimated for the corresponding levels of the process parameters coded in the region [-1 – 1]. The linear models giving the relation between the dimensionless coded units (x_i) and the natural units (z_i) are also presented in table 2. The estimated regression models can be used for both the EB power (z_1), kW, and the refining temperature *T*, K, ($T = 1100 + 40z_1$).

Table 2. Experimental regions for EBM process parameters and coding linear models.

Parameter	$z_{i,min}$	$z_{i,max}$	$x_i = f(z_i)$	$z_i = f(x_i)$
EB Power – z_1, kW	7.5	15	$x_1 = (z_1 - 11.25)/3.75$	$z_1 = 3.75x_1 + 11.25$
Refining time – z_2, min	3	30	$x_2 = (z_2 - 16.5)/13.5$	$z_2 = 13.5x_2 + 16.5$

Table 3. Regression models.

Output	Regression model	R^2	R^2_{adj}
C_{Sn}, ppm	$y_1 = 3.1214 - 4.5675x_1 - 0.4290x_2 + 1.9500x_1^2 + 0.4882x_1x_2$	0.9994	0.9992
C_{Sb}, ppm	$y_2 = 14.7457 - 10.3066x_1 - 0.7970x_2 - 0.0300x_1^2 - 0.0189x_2^2 + 0.8004x_1x_2$	0.9999	0.9999
C_{Pb}, ppm	$y_3 = 19.0650 - 14.3499x_1 - 1.3504x_2 + 0.1349x_1^2 + 1.2019x_1x_2$	0.9999	0.9999
C_{Bi}, ppm	$y_4 = 1.7909 + 0.4017x_2 - 0.0401x_2^2 + 0.1519x_1x_2$	0.9908	0.9883
C_{Ni}, ppm	$y_5 = 43.7803 - 3.0469x_1 - 1.1960x_2 + 0.0300x_1^2 + 0.9468x_1x_2$	0.9999	0.9999
C_{Ag}, ppm	$y_6 = 7.6409 - 0.1443x_1 + 0.1100x_2 + 0.5026x_1x_2$	0.9938	0.9922
C_{As}, ppm	$y_7 = 9.3795 - 1.8850x_1 - 0.2433x_2 + 0.2041x_1x_2$	0.9967	0.9958
C_{Zn}, ppm	$y_8 = 9.1077 + 12.7899x_2 + 28.9880x_2^2 + 39.2024x_1x_2^2$	0.9166	0.8750
C_{O2}, ppm	$y_9 = 18.8320 - 70.6268x_2 + 135.7644x_2^2 - 47.3527x_1x_2^2$	0.8351	0.7902
η, %	$y_{10} = 92.7034 + 3.3257x_1 + 3.2007x_2 - 9.1443x_2^2$	0.8752	0.8412
W_{loss}, g	$y_{11} = 8.6312 + 7.6359x_1 + 7.3461x_2 + 6.5154x_1x_2$	0.9995	0.9994

Based on the results in table 3, it can be seen that the estimated regression models are good and can be used for EBM process investigation and multicriterial optimization. The estimated models should be used within the experimental regions for the process parameters – EB power and refining time (table 2).

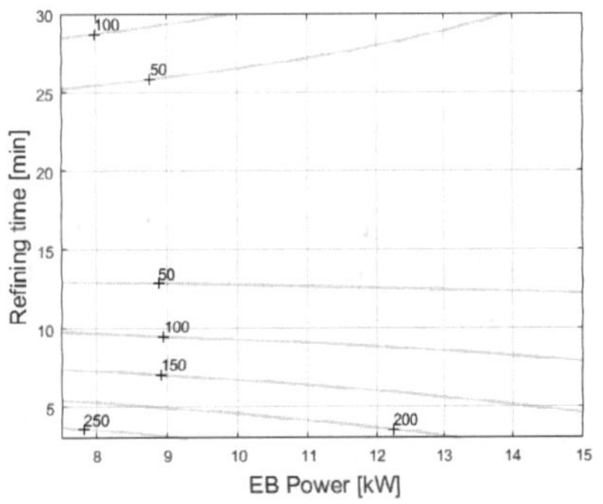

Figure 1. Contour plot of the O_2 concentration (y_9) depending on the EB power variation (z_1) and the refining time (z_2).

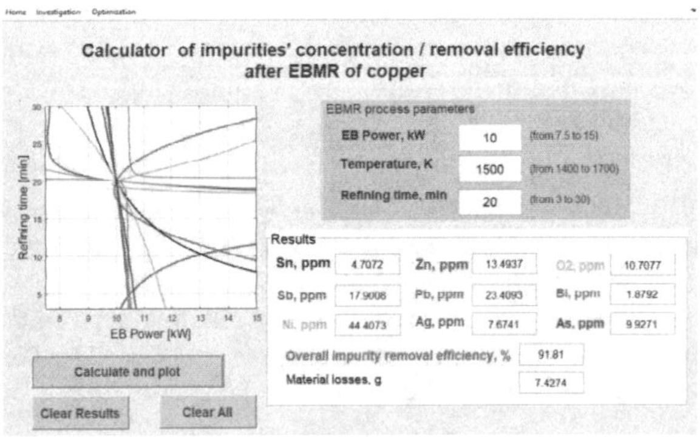

Figure 2. GUI module – Calculator.

Figure 1 is a contour plot of the residual O_2 concentration as a function of the EB power (z_1) and the refining time (z_2) using regression model y_9 (table 3). As seen, the refining time value is decisive for the residual O_2 concentration, while the influence of the EB power is weaker. It can also be seen that the optimal refining time values for minimizing the residual O_2 concentration are within the range of 17 – 22 min.

3.2. Graphical user interface development

The graphical user interface (GUI) based on the estimated regression models (table 3) was developed in Matlab environment (Guide tool). It consists of three modules for calculation, investigation and optimization.

3.2.1. Calculator GUI module (figure 2) provides the possibility of setting the process parameters values – EB power or refining temperature and refining time – and then estimate the values of the investigated impurities concentrations, as well as the values of the material losses and the overall impurity removal efficiency. The set regime point can be observed graphically as the intersection point of the corresponding contour lines of each of the EBMR quality characteristics presented as functions of the technological process parameters – the EB power and the refining time. Figure 2 presents the results in the case of an EBMR process at e-beam power of 10 kW (temperature 1500 K) and refining time of 20 min.

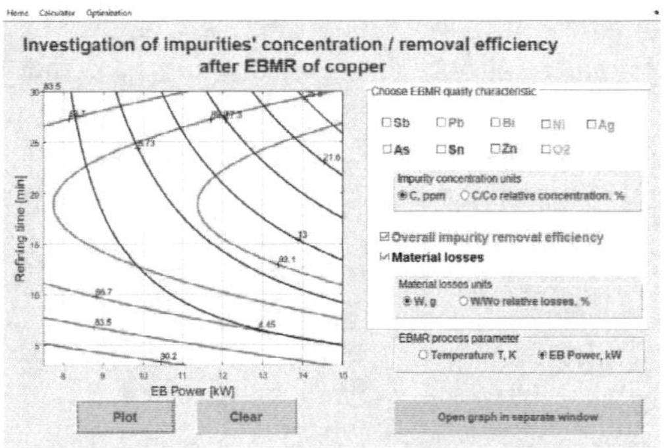

Figure 3. GUI module – Investigation.

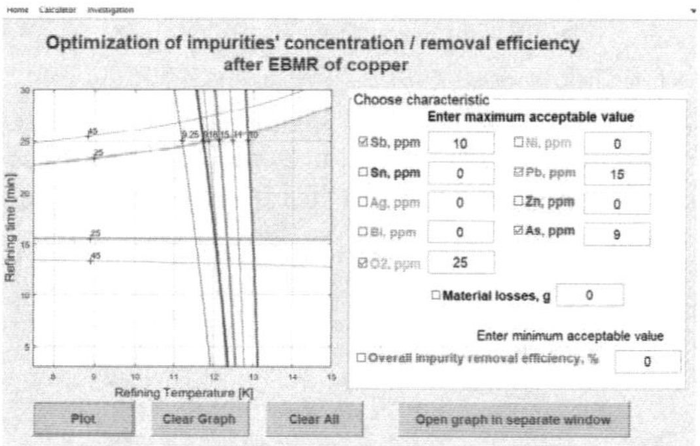

Figure 4. GUI module – Optimization.

3.2.2. Investigation GUI module (figure 3) provides a graphical presentation by contour plots of the selected quality characteristics as functions of the EBMR process parameters. Options are provided for choosing the impurities' concentrations units – ppm or relative residual concentration C/C_0, %, the material losses units – grams or relative losses W/W_0, %, as well as the EBMR process parameter – EB power or refining temperature. Figure 3 shows superimposed contour plots illustrating the variation of the material losses and the overall impurity removal efficiency. It is seen that the material losses increase with the increase of the EB power or/and the increase of the refining time. On the other hand, the overall impurity removal efficiency increases with the increase of the EB power. It reaches maximal values for a heating time in the range of 17 – 22 min; for longer refining times (more than 22 min), the values of the overall impurities' removal efficiency decrease.

3.2.3. Optimization GUI module (figure 4) provides a possibility to define optimal process parameter regions in view of fulfilling a certain set of producer requirements (constraints) to the quality characteristics considered. The multicriterial optimization approach used is graphical optimization. The optimal region is obtained by superimposing the one-sided constraints on the selected maximal acceptable concentrations for the impurities and the material losses with the minimal acceptable value for removal efficiency. Thus, the optimal region is obtained for selecting the EBMR process parameters. In figure 4, the following constraints are set: $C_{O2} \leq 25$ ppm, $C_{Sb} \leq 10$ ppm, $C_{Pb} \leq 15$ ppm and $C_{As} \leq$ ppm. The optimal process parameter region, where all constraints are fulfilled, is highlighted in yellow. The concentrations of other impurities, the material losses or the overall impurities' removal efficiency can be used as additional criteria to choose a specific optimal regime within the optimal region found.

The Investigation (figure 3) and Optimization (figure 4) modules provide greater convenience by allowing the contour plots to be viewed in full screen by pressing the "Open graph in separate window" button. The transition between the GUI modules is possible from the GUI menu or from the initial window. The GUI can be easily included as a module of a larger user software that contains results from other EBM experiments on copper or other pure materials.

3.3. Other multicriterial optimization approaches

Other multicriterial optimization methods can also be implemented – desirability function, loss function, distances function, etc. These methodologies define one overall criterion (function) that is minimized or maximized over the experimental region.

The maximal value of the impurities' removal efficiency obtained by implementing the regression equation in table 3 (y_{10}) and a generic optimization algorithm is 96.31 % for EB power $z_1 = 15$ kW ($T = 1700$ K) and refining time $z_2 = 18.86$ min. For these process parameters, the residual O_2 concentration is estimated to be $y_9 = 9$ ppm. For comparison, in [2], based on the experimental data presented, the highest purity of Cu (99.991 %) and an oxygen content of 20 ppm were reached at 1700 K for a melting time of 10 min, with the removal efficiency being 94.7 %.

Pareto-optimal solutions were obtained by implementation of a generic optimization algorithm [5] for simultaneously minimizing the residual impurities' concentration – C_{Sb}, C_{Pb}, C_{As} and C_{O2}, and the material weight losses W_{loss}, while maximizing impurities' removal efficiency η. Some of the results obtained are presented in table 4. To find a compromise, an optimization strategy should be implemented involving estimation of an overall function (e.g., Loss function), expert opinions or additional criteria.

Table 4. Pareto-optimal solutions.

	z_1, kW	z_2, min	C_{Sb}, ppm	C_{Pb}, ppm	C_{As}, ppm	C_{O2}, ppm	η, %	W_{loss}, g
1	15.00	25.88	4.4	4,7	7.5	12.5	93.8	25.9
2	8.85	13.06	21.7	28.9	10.7	47.6	89.2	2.9
3	15.00	22.80	4.4	4.8	7.5	5.1	95.5	22.7
4	10.60	3.00	17.4	23.1	10	233.4	80	1.09
5	10.80	8.49	16.5	21.7	9.8	110.5	87.2	3.82

4. Conclusions

Applying mathematical modeling to the e-beam melting technology is an important part of the refining processes aimed at obtaining high-purity metals, as well as an effective tool for increasing the efficiency and optimizing the technological process.

In this work, regression models are estimated and optimal EBMR technological regimes are obtained by implementing multicriterial parameter optimization approaches aiming to fulfill simultaneously the requirements for minimal residual contents of the investigated impurities, maximal refining rates and overall removal efficiency, and minimal material losses. A graphical user interface was designed and developed in view of supporting further investigations, training young specialists and assisting the EBM operators in choosing appropriate processing parameter settings. The graphical user interface allows one to calculate the values of the quality characteristics, investigate the dependencies of the quality characteristics on the EBMR process parameters thus achieving multicriterial optimization of the refined copper material quality.

Acknowledgements

The work was supported in part by the Bulgarian National Science Fund under projects KP-06-N27/18 and DN17/9.

References

[1] Mladenov G, Koleva E, Vutova K and Vassileva V 2011 *in book Practical Aspects and Applications of Electron Beam Irradiation* ed M Nemtanu and M Brasoveanu (Transworld Research Network India) chapter 3 pp 43–93
[2] Vassileva V and Vutova K 2020 *IOP Publ. J. of Phys.: Conf. Series* **492** 012014
[3] Vutova K, Vassileva V, Koleva E, Munirathnam N, Amalnerkar D and Tanaka T 2016 *Metals* **6** (11) 287
[4] Vasileva V, Vutova K, Georgiev L, Nikolov T and Mladenov G 1992 *EBMR-State of the Art*, ed R Bakish (Bakish Materials Corporation Englewood NJ) p 233–242
[5] QstatLab http://www.qstatlab.co.uk/eng/index.html

Multicriterial optimization strategies for electron beam welding processes

E Koleva[1,2,4], L Koleva[1], Dm Trushnikov[3], G Kolev[1] and Z Petrova[1]

[1]University of Chemical Technology and Metallurgy – Sofia, 8 Kliment Ohridski Blvd., Sofia 1756, Bulgaria
[2]Institute of electronics, Bulgarian Academy of Sciences, 72 Tsarigradsko Chaussee, Sofia 1784, Bulgaria
[3]Perm National Research Polytechnic University, Perm, Russian Federation

E-mail: eligeorg@uctm.edu

Abstract. The electron beam welding process is analyzed of high-strength steel type 15Cr5Mo samples with deflection oscillations along the zone of interaction. The geometries of the molten and the heat affected zones are studied in order to investigate the influence of the deflection oscillations parameters and to improve the seam quality. A robust engineering approach is implemented for the case of production conditions considering the errors in the process parameter settings. The problem of quality improvement through fulfilling pre-set technological and quality requirements is solved by scalarization of the vector criteria by implementing a reference point strategy. Other multicriterial optimization strategies are also proposed.

1. Introduction

Deflection oscillations during the process of electron beam welding (EBW) are used to avoid spiking in the root of the weld, as well as the appearance of globules, which tend to form a ridge along the top weld surface at high welding speeds; to standardize the penetration depth; to reduce a significant fraction of the weld porosity; and to create a more appropriate weld structure and mechanical properties of the electron beam welded joints [1]. The beam deflection oscillations change the electron bombardment distribution on the keyhole walls (typically formed during the EBW process), as well as the material and thermal flows in the molten weld pool; as a consequence, the keyhole shape is strongly affected and the weld profile undergoes limited changes [1].

The geometry of the molten and the heat affected zones of special steels, like the high-strength steel (type 15Cr5Mo) considered in this paper, as well as of other metals and alloys obtained after EBW processing, depend strongly on the process parameters: welding speed, electron beam (EB) power, distances from the magnetic lens of the EB gun to the sample surface and to the EB focusing plane [2], the EB emittance and brightness [3], as well as on the way of implementing the EB deflection oscillations – the shape and direction of the oscillations, and their amplitudes and frequencies [1]. The shape and type of the EB deflection oscillations are determined by the coil current deflection law (linear, sinusoidal, etc.). The focus position is controlled by varying the focusing current I_f; the optimal focus position in EBW in terms of maximizing the weld depth (for stainless

[4] To whom any correspondence should be addressed.

Content from this work may be used under the terms of the Creative Commons Attribution 3.0 licence. Any further distribution of this work must maintain attribution to the author(s) and the title of the work, journal citation and DOI.
Published under licence by IOP Publishing Ltd

Journal of Physics: Conference Series **2240** (2022) 012038 doi:10.1088/1742-6596/2240/1/012038

steel [2]) is at a certain distance below the sample surface. The model-based robust approach for improving the quality of the process [4, 5] can be successfully applied to different industrial processes. For each of the quality performance characteristics and using their regression models, two other models are estimated – for their mean values and variances. The quality improvement is performed using a given overall criterion, or simply by minimizing the performance characteristics variances while keeping their mean values close to their target values.

In this paper, we report on studies on the geometries of the molten and the heat affected zones in order to investigate the influence of the deflection oscillations parameters and to improve the seam quality. Robust engineering approach is implemented for the case of production conditions considering the errors in the process parameter settings. The problem of improving the quality by fulfilling pre-set technological and quality requirements is solved by scalarization of the vector criteria by implementing a reference point strategy. Other multicriterial optimization strategies are also proposed.

2. Experimental results

Special purpose steel alloy (type 15Cr5Mo) seamless pipes find wide applications in heat-resistant steel boilers, gas and oil pipelines, in machinery industry, to name a but few areas. Heat exchange equipment and, in particular, shell-and-tube heat exchangers are currently widely used in the petrochemical and related industries [6].

We analyzed the experimental results (30 experimental runs [1]) of electron beam welding of high-strength steel type 15Cr5Mo samples with deflection oscillations along the zone of interaction at a beam current of 45 mA, an electron energy of 60 keV, a welding speed of 0.5 cm/s and a distance from the magnetic lens of the electron gun to the sample surface of 10 cm. The influence was investigated of varying the following process parameters: focusing current from 820 mA to 850 mA (p_1'); EB deflection oscillations frequencies (p_2') from 90 Hz to 1400 Hz; and amplitudes (p_3') on four levels – 0.7 mm, 1 mm, 1.5 mm and 2.7 mm. The geometrical characteristics considered of the weld cross-sections and the heat affected zones (figure 1) were: the depth of the weld cross-section y_1, mm; the weld width at half the weld depth y_2, mm; the top weld width y_3, mm; the weld width together with the heat affected zone y_4, mm.

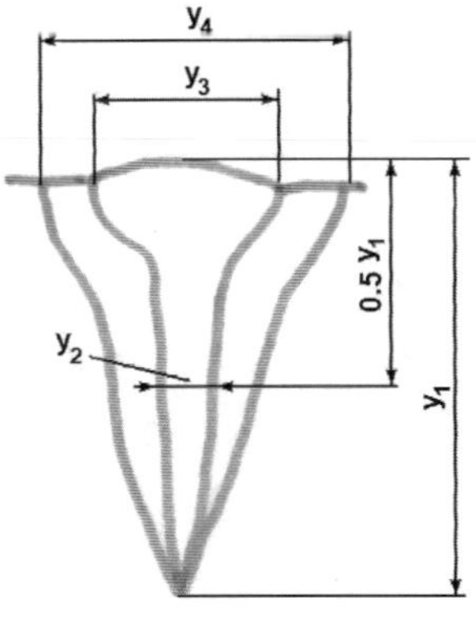

Figure 1. Geometrical characteristics of weld cross-section.

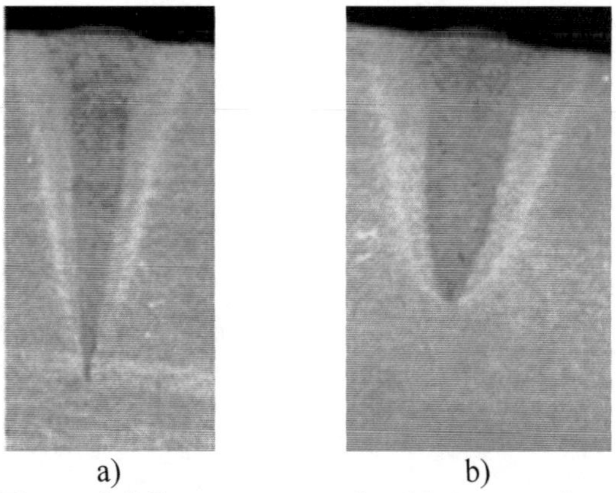

a) b)

Figure 2. Microstructures of weld cross-sections.

Figure 2 presents the microstructures of weld cross-sections obtained at: a) focusing current of 843 mA; deflection oscillations amplitude 1.5 mm and frequency 725 Hz; weld depth $y_1 = 12.2$ mm; and widths $y_2 = 1.42$ mm, $y_3 = 3.66$ mm, $y_4 = 6.41$ mm; b) focusing current 828 mA; deflection oscillations amplitude 2.7 mm and frequency 1120 Hz; weld depth $y_1 = 8.6$ mm; and widths $y_2 = 2.22$ mm, $y_3 = 4.16$ mm, $y_4 = 6.94$ mm.

3. Robust engineering approach

The regression models for the geometrical characteristics (figure 1) were estimated as functions of the process parameters (in coded values) – the focusing current (p_1) and the frequency (p_2) and amplitude (p_3) of the e-beam deflection oscillations. They are presented in table 1 for the process parameters levels coded in the region [-1 – 1], together with the corresponding multiple correlation coefficients (measures for the models accuracy).

The models are considered good enough to be implemented for investigation, estimation of robust models and optimization. The robust engineering approach is implemented for the case of production conditions considering the errors in setting the process parameter. Two models are estimated for each of the output weld quality characteristics – for their mean values $\tilde{y}_i$ and their variances σ_i^2 [3, 4], using the regression models in table 1. The variance of the process parameters under production conditions are estimated by their tolerance limits given in table 2. Contour plots for the mean values and the corresponding variances for the weld depth and the top weld are presented in figure 3. They have the biggest variances of the set conditions. The process parameters regions where maximal weld depth and minimal weld width are obtained can also be defined for the cases considered of production conditions.

Table 1. Regression models for the geometrical parameters of the molten and the heat affected zones.

Output	Regression model	R
y_1	$10.64 + 2.47p_1 - 1.92p_1^2 + 0.89p_3^2 - 1.36p_1^3 - 1.49p_2^3 + 1.97p_1p_2 - 0.93p_2p_3 - 0.59p_1p_3 - 1.56p_1^2p_2 + 2.16p_1p_2^2 + 1.61p_1p_2p_3 - 0.94p_2^2p_3$	0.95
y_2	$1.85 - 0.75p_1 + 0.31p_2^2 + 0.61p_1^3 - 0.47p_2^3 + 1.02p_1^2p_2 + 0.27p_2p_3$	0.85
y_3	$3.48 - 2.47p_3 + 1.81p_1^2 + 0.76p_3^2 + 1.78p_3^3 + 0.67p_1p_2p_3 - 1.10p_1p_3^2 + 2.46p_1^2p_3 - 0.70p_1p_3$	0.79
y_4	$5.10 + 0.49p_2 - 4.44p_3 + 1.75p_1^2 + 0.48p_2^2 + 1.17p_3^2 + 4.66p_3^3 - 0.66p_1p_3 - 0.76p_2p_3^2 + 0.85p_1^2p_2$	0.88

Table 2. Tolerance limits and coded variances of the process parameters in production conditions.

Parameter	$p'_{i,min}$	$p'_{i,max}$	Tolerance limits	Coded variance
p_1', mA	820	850	$p_1' \pm 3\%\ p_1'$	$\sigma_1^2 = 0.30988 + 0.01113p_1 + 0.0001p_1^2$
p_2', Hz	90	1400	$p_2' \pm 3\%\ p_2'$	$\sigma_2^2 = 0.00013 + 0.00023p_2 + 0.0001p_2^2$
p_3', mm	0.7	2.7	$p_3' \pm 0.1$	$\sigma_3^2 = 0.0011$

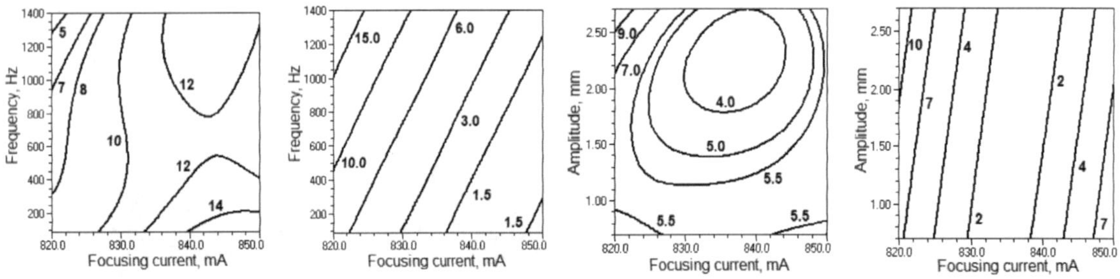

a) $\tilde{y}_1(p_1', p_2')$ for p_3' = 0.7 mm. b) $\sigma_1^2(p_1', p_2')$ for p_3' = 0.7 mm. c) $\tilde{y}_3(p_1', p_3')$ for p_2' = 398 Hz. d) $\sigma_3^2(p_1', p_3')$ for p_2' = 398 Hz.

Figure 3. Contour plots for the mean values $\tilde{y}_i$ and the corresponding variances σ_i^2.

4. EBW multicriterial optimization strategies

The defined multicriterial optimization strategies for the quality improvement of the EBW process are model-based and refer to production conditions in the presence of errors in the process parameter settings.

4.1. Robust optimization

The estimated models for the mean values and the variances of the geometry characteristics can be used for multicriterial optimization by implementing the overall robust desirability function approach [7]. The graphical multicriterial optimization is applicable when acceptable constraints to the characteristics are set and the optimal regions for the process parameters are obtained by superimposing the contour plots of different characteristics [7]. The variances minimization while keeping their mean values close to their target values aims at improving the variations and the repeatability of the quality (geometry) characteristics, thus improving the welded joints quality.

The following optimization task is set: find the minimal variance of the weld depth $\sigma_1^2(\boldsymbol{p}) \rightarrow \min$, while the technological requirements for the mean depth is $\tilde{y}_1(\boldsymbol{p}) \geq 12$ mm and for the weld top width is $\tilde{y}_3(\boldsymbol{p}) \leq 4$ mm. A generic algorithm [8] is applied to find the optimal technological parameters: focusing current $p_{1,opt'}{}' = 837.9$ mA, deflection oscillation frequency $p_{2,opt'}{}' = 226.7$ Hz and amplitude $p_{3,opt'}{}' = 2.7$ mm. The optimal values for the quality characteristics are: - $\sigma_{1,min}^2(\boldsymbol{p}) = 1.09$ mm^2, $\tilde{y}_1(\boldsymbol{p}) = 12$ mm; $\tilde{y}_3(\boldsymbol{p}) = 4$ mm.

4.2. Optimization by additional criteria

The following optimization criteria f are proposed to obtain an optimal geometry of the welded seams: the area of the heat affected zone (HAZ) d_1, mm^2, being the difference between the whole cross-section area S of the weld and HAZ minus the area of the weld S_w; the difference between the weld top width (y_3) and the width at the half weld depth (y_2) d_2, mm; the ratio of the weld depth to the weld top width (y_1/y_3) d_3; the difference between the width of the weld together with the HAZ and the top weld width ($y_4 - y_3$) d_4. These criteria are related to minimizing the HAZ, obtaining welds with less widening at the upper part of the weld, obtaining deep and narrow welds, as well as reducing the width of the HAZ on both sides of the molten zone at the top of the weld. Regression equation for the defined criteria is estimated as functions of the process parameters considered (in coded values). They are presented in table 3, together with the corresponding multiple correlation coefficients.

Table 3. Regression models for the formulated geometry optimization criteria.

Output	Regression model	*R*
d_1	$15.20 - 6.95p_1^4 - 14.06p_1^2p_2 + 16.75p_2p_3 + 10.09p_1p_2p_3 + 2.74p_1p_3^2 + 8.52p_1^2p_3^2 - 21.04p_2p_3^3$	0.85
d_2	$2.11 + 0.47p_1 + 2.02p_1^2 - 0.58p_3^3 - 0.73p_1p_3 + 1.88p_1^2p_3 + 0.50p_1p_2p_3 - 0.82p_1p_3^3$	0.84
d_3	$2.63 + 1.33p_1 + 0.25p_3 - 0.94p_1^2 - 1.07p_1^3 - 0.43p_2^3 - 0.49p_1^2p_3$	0.84
d_4	$2.26 + 0.27p_2^3 + 0.91p_3^3 - 1.79p_1^2p_3 - 0.81p_2^2p_3 - 0.59p_1p_2p_3 + 0.89p_1p_3^2$	0.75

Table 4. Compromise Pareto-optimal solution selection by additional criteria.

№	p_1' mA	p_2' Hz	p_3' mm	$\tilde{y}_1$ mm	σ_1^2 mm^2	$\tilde{y}_2$ mm	σ_2^2 mm^2	$\tilde{y}_3$ mm	σ_3^2 mm^2	$\tilde{y}_4$ mm	σ_4^2 mm^2	d_1 mm^2	d_2 mm	d_3 -	d_4 mm
1	848.1	1393.6	0.7	12.75	2.29	2.26	0.62	4.29	11.36	9.07	1.99	**5.03**	2.69	2.08	5.05
2	840.9	90.3	2.7	12.58	1.39	2.26	0.29	3.99	1.52	7.39	0.35	19.89	**1.51**	3.54	2.39
3	840.9	90.3	2.7	12.58	1.39	2.26	0.29	3.99	1.52	7.40	0.35	19.89	1.51	**3.54**	2.39
4	833.2	810.6	0.7	10.75	5.13	2.24	0.61	5.75	1.47	6.25	0.59	15.52	2.65	2.21	**1.27**

The multicriterial optimization task for the characteristics mean values $\tilde{y}_i$ and their variances σ_i^2 was set as: $\tilde{y}_1(\boldsymbol{p}) \rightarrow \max$; $\tilde{y}_2(\boldsymbol{p}) \rightarrow \min$, $\tilde{y}_3(\boldsymbol{p}) \rightarrow \min$, $\tilde{y}_4(\boldsymbol{p}) \rightarrow \min$, $\sigma_1^2(\boldsymbol{p}) \rightarrow \min$, $\sigma_2^2(\boldsymbol{p}) \rightarrow \min$, $\sigma_3^2(\boldsymbol{p}) \rightarrow \min$ and $\sigma_4^2(\boldsymbol{p}) \rightarrow \min$. In order to fulfill all these requirements simultaneously, compromise solutions were found by implementing a generic algorithm. Fifty Pareto-optimal solutions were obtained [8]. All these solutions are subject to the property that if two random solutions are compared at least one of the quality characteristics will have a worse value than for the other solution. Therefore, selecting among the Pareto-optimal solutions that fulfil each of the requirements to a certain extent could be a difficult task. The approach proposed here is based on the use of the formulated additional geometry optimization criteria d_i. Their values are calculated for each Pareto-optimal solution and the solutions that are selected and presented in table 4 are those which individually minimize the criteria d_1, d_2 or d_4 or maximize d_3.

4.3. Optimization under uncertainty with discreet number of variants

The most widely used strategies for multicriteria optimization are the strategies that apply a reference point approach [8]. These methods are part of the wide class of scalarizing methods. Each characteristic is given a referenced value (referenced point). The best values (minimal or maximal) for each characteristic from all Pareto-optimal solutions are assigned to the referenced values q_j^*, (j = 1, 2, …, m). Thus, "optimistic" strategy should be applied, based on the overall Loss function [9]:

$$F_L^o(\boldsymbol{p}) = \frac{1}{m}\sum_{j=1}^{m}\left[\Delta q_j(\boldsymbol{p})\right]^2 = \frac{1}{m}\sum_{j=1}^{m}\left(\frac{q_j^*-q_j(\boldsymbol{p})}{q_{j,max}-q_{j,min}}\right)^2 \to \min_{p}.$$

where $q_j(\boldsymbol{p})$ is uncompromised solution for the j-th quality characteristic within the experimental region of the process parameters (p_i), $q_{j,max}$ and $q_{j,min}$ are accepted as the maximal and minimal values of each characteristic. They were determined from the Pareto-optimal solutions and presented in table 5. The three best optimal solutions (process parameter values p_i, mean values $\tilde{y}_i$ and the variances σ_i^2 of the geometrical characteristics and the additional criteria d_i that minimize the overall Loss function $F_L^o(\boldsymbol{p})$ are presented in table 6.

Table 5. Goals, reference values and the maximal and minimal characteristic values $q_j(\boldsymbol{p})$.

Criterion	$\tilde{y}_1$ mm	σ_1^2 mm^2	$\tilde{y}_2$ mm	σ_2^2 mm^2	$\tilde{y}_3$ mm	σ_3^2 mm^2	$\tilde{y}_4$ mm	σ_4^2 mm^2	d_1 mm^2	d_2 mm	d_3 -	d_4 mm
Goal	max ↑	min ↓	min ↓	min ↓	min ↓	min ↓	min ↓	min ↓	min ↓	min ↓	max ↑	min ↓
Ref.	15.0	1.11	1.4	0.30	3.3	1.43	4.0	0.32	5.01	1.4	3.5	1.01
$q_{j,max}$	15.0	8.13	3.3	1.95	7.6	12.44	10.0	2.29	40.17	4.7	3.5	4.54
$q_{j,min}$	10.2	1.11	1.4	0.30	3.3	1.43	4.0	0.32	5.01	1.4	1.6	1.01

Table 6. Optimal solution by overall Loss function.

	F_L^o	p_1'	p_2'	p_3'	$\tilde{y}_1$	σ_1^2	$\tilde{y}_2$	σ_2^2	$\tilde{y}_3$	σ_3^2	$\tilde{y}_4$	σ_4^2	d_1	d_2	d_3	d_4
1	0.12	838.2	95.6	2.34	12.2	1.13	2.6	0.33	3.3	1.43	4.5	0.32	9.72	1.97	3.42	1.8
2	0.13	839.2	163.9	2.53	12.0	1.15	2.3	0.31	3.5	1.44	5.5	0.33	13.31	1.77	3.38	2.2
3	0.15	837.8	789.0	2.17	10.7	2.04	2.1	0.34	3.4	1.55	4.1	0.32	15.8	2.14	2.94	2.4

5. Conclusions

Experimental results of EBW of high-strength steel type 15Cr5Mo samples with deflection oscillations along the zone of interaction are analyzed by implementing a model-based robust approach. The geometry characteristics of the molten and the heat affected zones are studied by estimating two models – for the mean value and the variance of each characteristic. Multicriterial optimization strategies are proposed and implemented aiming at improving the quality of the joints produced by EBW by improving the performance of the quality characteristics, increasing the repeatability of their geometry characteristics while simultaneously fulfilling additional technological requirements.

Acknowledgements

The research was supported by the Bulgarian National Science Fund under projects KP-06-N27/18 and KP-06-N27-8/2018.

References

[1] Trushnikov D, Koleva E, Mladenov G and Belenkiy V 2013 *J. of Mat. Process. Technol.* **213** 1623–1634

[2] Koleva E 2001 *Vacuum* **62** 151–157

[3] Koleva E, Koleva L, Todorov D and Mladenov G 2016 *IOP J. of Phys.: Conf. Ser.* **700**(1) 012013
[4] Koleva E and Vuchkov I 2005 *Vacuum* **77** 423–428
[5] Vuchkov I and Boyadjieva L 2001 *Quality Improvement with Design of Experiment* ed A Keller (Kluwer Academic Publishers: The Netherlands)
[6] Rizvanov R, Mulikov D, Karetnikov D, Fairushin A and Tokarev A 2018 *IOP Conf. Ser.: Mat. Sci. and Eng.* **317** 012077
[7] Koleva E, Koleva L, Nemţanu M and Braşoveanu M 2017 *MTM* **12** 575–578
[8] *QstatLab* http://www.qstatlab.co.uk/eng/index.html
[9] Stoyanov S 2009 *Bioautomation* **13**(2) 1–18

Journal of Physics: Conference Series **2240** (2022) 012039 doi:10.1088/1742-6596/2240/1/012039

Accuracy comparison of intraoral *versus* laboratory scanners used in the contemporary dental practice

H Galeva[1,3], T Uzunov[1], Y Sofronov[2] and G Todorov[2]

[1]Faculty of Dental Medicine, Medical University of Sofia, Sofia, Bulgaria
[2]Faculty of Industrial Technologies, Technical University of Sofia, Sofia, Bulgaria

E-mail: dr.hrgaleva@gmail.com

Abstract. The aim of this *in vitro* study was to compare the quality of the data obtained by intraoral and by laboratory 3D scanners. An artificial jaw with an acrylic tooth prepared for a full metal ceramic crown was used as a master model of the prosthetic field. A reference model was created by scanning this field by a coordinate measuring machine (CCM). The acrylic master model was scanned by four scanners using three different methods and six digital models were obtained. They were divided in three groups – direct intraoral scanning (DIS), laboratory stone scanning (LSS) and laboratory conventional impression scanning (LCIS). Each scan was saved in the STL file format. Using a computer program, each of the six digital models were compared with the reference model created by the CCM and the results were displayed as color maps. The minimal acceptable deviation was defined as the deviation from the digital standard within the ± 0.02 mm limits. Thus, the LSS group exhibited an average of 13.5% points within these limits, followed by the LCIS group with 17.25%. The result for the DIS group was 44.75%.

1. Introduction

Nowadays, the digital technologies are part of everyone's life and are used in every field. The dental medicine is no exception. The CAD/CAM technology becomes more and more popular in its daily practice. This is why some more advanced dental schools even include it as a subject in their curricula [1].

This technology is widely used in the prosthetic dentistry and the orthodontics. The first step for a successful outcome is taking an accurate impression. In addition to conventional impressions, various digital ones already exist – optical, contact, and ultrasound scanners can be used [2]. Numerous studies have examined and compared their accuracy [3, 4, 5].

The classical physical impressions are gradually being replaced by more modern digital ones. Optical digital impressions can be obtained by direct intraoral scanning or indirect laboratory scanning. There are two types of laboratory scanning – impression and stone model scanning. However, the laboratory scanners' accuracy has already been proved [6]. Many authors have compared the intraoral scanners and the conventional impression techniques. According to Ferrini et al., the intraoral scanners could be an alternative to the conventional impressions. They also compared different intraoral scanners and recommended the PlanScan as the most accurate one [7]. Schlenz et al. have focused on the interdental areas and have also found that intraoral impressions

[3] To whom any correspondence should be addressed.

Content from this work may be used under the terms of the Creative Commons Attribution 3.0 licence. Any further distribution of this work must maintain attribution to the author(s) and the title of the work, journal citation and DOI.
Published under licence by IOP Publishing Ltd

yield better results than the analog ones at these specific areas [8]. In a similar research, Sharma et al. have compared the conventional impression with different digital ones. Their results gave preference to the digital techniques [9].

Generally, many studies have highlighted the intraoral scanners advantages, such as reduced patients' discomfort, real-time visualization, easy replication, rapid communication etc. [10, 11, 12]. Other authors have not only compared the intraoral scanning with conventional techniques, but also compared the performance of the different types of laboratory scanning. According to Shimizu et al. [13], there is significant difference between the intraoral and extraoral scanners in what concerns the scanning trueness. Some researchers have evaluated the importance of surface topography in relation to the scanning accuracy. Lee et al. recommended the extraoral scanners for full-arch scanning [14].

In contrast, Tomita et al. have found that the intraoral scans could be even more accurate compared to the laboratory scans. [15] The results achieved by Sason et al. and Malaguti et al. confirmed this claim [16, 17].

2. Materials and methods

An experimental prosthetic field was created by preparing an artificial acrylic tooth for a full metal ceramic crown according to the standard guidelines using NTI-Kahla GmbH (Germany) rotary tools and continuous water-cooling. The artificial jaw was fixed on the coordinate measuring machine's (Optiv Performance 664, Germany) working table. A touch probe – a spherical ruby ball with 1-mm diameter, collected the data for the prepared tooth shape, which was used to create a digital model. This model served as a digital reference model and was used to compare the performance of the digital models produced by the other scanners (intraoral and laboratory scanners).

Two different intraoral scanners making use of the scan principle of active stereo imaging (3D iOS and Zhf) were employed to scan the prosthetic field. The scanning was carried out according to the manufacturer's recommendations – starting from the occlusal surface for the full dental arch, the process continued on the lingual surfaces and finished on the vestibular surface.

Using a standard impression trays and A-silicon (Zhermack, Germany, putty and light body), two conventional impressions were taken of the same prosthetic field by means of the two-layers two-stages technique (the most commonly used one in the contemporary dental practice). The first impression was used to prepare a stone model with removable dies.

The conventional intraoral impression and the stone model were scanned (figure 1) by two different laboratory scanners with structured blue light (Ceramill map 400 and Zhf).

Figure 1. Stone model scanning.

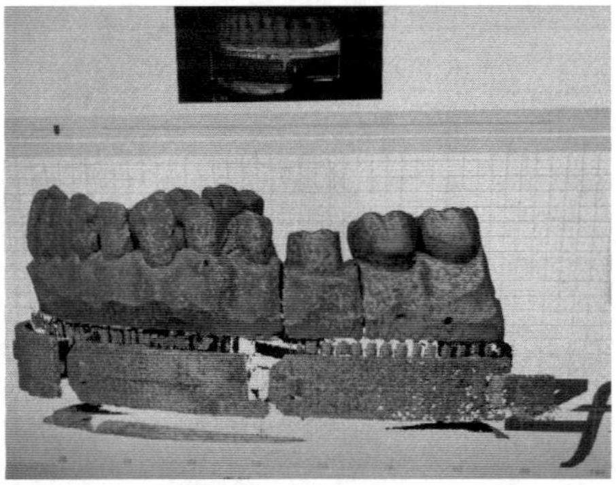

Figure 2. Digital image obtained by laboratory scanner.

Seven prosthetic's field digital images were obtained – one by the coordinate measuring machine (digital standard), and six by the optical scanners. The data was divided in three groups according to the scanning method: direct intraoral scanning (DIS), laboratory conventional impression scanning

(LCIS), and laboratory stone scanning (LSS). All devices except the coordinate measuring machine use structured white light and triangulation as a method of producing a digital image (figure 2).

The digital images from each scanner were compared with the digital reference standard provided by the CMM. Using the Reshaper 3D software, every scan was superimposed on the reference model (figure 3) and color maps were created to display the deviations from the standard (figure 4).

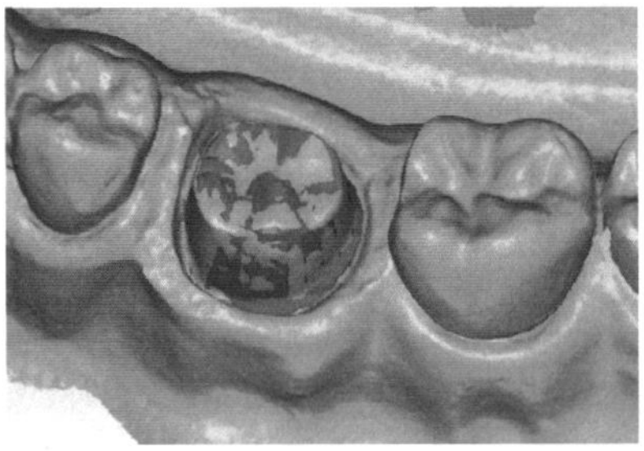

Figure 3. Superimposed mesh obtained by two scanners.

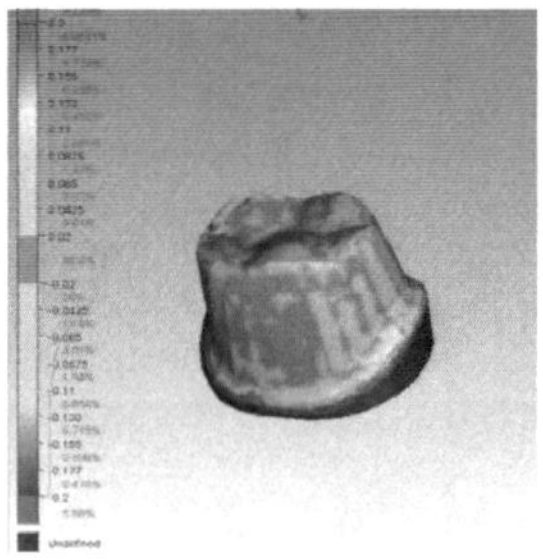

Figure 4. Color map.

3. Results

The Bestfit algorithm was chosen to compare the digital images, which is achieved by aligning the models and seeking the least-squares deviations between the triangles that construct the studied object's images. The color maps exhibiting the differences consist of 19 areas and the differences (in percent) are shown on a color-coded scale on the side. Results within ± 0.02 mm were considered as optimal. For the LSS group, 13.5% of the data fell within these limits, 17% for the LCIS group, and 45% for the DIS group (figure 5).

The results are summarized in table 1 and the diagram below.

Table 1. Deviation from the digital reference standard, in mm.

Deviation from the digital standard in mm / Type of scanning	Group LSS (Ceramill map 400)	Group LSS (Zhf)	Group LCIS (Ceramill map 400)	Group LCIS (Zhf)	Group DIS (3D Ios)	Group DIS (Zhf)
Up to +0.2	2.15	1.52	2.06	1.71	0.22	1.26
+0.2 to +0.17	1.58	1.34	0.59	0.35	0.08	0.52
+0.17 to +0.15	2.19	2.31	0.79	0.72	0.11	0.66
+0.15 to +0.13	3.81	3.25	1.53	0.91	0.25	0.63
+0.13 to +0.11	4.63	4.24	3.21	1.78	0.43	0.71
+0.11 to +0.08	6.43	4.81	4.6	4.3	0.88	1.04
+0.08 to +0.06	5.41	8.21	7.01	6.74	1.47	4.39
+0.06 to +0.04	5.46	7.87	8.08	8.24	2.37	9.97
+0.04 to +0.02	7.5	7.94	8.88	8.19	9.51	18.4
+0.02 to -0.02	14.2	12.6	18.6	15.9	38.6	50.9
-0.02 to -0.04	9.48	9.25	10.2	11	20	5.47
-0.04 to -0.06	9.38	8.52	9.5	12.3	11.5	2.4
-0.06 to -0.08	8.88	9.73	8.42	9.18	4.01	1.07
-0.08 to -0.11	7.45	6.44	6.19	6.32	1.94	0.88
-0.11 to -0.13	4.81	4.92	4.08	5.61	0.85	0.71
-0.13 to -0.15	3.8	3.6	3.44	4.21	0.71	0.41
-0.15 to -0.17	1.27	1.61	2.3	1.8	0.69	0.27
-0.17 to -0.2	0.67	0.89	0.48	0.53	0.47	0.22
Up to -0.2	0.89	0.97	0.02	0.22	5.88	0.07

4. Discussion

Restoring the harmony in the patient's occlusion necessitates in the first place that the data on the prosthetic field should be transferred correctly to the technician's laboratory. This why the impression's accuracy is of paramount importance. Also, increasing the number of steps during the information transfer increases the risk of errors and deviations, which is confirmed by the results obtained by the present *in vitro* study. However, it was conducted under perfect conditions that cannot be achieved in a clinical environment – certain factors, such as saliva, blood, patient's movements etc. are missing.

Further, it is difficult to eliminate completely the volume changes of the dental materials used – impression materials and stone. This may be an additional reason why the intraoral scanning impressions are more accurate than the laboratory scans, the latter requiring a physical model of the prosthetic field.

However, all type of digital impressions obtained during this study are clinically acceptable.

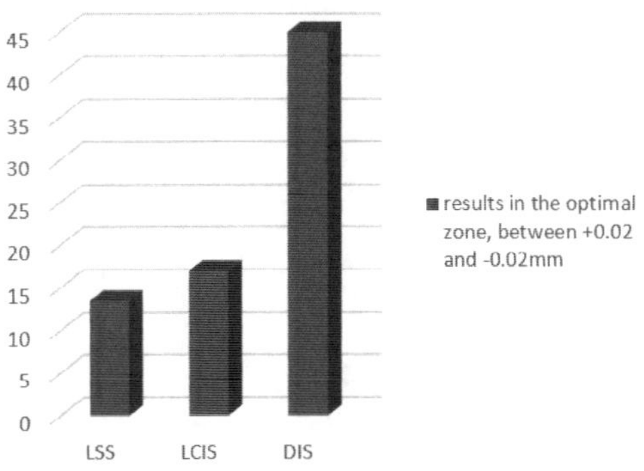

Figure 5. Results in the optimal match zone for the different scanning methods groups.

5. Conclusions

Based on the current *in vitro* study, one could draw the following conclusions:

Of the different scanning methods for clinical dentistry, the intraoral scanners showed the highest accuracy.

The laboratory scanners' accuracy depends on the scanned objects. The results are better when a conventional silicon impression is scanned compared with the stone model scanning.

However, the three scanning methods are acceptable for clinical purposes.

References

[1] El-Kerdani T 2016 *MedEdPortal* **12** 10487
[2] Praca L, Pekam F C, Rego R O, Radermacher K, Wolfart S and Marotti J 2018 *Dent. Mat.* **34** e280-e288
[3] Joda T, Zarone F and Ferrari M 2017 *BMC Oral Health* **17** 124
[4] Svanborg P 2020 *Biomat. Invest. In Dentistry* 7(1) 9–15
[5] Shopova D, Bozkova T, Slavchev D and Musurlieva N 2019 *KNOWLEDGE-Int. Journ.* **35.4** 1087–1093
[6] Galeva H, Uzunov T, Sofronov Y and Todorov G 2020 *J. Phys.: Conf. Ser.* **1492** 012017
[7] Ferrini F, Sannino G, Chiola C, Cappare P, Gastaldi G and Gherlone E F 2019 *Int. J. Environ. Res. Public Health* **16** 544
[8] Schlenz M A, Schubert V, Schmidt A, Wostmann B, Ruf S and Klaus K 2020 *Int. J. Environ. Res. Public Health* **17** 4725
[9] Sharma A, Abraham D, Gupita A, Singh A and Sharma N 2020 *Cont. Clin. Dentistry* **11(3)** 237–244
[10] Takeuchi Y, Koizumi H, Furuchi M, Sato Y, Ohkubo C and Matsumura H 2018 *J. of Oral Science* **60** (1) 1–7
[11] Suese K 2020 *Dent. Mat. J.* **39** (1) 52–56
[12] Gogushev K, Abadzhiev M, Georgieva K and Denkov I 2017 *Varna Medical Forum* 10.14748?vmf.v6i1.1925
[13] Shimizu S, Shinya A, Kuroda S and Gomi H 2017 *Dent. Mat. J.* **36** (4) 402–407
[14] Lee S, Kim S W, Lee J J and Cheong W 2020 *Dent. J.* **52** 10.3390/di8020052

[15] Tomita Y, Uechi J, Konno M, Sasanoto S, Iijima M and Mizoguchi I 2018 *Dent. Mat. J.* **37** (4) 628-633
[16] Sason G K, Mistry G, Tabassum R and Shetty O 2021 *The J. of Indian Prosth. Society* **18** 108-116
[17] Malaguti G, Rossi R,Marziali B, Esposito A,Bruno G, Dariol C and Di Fiore A 2016 *Oral and Impl.* **9** 21-27

Optimization of titanium and titanium alloy surface properties by ultra-short laser processing for improved antibacterial characteristics

D M Aceti[1,5], A Daskalova[1], L Angelova[1], E Filipov[1], L Sotelo[2,3], A Andreeva[4], A Trifonov[4] and I Buchvarov[4]

[1]Institute of Electronics, Bulgarian Academy of Sciences, 72 Tsarigradsko Chaussee, 1784, Sofia, Bulgaria
[2]Friedrich-Alexander-Universität Erlangen-Nürnberg, 91058 Erlangen, Germany
[3]Innovation-Institut für Nanotechnologie und korrelative Mikroskopie gGmbH, 91301 Forchheim, Germany
[4]Faculty of Physics, St. Kl. Ohridski University of Sofia, 5 J. Bourchier Blvd. 1164 Sofia, Bulgaria

E-mail: acetidm@gmail.com

Abstract. The aim of the current study is to improve in a one step process the properties of Ti and Ti alloy surfaces by enhancing their bioactivity in order to provide better conditions for microbial rejection. We propose to alter the biomaterial characteristics by a method alternative to the chemical ones, namely, non-contact processing of the surface by ultra-short laser pulses. The laser-induced modification results in a surface with different topographic features and an increased presence of oxides. We performed hierarchical laser patterning of the surface inducing the formation of areas covered by nanostructures, or laser-induced periodic surface structures (LIPSS), alternating with areas covered by micropillars in their turn surmounted by LIPSS. The increased roughness achieved due to the presence of micropillars, together with a marked presence of oxides, has been proven by several studies to enhance the biocompatibility of the material by improving the surface wettability and, furthermore, promoting the cells adhesion and osseointegration, while reducing the adhesion of pathogens.
The laser processed surface, consisting of a pattern of parallel lines, showed improved and anisotropic wettability. The water contact angle value decreased by ~10° in a direction orthogonal to the lines of the pattern, and by ~40° in a parallel direction.

1. Introduction

The application of diverse types of biomaterials for the creation of prosthetic implants needs to fulfil several requirements: mechanical and biological compatibility with the host tissues and resistance to wear and corrosion in biological environments. Titanium-based materials meet all the requirements to be an optimal candidate as an orthopedic implant material and have been proved to be superior compared to other metallic materials [1]. These materials are among the most widely used ones for bone prosthetic implants owing to their high strength and toughness, although their main advantage resides in the lightweight and in the lower Young modulus, compared to other metals used for bone implants as cobalt-

[5] To whom any correspondence should be addressed.

Content from this work may be used under the terms of the Creative Commons Attribution 3.0 licence. Any further distribution of this work must maintain attribution to the author(s) and the title of the work, journal citation and DOI.
Published under licence by IOP Publishing Ltd

chrome alloys, stainless steel, or tantalum. The Young modulus is closer to the bone elastic modulus so that the stress shielding effect is less likely to occur. Moreover, titanium has been proven to be a bio-inert and corrosion resistant material. Under biological condition, it undergoes passivation by forming a stable layer (~10 nm) of TiO_2.

The surface morphology and topography of a material play a key role in its behavior in a biological environment in terms of interaction with cells and bacteria. Critical requirements are promoting cell proliferation [2] and differentiation [3] while avoiding bacteria adhesion [4] and biofilm formation [5].

The approach proposed by us uses Ti:sapphire laser processing to obtain a pattern consisting of a combination of micro- and nanostructures on the surface of commercially pure titanium (cp-Ti) and titanium grade 6 (Ti-6Al-4V). Laser processing appears to be preferable compared to other surface modification processes [6] (e.g. sand blasting, machining, plasma spray, etc.), as it does not require additional materials for coating and also eliminates the use of chemicals (e.g., acid etching) to form surface features. In addition, the use of ultra-short pulses makes the process more precise and cleaner, with less debris formation [7], while thermal damage is avoided [8].

The process of interaction between the laser pulses and the surface of Ti and Ti alloy leads to the formation of laser-induced periodic surface structures (LIPSS), a topographical structure that has been shown to eliminate bacterial adhesion [9], while positively affecting the cellular response in terms of alignment and differentiation [10]. The appearance of new surface features comes together with a modification of the surface chemistry leading to the formation of an amorphous layer of TiO_2 [11].

LIPSS has also been proven, under particular conditions, to reduce the coefficient of friction and wear of laser-treated surface materials [12]; the use of LIPSS with such an aim in a biological environment is yet to be well explored. Cunha et al. demonstrated [13] the effectiveness of LIPSS on Ti against proliferation of *Staphylococcus aureus*. The group of Lee et al. found that LIPSS induces cells alignment and increases their metabolism [14], while Gnilitskyi et al. observed improved cell adhesion and proliferation [15] due to the presence of LIPSS.

Although these structures are usually hydrophobic, the presence of microscale roughness, combined with LIPSS, provides an improved wettability [16] that facilitates the cells adhesion and differentiation [17], as well as the osseointegration due to the better interlocking between the material surface and the actual bone [18] compared to a smooth surface.

In this paper, we describe the formation of specific structures obtained by ultrashort laser processing and the corresponding physico-chemical properties. The ratio between the areas covered by the LIPSS only and by a combination of LIPSS and micro-pillars, as well as the pillars size, can be tuned by varying the processing parameters, namely, hatch distance and laser scan speed and power.

2. Material and methods

Commercially available pure titanium and titanium-6 aluminium-4 vanadium plates with sizes 20 mm×20 mm×2 mm and 10 mm×10 mm×1 mm were patterned employing a Ti:sapphire laser system generating pulses with duration $\tau < 130$ fs at the central wavelength of 800 nm and a repetition rate of 1 kHz with a peak power above 1 W.

The plates were positioned on a two-axis motorized translation stage controlled by LabVIEW software. The laser beam was focused using an achromatic bi-convex lens with a 1-cm diameter and 10-cm focal length. The beam diameter before the lens had a 7-mm diameter; a chopper was used to change the repetition rate to 500 Hz.

The patterned surfaces were obtained using a variety of sets of parameters – translation stage scan speed ranging from 0.125 mm/s to 16 mm/s, pulse energy fixed at 0.06 mJ and 0.12 mJ, distance between adjacent lines varying between 100 μm and 150 μm.

The material was processed in air at room temperature. The processed samples were ultrasonically cleaned in deionized water for 10 minutes; this procedure was repeated four times, and then the samples were rinsed in water and ethanol to eliminate any residual produced in the process.

The morphology obtained via femtosecond laser processing was observed and studied using atomic force microscopy (AFM) (MFP-3D Asylum Research, Oxford Instruments, UK), an optical profilometer

(Zeta-20, Zeta Instruments, USA), and scanning electron microscopy (SEM) (TM4000, Hitachi); the wettability tests were performed using a 10-µl drop of deionized water and the contact angle was evaluated by using ImageJ software for picture analysis. A Raman spectrometer (LabRAM HR Evolution, Horiva) was used to investigate the chemical composition of the surface before and after laser processing using a 10× objective and a 532-nm laser with a power of 1.68 W.

3. Results

The femtosecond laser treatment induced the formation of diverse topographic features depending on the process parameters. Similar results were observed for both cp-Ti and Ti-6Al-4V. The laser exposure induced the formation of LIPSS with a periodicity of ~600 nm in all cases.

At the scan speed of ~1 mm/s, grooves were formed on both materials in the center of the lines, corresponding to the area subject to a higher energy density; spikes were seen at lower values of the scan speed. Both topographies are surmounted by LIPSS, as shown in figure 1 (a, b). Higher scan speed values allow the formation of LIPSS only, see figure 1 (c, d).

a) **b)** **c)** **d)**

Figure 1. Surface morphology obtained by ultrashort laser processing, **a)** SEM micrograph, general appearance of the sample surface, laser power 60 mW, scan speed 0.25 mm/s, scale bar 100 µm; **b)** SEM micrograph, detail of the edge between the two kinds of morphology, laser power 60 mW, scan speed 0.25 mm/s, scale bar 5 µm; **c)** SEM micrograph of LIPSS only morphology, laser power 3 mW, scan speed 4 mm/s, scale bar 1 µm **d)** AFM topographic image of LIPSS, laser power 60 mW, scan speed 4 mm/s, the inset shows the 2D-FFT image highlighting the LIPSS period, scale bar 5 µm.

Using a laser power of 60 mW, scan speeds of 0.25 mm/s and 0.125 mm/s, and 100-µm distance between lines for cp-Ti and 150 µm for Ti-6Al-4V, half of the obtained surfaces is covered by LIPSS only, and half, by micropillars surmounted alternately by LIPSS. This composite morphology affects the properties of the surface. The first effect, due to the striped pattern, is the anisotropy in the wetting behavior in accordance with the direction of the laser-modeled lines. Generally, the drop spreads on the surface more easily in the direction parallel to the created lines, and less so along the perpendicular one. In the case of only LIPSS covering the surface, the hydrophobicity was improved; in contrast, the presence of micron-sized features improved the wettability of the surface. The water contact angle value is measured by observing the sample both along and across the laser-induced pattern lines; its value decreased by ~25° and ~10° respectively.

As seen in Figure 2, a higher pulse energy value is required to obtain a dense and homogeneous presence of pillars, while varying the scan speed affects the size of these structures. The longer exposure allows the pillars to become larger and eventually merge.

The Raman signal for the untreated material is of low intensity (see Figure 3), as expected for Raman scattering of metals; the presence of oxides becomes more evident while increasing the degree of modification of the surface. The absence of well-defined peaks is attributed to the formation of amorphous TiO_2; however, the presence of other kind of oxides, such as Ti_2O_3, cannot be excluded.

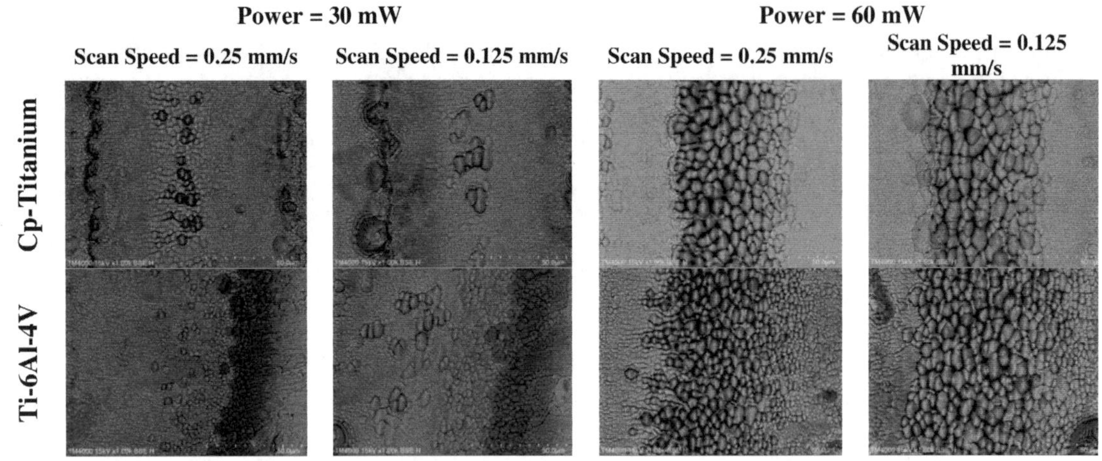

Figure 2. SEM micrographs of composite morphology obtained via fs laser processing on cp-Titanium and on Ti-6Al-4V with different process parameters. Scale bars are 50 μm.

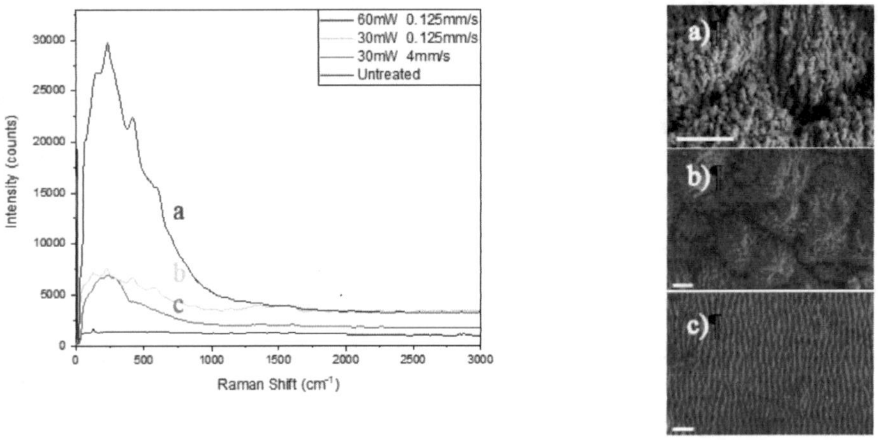

Figure 3. Raman spectra and the corresponding SEM micrographs of three different kinds of morphology. a) micropillars surmounted by LIPPS (blue line); b) incomplete formation of pillars covered by LIPSS (green line); surface covered by LIPPS only (red line). Scale bars are 2 μm.

4. Conclusions

Our study demonstrated that the ultrashort laser processing of a Ti surface results in a hierarchical morphology consisting of alternation of LIPSS only and of micropillars surmounted by LIPSS; at the same time, the formation of oxides increases, thus making the surface more hydrophilic. All these features are known to improve the biocompatibility and disadvantage the proliferation of bacteria.

The laser induced surface structuring provides a potentially favorable environment for the cells proliferation, while disfavoring the bacterial colonization. The hierarchical morphology can positively affect the cells adhesion and simultaneously lead to repelling or trapping of bacteria and preventing the formation of a biofilm. Our findings suggest that this approach can be widely applied to surface treatment of prosthetic materials, since no extreme conditions (e.g. vacuum) or additional chemicals (e.g. acids) are required and the process can be tailored according to specific needs.

Acknowledgments

This research has received funding from the European Union's H2020 research and innovation program under the Marie Sklodowska-Curie Grant Agreement No. 861138. The authors would like to acknowledge the Bulgarian National Science Fund (NSF) under grant number No. KP-06-H48/6 (2020–2023) and project H2020 FET Open METAFAST Grant Agreement No. 899673.

References

[1] Geetha M, Singh A K, Asokamani R and Gogia A K 2009 *Progress in materials science* **54** (3) 397–425
[2] Ponsonnet L, Reybier K, Jaffrezic N, Comte V, Lagneau C, Lissac M and Martelet C 2003 *Materials Science and Engineering:* C **23** (4) 551–560
[3] Raines A L, Olivares-Navarrete R, Wieland M, Cochran D L, Schwartz Z and Boyan B D 2010 *Biomaterials* **31** (18) 4909–4917
[4] Cheng Y, Feng G and Moraru C I 2019 *Frontiers in microbiology* **10** 191
[5] Jaggessar A, Shahali H, Mathew A and Yarlagadda P K D V 2017 *Journal of nanobiotechnology* **15** (1) 1–20
[6] Gaggl A, Schultes G, Müller W D and Kärcher H 2000 *Biomaterials* **21** (10) 1067–1073
[7] Chichkov B N, Momma C, Nolte S, von Alvensleben F and Tünnermann A 1996 *Applied physics* A **63** (2) 109–115
[8] Phillips K C, Gandhi H H, Mazur E and Sundaram S K 2015 *Advances in Optics and Photonics* **7** (4) 684–712
[9] Lutey A H A, Gemini L, Romoli L, Lazzini G, Fuso F, Faucon M and Kling R 2018 *Scientific reports* **8** (1) 1–10
[10] Dumas V *et al.* 2015 *Biomedical Materials* **10** (5) 055002
[11] Landis E C, Phillips K C, Mazur E and Friend C M 2012 *Journal of Applied Physics* **112** (6) 063108
[12] Bonse J, Koter R, Hartelt M, Spaltmann D, Pentzien S, Höhm S, Rosenfeld A and Krüger J 2014 *Applied physics* A **117** (1) 103–110
[13] Cunha, Alexandre, Elie A-M, Plawinski L, Serro A P, Botelho do Rego A M, Almeida A, Urdaci M C, Durrieu M-C and Vilar R 2016 *Applied Surface Science* **360** 485–493
[14] Lee B E J, Exir H, Weck A and Grandfield K 2018 *Applied Surface Science* **441** 1034–1042
[15] Gnilitskyi I, Pogorielov M, Viter R, Ferraria A M, Carapeto A P, Oleshko O, Orazi L and Mishchenko O 2019 *Nanomedicine: Nanotechnology, Biology and Medicine* **21** 102036
[16] Klos A, Sedao X, Itina T E, Helfenstein-Didier C, Donnet C, Peyroche S, Vico L, Guignandon A and Dumas V 2020 *Nanomaterials* **10** (5) 864
[17] Pfleging W, Kumari R, Besser H, Scharnweber T and Majumdar J D 2015 *Applied Surface Science* **355** 104–111
[18] Shah F A, Simonsson H, Palmquist A and Thomsen P 2016 *PloS one* **11** (6) e0157504

Altering the surface morphology and wettability of chitosan/graphene coatings by femtosecond and nanosecond laser processing

E Filipov[1,4], L Angelova[1], D Aceti[1], V Marinova[2], D Karashanova[2], A Trifonov[3], I Buchvarov[3] and A Daskalova[1]

[1]Institute of Electronics, Bulgarian Academy of Sciences, 72 Tsarigradsko Chaussee Blvd., 1784, Sofia, Bulgaria
[2]Institute of Optical Materials and Technologies, Bulgarian Academy of Sciences, Sofia 1113, Bulgaria
[3]Department of Physics, St. Kliment Ohridski University of Sofia, 5 J. Bourchier Blvd., Sofia 1164, Bulgaria

E-mail: emil.filipov95@gmail.com

Abstract. A major drawback of the currently used orthopedic implants is the formation of a bacterial biofilm on their surface after implantation. A potential strategy for overcoming this problem is developing micro- and nano-structured motifs on the implant's surface to prevent bacterial attachment. Additionally, the use of materials with inherent antimicrobial properties could further stall the aggregation of bacteria. This paper demonstrates an approach to obtaining a modified surface topography on chitosan-graphene 2D films via femtosecond and nanosecond laser treatment. Atomic force microscopy (AFM) revealed that laser irradiation leads to a strong surface roughness increase due to formation of non-ordered topography features with peak-to-valley distance ranging between 200 nm and 1 μm. Interestingly, sharp single protrusions (~200 nm) were observed on the untreated samples. Transmission electron microscopy analysis confirmed the presence of graphene monolayers within the composite films. Wettability tests showed that nanosecond laser ablation of the samples resulted in super hydrophilic surface properties, which repel bacteria *per se*. Based on our findings, we hypothesized that the topography motifs of both laser treated and untreated samples could potentially impede bacterial attachment via different mechanisms – reduction of contact points for adhesion or restriction of motility.

1. Introduction

Insertion of medical implants at sites of injury in the body is one of the most common medical interventions occurring on a daily basis. Although saving patients' lives and facilitating the recovery of injuries, implants could become the reason for the development of a bacterial infection at the site of implantation. In the USA, it has been assessed that about 25% of all infections in healthcare are associated with biomedical implants, with the most common pathogens being *S. aureus, S. epidermidis* and *E. coli* [1], [2]. Two events have been outlined as crucial for the development of an implant-related

[4] To whom any correspondence should be addressed.

Content from this work may be used under the terms of the Creative Commons Attribution 3.0 licence. Any further distribution of this work must maintain attribution to the author(s) and the title of the work, journal citation and DOI.
Published under licence by IOP Publishing Ltd

infection – the initial attachment of cells to the implant's surface and the gradual secretion of a biofilm that serves as a protection against the host's immune response and antimicrobial agents [1].

A few strategies for eliminating possible microbial-implant interactions have been under investigation: i) coating an implant surface with a thin polymer layer that could alter the surface chemistry of the implant thus preventing bacterial attachment [2]; ii) adding a layer of antimicrobial agents (e.g. chlorohexedeine) onto the surface of the implant that are gradually released in the cellular environment [2]; iii) developing bactericidal properties by changing the topography of the implants surface on micro- or nano-scale, thus impeding bacterial adhesion [3].

The surface topography and roughness are crucial factors in the initial contact between the material and the cell. Forming defined superficial patterns and surface roughness on different scales allows one to tune the physiochemical properties of a given material [3]. For example, different morphologies could turn a surface from strongly hydrophilic to super hydrophobic and vice versa. One method for producing precise topographical patterns that could be implemented in the development of bactericidal surfaces is laser processing by ultra-short laser pulses. Ultra-short femtosecond laser pulses allow a highly precise surface structuring on both micro- and nano-scales. The laser-matter interaction involving nanosecond pulses leads to a build-up of thermal energy within the sample that is transferred to the lattice [4]. Thus, a heat-affected zone is formed resulting in unwanted collateral damage to the material and ejection of melted material, which could affect the adjacent surface structures. With ultra-short pulsed lasers, the time period for heat transfer from the free electrons to the lattice electrons is much longer (~10 ps) than the pulse itself (range of femtoseconds) [5]. Therefore, no-heat affected zone is created and no collateral damage is observed. This allows for a more precise and clear material processing creating surface topographical models with defined dimensions.

The effects of bactericidal surfaces could be improved if the material itself possesses inherent antibacterial properties. Due to its cationic nature, chitosan (a deacetylated form of chitin) can interact with anionic components of the bacterial cell wall, which could potentially lead to disruption and bacterial death [6]. Chitosan has been investigated for its bone-promoting properties as it could stimulate both the adhesion of bone cells and their capability to form extracellular matrix that gives rise to mineralized bone tissue [7]. In tissue engineering, the polysaccharide has been used as a material for the fabrication of scaffolds for damaged tissue regeneration. A common approach is to combine a few materials in one scaffold so that they complement each other's properties. Incorporation of graphene nanoparticles within a chitosan-based scaffold would further improve the mechanical stability, antimicrobial properties and the biocompatibility of the construct [8]. Graphene is a two-dimensional layer of carbon atoms with exceptional heat and electrical conductivity and good elastic modulus and optical transmittance [8]. As an antimicrobial agent, different studies have come to the conclusion that due to its physiochemical properties it could lead to disruptions in the bacterial cell wall [9], [10].

In this paper, we demonstrate the application of femtosecond and nanosecond laser processing to inducing morphological changes at nano and micro level on the surface of two-dimensional chitosan-graphene coatings. By obtaining different topographical patterns and roughness, we aim at producing a bactericidal surface with a potential application in bone tissue engineering.

2. Materials and methods

2.1 Fabrication of 2D chitosan-graphene films

Graphene nanopowder (3 wt %) (AO-2: 8 nm Flakes-25 g, "Graphene Supermarket") was dispersed in deionized water by sonication. Separately, chitosan (medium molecular weight, Sigma Aldrich) was dissolved in 0.5 M acetic acid. Aqueous graphene dispersion was mixed with chitosan solution to obtain a 0.5 % v/v final solution. The mixture was then split in two groups as one was stirred magnetically for 3 hours and the other overnight. The chitosan-graphene solution was poured onto cover glass to form a thin film and left to dry.

2.2 Processing of chitosan-graphene 2D films with a femtosecond and a nanosecond laser
The dried films were irradiated by a Ti:sapphire mode-locked laser (Quantronix-Integra-C System, Hamden, CT, USA) with a pulse duration of 130 fs at a repetition rate of 1 kHz and a central wavelength $\lambda = 800$ nm. By using an optical chopper, the repetition rate was reduced to 500 Hz for all experiments. Samples for irradiation were mounted on a vertical XZ translation stage (positioned perpendicularly with respect to the laser beam direction). The number of pulses (N) delivered to the laser-treated zone was defined by the speed of the software-controlled translation stage. The number of applied laser pulses varied between 2 and 10 and the laser fluence, between 0.166 ~ 0.248 J/cm^2.

Another set of dried composite layers was modified by the fourth harmonic ($\lambda = 266$ nm) of a Q-switched Nd:YAG nanosecond laser system (LOTIS TII, Belarus) with a pulse duration of 18 ns, a pulse repetition rate of 1 ~ 10 Hz. The laser fluence was varied in the range 2.02 ~ 6.9 J/cm^2.

2.3 Morphological analysis of textured coatings via microscopic methods
The samples were observed by a light microscope (Optika, Italy) for initial identification of surface modifications. Scanning electron microscopy (SEM) (TM4000, Hitachi) was performed for detailed investigation of the morphology of the obtained surface modifications. Samples were sputter-coated by a gold layer (~8 nm). In order to assess the presence and distribution of graphene nanoparticles within the composite material, transmission electron microscopy (TEM) (JEOL JEM 2100) was applied. Topography patterns and changes in the surface roughness of the processed samples were examined via AFM (MFP-3D, Asylum Research, Oxford Instruments, UK). Root-mean-square (RMS) values were taken into account to assess surface roughness. Due to the unknown morphology of the processed samples, the AFM was set to scan a 5×5 μm area in order to avoid "dragging" of the material by the tip of the cantilever.

2.4 Assessment of surface properties of treated samples via water contact angle (WCA) measurement
For this test, a 1-μl droplet of deionized water was pipetted onto a non-irradiated or an irradiated area on the samples. The angle evolution was followed for 10 s after the initial contact between the droplet and the material. The WCA was measured by using the ImageJ software with a contact angle plug-in.

3. Results

3.1 Obtaining surface roughness at nanoscale on both femtosecond and nanosecond laser irradiated chitosan-graphene coatings
Surface roughness and topography play an important role in adhesion of bacterial cells since it could either facilitate or impede the process [11]. The AFM analysis showed that the surface topography of coatings treated with the nanosecond laser consisted of sharp peaks reaching nearly 400 nm and surrounded by lower smoother protrusions (figure 1, b). The femtosecond laser texturing formed valleys as low as 200 nm and hillocks reaching 1 μm (figure 1, c). This morphology could potentially

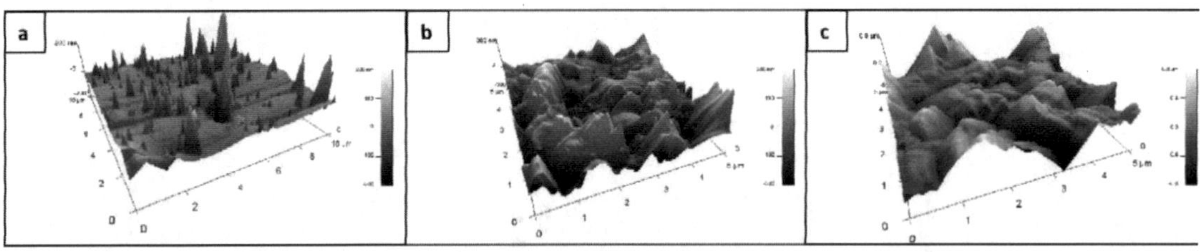

Figure 1. AFM images of a chitosan-graphene layers before (a), after nanosecond (b) ($\lambda = 266$ nm, $F = 6.32$ J/cm^2, $N = 1$) and femtosecond (c) ($\lambda = 800$ nm, $F = 0.207$ J/cm^2, N = 2) laser texturing; non-irradiated control surface (c). RMS values: (a) 107 nm; (b) 190 nm; (c) 572 nm.

decrease the number of contact points for the bacterial cells; however, the larger valleys (~2.4 μm in width) could be suitable for cells to fit and aggregate, the diameter of *E. coli* and *S. aureus* being 1 ~ 2 μm [12], [13]). Single sharp peaks (~50 nm) were visualized on the control samples and attributed to the naturally rough surface of the composite layers (figure 1 a). The average distance between peaks was ca. 1 μm; hence, we hypothesize that the distribution of these peaks could affect the adhesion or restrict the motility of the bacterial cells on the surface, as shown in bacterial entrapment experiments [14].

3.2 Preservation of graphene monolayers within chitosan-graphene composite

The TEM results proved the presence of separate graphene monolayers in the samples that were prepared by an overnight stirring of the materials and then irradiated by the nanosecond laser ($\lambda = 266$ nm, $F = 6.32$ J/cm^2, $N = 1$) (figure 2 a, b). In contrast, clusters of nanoparticles had formed in the coatings that were magnetically stirred for 3 hours prior to film formation and were ablated by the femtosecond laser ($\lambda = 800$ nm, 130 fs, $F = 0.166$ J/cm^2, $N = 10$) (figure 2 c, d). Further analyses and experiments are required in order to clearly distinguish whether the fabrication method or the type of laser texturing exert stronger effects on the nanoparticles distribution.

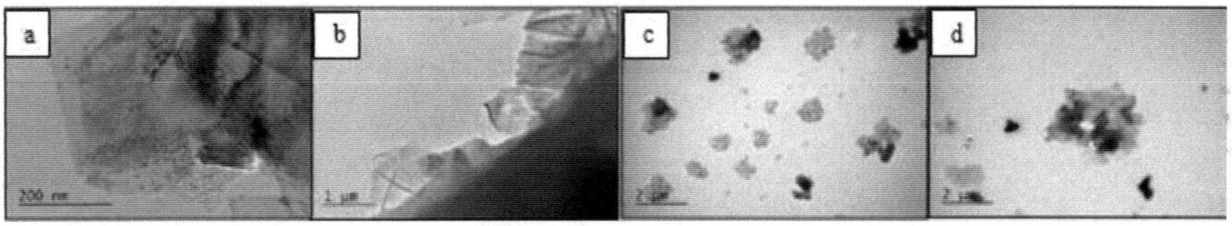

Figure 2. TEM images of graphene nanoparticles distributed within chitosan after nanosecond laser ablation (a, b) ($\lambda = 266$ nm, $\tau = 18$ ns, $F = 6.32$ J/cm^2, $N = 1$) and femtosecond laser ablation (c, d) ($\lambda = 800$ nm, $\tau = 130$ fs, $F = 0.166$ J/cm^2, $N = 10$).

3.3. Wettability assessment

The materials' wettability properties are of extreme importance for the interaction between the material *per se* and the bacterial or human cell [15]. Chitosan-graphene coatings treated with the femtosecond laser exhibited an increased hydrophilicity when compared to untreated areas of the sample (figure 3). Compared to these results, the contact of the water droplet and the dried composite layers processed with the nanosecond laser led to complete wetting of the modified surface. Surfaces with extremely high degree of wettability have been investigated to prevent attachment of bacterial cells [16].

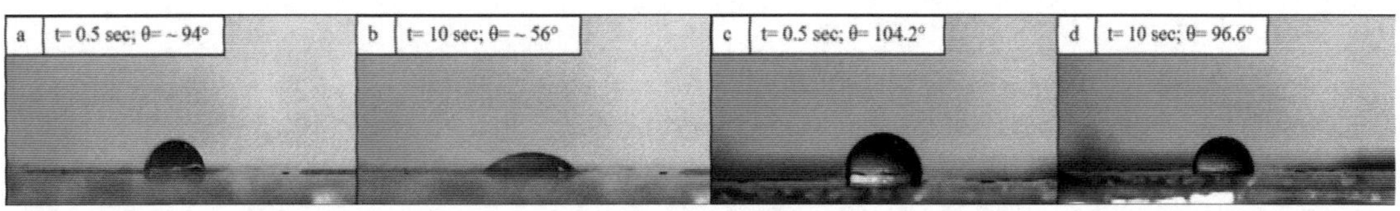

Figure 3. WCA assessment of chitosan-graphene coating treated by a femtosecond laser (a, b); control samples (c, d). Femtosecond laser texturing at $\lambda = 800$ nm, $F = 0.166$ J/cm^2, $N = 2$.

4. Conclusions

Using femtosecond or nanosecond laser texturing of chitosan-graphene coatings, we obtained distinct morphologies at a nanoscale. We hypothesize that both the laser treated and the untreated surfaces of the coatings could impact bacterial adhesion. Furthermore, the application of both types of laser pulses lead to an increase in the wettability and to a complete wetting of the surface of the nanosecond laser treated composite layers. Such properties changes would further improve the inherent antibacterial properties of the chitosan-graphene composite and widen its application in tissue engineering.

Acknowledgements
This project has received funding from the European Union's H2020 research and innovation program under the Marie Sklodowska-Curie Grant Agreement No. 861138. Authors would like to acknowledge the support under the Bulgarian National Science Fund project KP-06-H48/6 (2020–2023), "Development of hybrid functional micro/nanoporous biomaterial scaffolds by ultra-fast laser modification" and under H2020 FET Open METAFAST Grant Agreement No. 899673.

References
[1] Arciola C R, Campoccia D and Montanaro L 2018 *Nat. Rev. Microbiol.* **16** 7 397–409
[2] VanEpps J S and Younger J G 2016 *Shock* **46** 6 597–608
[3] Decuzzi P and Ferrari M 2010 *Biomaterials* **31** 1 173–179
[4] Hamad A H 2016 *High Energy and Short Pulse Lasers* ed R Viskup InTech
[5] Hergenröder R, Samek O and Hommes V 2006 *Mass Spectrom. Rev.* **25** 4 551–572
[6] Levengood S K L and Zhang M 2014 *J. Mater. Chem.* B **2** 21 3161
[7] Mathews S, Gupta P K, Bhonde R and Totey S 2011 *Cell Prolif.* **44** 6 537–549
[8] Geetha Bai R, Muthoosamy K, Manickam S and Hilal-Alnaqbi A 2019 *Int. J. Nanomedicine* **14** 5753–83
[9] Dallavalle M, Calvaresi M, Bottoni A, Melle-Franco M and Zerbetto F 2015 *ACS Appl. Mater. Interfaces* **7** 7 4406–14
[10] Perreault F, Fonseca de Faria A and Elimelech M 2015 *Chem. Soc. Rev* **44** 16 5861–96
[11] Hsu L C, Fang J, Borca-Tasciuc D A, Worobo R W and Moraru C I 2013 *Appl. Environ. Microbiol.* **79** 8 2703–12
[12] Gangan M S and Athale C A 2017 *R. Soc. open sci.* **4** 2 160417
[13] Harris L, Foster S and Richards R 2002 *Eur. Cell Mater.* **4** 39–60
[14] Ge X *et al.* 2019 *Bioact. Mater.* **4** 346–357
[15] Zheng S, Bawazir M, Dhall A, Kim H-E, He L, Heo J and Hwang G 2021 *Front. Bioeng. Biotechnol.* **9** 643722
[16] Qin X-H, Senturk B, Valentin J, Malheiro V, Fortunato G, Ren Q, Rottmar M and Maniura-Weber K 2019 *Langmuir* **35** 5 1882–94

Ultra-fast laser modification of poly-lactic acid (PLA) – towards enhanced biocompatibility

L Angelova[1,4], A Daskalova[1], R Mincheva[2], X Carette[2], A Trifonov[3], E Filipov[1], D Aceti[1] and I Buchvarov[3]

[1]Institute of Electronics, Bulgarian Academy of Sciences, 72, Tsarigradsko Chaussee Blvd., 1784 Sofia, Bulgaria
[2]Laboratory of Polymeric and Composite Materials (LPCM), University of Mons, Place du Parc, 23, B-7000, Mons, Belgium
[3]Faculty of Physics, St. Kliment Ohridski University of Sofia, 5 J. Bourchier Blvd., 1164 Sofia, Bulgaria

E-mail: lily1986@abv.bg

Abstract. In this study, the interaction was investigated of femtosecond laser radiation (pulse duration $\tau = 130$ fs, wavelength $\lambda = 800$ nm) with poly-lactic acid (PLA) 2D scaffolds. Two sets of laser fluences (*F*) and scanning speed (*V*) values were applied to PLA matrices – $F = 1.66$ J/cm^2, $V = 3.8$ mm/s, and $F = 0.83$ J/cm^2, $V = 3.8$ mm/s. The PLA samples thickness, roughness, and water contact angle (WCA) were characterized before and after the laser structuring. The fs-induced microstructures were investigated by SEM, EDX, and FTIR analyses. Preliminary cell fibroblast studies were performed. The results obtained clearly show that a precise laser surface structuring could orient the cells matrix ingrowth and thus make PLA bone tissue engineering interbody future application more successful and adaptable to the personal needs of the recipients.

1. Introduction

Bone tissue engineering relies on temporary matrices that mimic the natural cell environment and provide a stable extracellular structure for their natural growth – the body regenerates naturally, as such platforms significantly affect the cellular behavior and, in this way, the overall osseointegration of implants in the body [1]. Functional bone tissue can be developed to replace the damaged one through the application of such "smart osteoimplants". Their optimal composition and structure play a significant role in meeting all cell needs for natural body ingrowth [2, 3]. Bio interfaces on the other hand provide the so needed extracellular matrix qualities for initial adhesion of "osteo-cells" and subsequent integration of permanent bone implants, making the whole process more stable and immune friendly [4]. Such temporal cell matrices must possess the qualities of natural bone, providing stability, strength, resistance, hierarchical porosity, and surface roughness, which are crucial for improving cell adhesion, proliferation, differentiation, and osteointegration [5, 6]. To achieve all these requirements, their additional structuring is obligatory.

Poly-lactic acid (PLA) is a biocompatible and biodegradable polyester of the lactic acid; it is nontoxic for the recipient, as it degrades to by-products natural to the body. This synthetic polymer is

[4] To whom any correspondence should be addressed.

Content from this work may be used under the terms of the Creative Commons Attribution 3.0 licence. Any further distribution of this work must maintain attribution to the author(s) and the title of the work, journal citation and DOI.
Published under licence by IOP Publishing Ltd

mechanically very stable, which makes it very appropriate for a basic bone graft material [7, 8]. PLA is approved by the US Food and Drug Administration for direct contact with biological fluids applications [9]; it interacts easily with the bone cells and as such a biomaterial can be used not only as a bone scaffold substrate [10-13], but also as contact surface substrate between the natural bone and the external permanent implant [13].

The ultrafast femtosecond laser structuring is a noncontact method, which provides a precise control over the surface roughness, wettability, and porosity of the treated biomaterial. It is an alternative and innovative approach allowing fine-tuning of the treatment process for functionalization of implant surfaces made of a wide range of different types of materials, including transparent biopolymers [14]. The femtosecond (fs) laser processing method can be applied to biomaterials surface functionalization with a high-level of precision without changing the quality composition of the matrix; due to the ultrashort time of interaction with the material and the precise control of ablation patterns, surface microstructuring is obtained without chemical structure alterations of the biomaterial processed [14]. By optimizing the parameters of the applied laser radiation, optimal cell living conditions could be achieved [15]. The study of Lee and co-authors, for example, demonstrates that fs laser processing can be used to increase the cells infiltration into 3D electrospun nanofibrous poly(l-lactide) scaffolds and, at the same time, facilitate the endothelial cell ingrowth [16].

In view of studying the interaction of femtosecond laser radiation with 2D PLA scaffolds, we explored two groups of femtosecond laser microstructured PLA matrices by SEM, EDX, and FTIR and compared them with an untreated control PLA surface. The thickness and roughness of the samples before and after laser structuring were characterized. The wettability change of the treated PLA matrices was also monitored. Preliminary fibroblast studies were performed to evaluate the viability of the cells for a period of seven days. The results obtained clearly show that precise laser surface structuring could be used for the functionalization of PLA-based temporal cell scaffolds.

2. Material and methods

2.1. PLA 2D samples preparation

The PLA samples were prepared using compression molding (Carver 4122 12-12H manual heated press) of the raw amorphous biopolymer compound (PLA 4060D, Nature Works) – the pellets were vacuum dried at 60°C/overnight, then molded according to the following procedure: 3-min contact at 180°C, several degassing cycles, and 2 min at 12 bars. The as prepared PLA 2D samples were cut into 2×2-cm squares and subjected to fs laser structuring.

2.2. Laser treatment

For femtosecond laser micro-processing of the prepared polymer 2D scaffolds, a Ti:sapphire laser was used (Quantronix-Integra-C) system with $\tau = 130$ fs pulse duration at a central wavelength of $\lambda = 800$ nm, $\nu = 0.5$ kHz repetition rate and a scanning speed $V = 3.8$ mm/s. Based on previous studies, two sets of laser fluences (F) were chosen for structuring the PLA matrices – $F = 1.66$ J/cm^2, (**group 1 fsPLA**) and $F = 0.83$ J/cm^2 (**group 2 fsPLA**). The samples were positioned perpendicular to the focused fs laser beam ($f = 10$ cm, $d = 50$ μm) on an XYZ translation stage (figure 1). The precise control of the experiment was performed by the LabView software. Each laser-processed PLA sample of both groups 1 and 2 was further analyzed with respect to the control – an untreated PLA 2D matrix (**group 3 cPLA**).

2.3. PLA samples analyses

The surface morphology and elemental composition of the 2D PLA samples (**group 1÷3**) were evaluated employing scanning electron microscopy (**SEM**) with energy dispersive X-Ray analysis (**EDX**) (TESCAN/LYRA/XMU), after covering the samples with a 10-nm gold (Au) layer.

Possible chemical bonds alterations and phase transformations after the laser treatment were evaluated using a Fourier-transform infrared (**FTIR**) spectrophotometer (IR Affinity-1, Shimadzu, Kyoto, Japan) – working range 4500÷500 cm^{-1} and resolution 4 cm^{-1}.

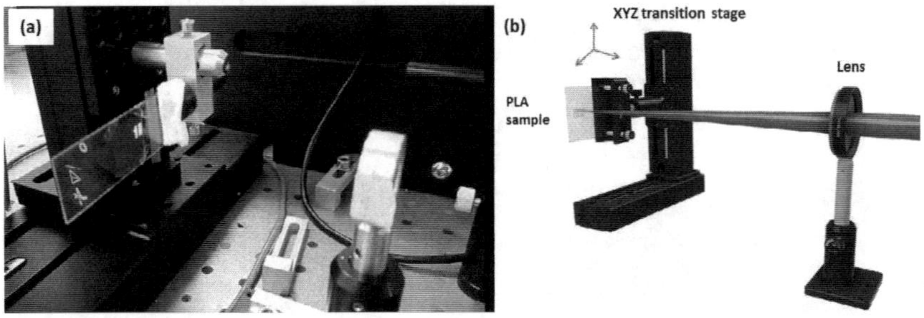

Figure 1. Photograph (a) and scheme (b) of the high-precision XYZ translation stage with the PLA sample positioned perpendicular to the fs laser beam focused by a 10-cm focal distance lens.

The PLA samples **thickness** (coating thickness gauge VA 8042 coating meter) and surface roughness (3D optical profiler, Zeta-20) were characterized before and after the laser structuring. The roughness parameters $\boldsymbol{R_a}$ **and** $\boldsymbol{S_a}$ were measured by means of 2D and 3D roughness analyses. Every value (in μm) was averaged over five thickness/roughness measurements of the fs-treated/control surface areas.

The wettability change of treated PLA matrices was monitored by water contact angle (**WCA**) measurements in air following a contact-angle goniometry method by a homemade installation – 1 μL of dH_2O was deposited by a micropipette (Dlab, ISO9001/13485-0.1-10 μL) fixed above the sample. The ImageJ software equipped with contact angle measurement plug-in was used for WCA evaluation for a period of seven seconds along and perpendicular to the fs grooves with respect to the control sample. Each WCA value was averaged over ten separate measurements.

2.4. Cell viability test

Preliminary cellular experiments were performed on group 1÷3 PLA 2D matrices – the samples from the three groups were first cleaned in 90% ethanol for two hours, then kept in a DMEM culture medium (with 10% FSA and 5% Streptomycin) overnight. Before cell seeding, the PLA samples were taken out of the medium and sterilized by UV light for one hour in a 6-well plate. A fibroblast stem cells stock (cell suspension with culture medium) was prepared at a concentration of 1×10^5 and an appropriate volume of cell stock was added to each well containing a scaffold (the volume was adjusted in accordance with the volume of the 6-well plate). The fibroblast stem cells viability was evaluated on day 1, day 3, and day 7 according to the CrystalViolet staining assay protocol. The stained cells (deep purple nuclei) were visualized by an optical microscope (4× and 10×). The absorbance was measured at 570 nm by an Epoch microplate spectrophotometer (BioTek, USA) against a background control as blank.

3. Results and discussion

As can be seen in the SEM images presented (figure 2 (a)), the fs laser treatment with the selected parameters leads to an ejection of material around the zones of interaction and to formation of surface grooves, thus enhancing the PLA topography micro-roughness in both groups 1 and 2, compared to the smooth surface of the control group 3 (figure3 (a)). Hierarchical porous frameworks can be visualized in the images taken at a higher magnification (5000×). The laser treatment proceeds without the formation of cracks or melted zones at the side of interaction, with no damage of the structure of the biomaterial.

According to the EDX elemental composition [wt. %] measurements (figure 2 (b)), no uncommon elements for PLA are observed after laser treatment. The change in the [C] and [O] concentrations (doubling the O/C ratio) can be assigned to surface oxidation resulting from the fs modification. On the other hand, the positions and shapes of all transmittance peaks in the FTIR spectra of the treated PLA samples (figure 2 (c)) remain at the correct positions with respect to the control PLA 2D matrix, while only the intensity of the characteristic for the PLA peaks (given on figure 2 (c)) is changed relatively to the control group, which does not point to any possible chemical bonds alterations and phase transformations in the structure of the PLA samples.

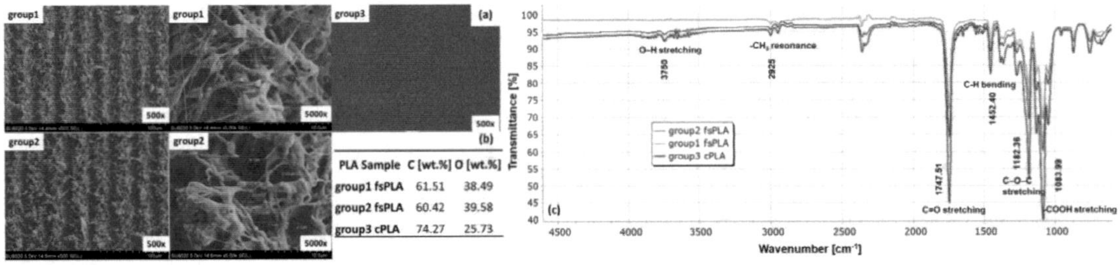

PLA Sample	C [wt.%]	O [wt.%]
group1 fsPLA	61.51	38.49
group2 fsPLA	60.42	39.58
group3 cPLA	74.27	25.73

Figure 2. (a) SEM images of group 1÷3 PLA matrices observed at 500× and 5000× magnification; **(b)** EDX elemental composition [wt. %] and **(c)** FTIR transmittance (%) spectra of the corresponding samples.

Although the grooves in group 1 and 2 look very similar on the SEM images provided in figure 2 (a), differences in the width and depth (R_a, S_a respectively) of the fs grooves are clearly observed in the 3D roughness true color composite images given in figure 3 (a) –increasing the fluence applied leads to deeper and narrower grooves and to thinner samples.

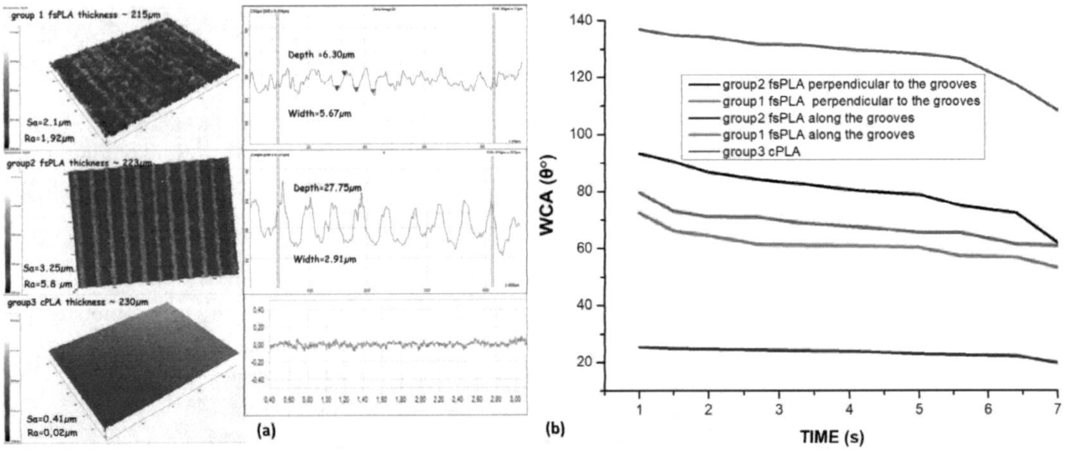

Figure 3. (a) 3D roughness true color composite images of group 1÷3 PLA samples; **(b)** WCA evaluation of group 1 fs and 2 fs PLA matrices performed along and perpendicular to the grooves created by the fs laser processing with respect to group 3 control surface.

A notable change in the wettability of the fs laser functionalized samples is observed in both ways of water drop application with respect to the control – figure 3 (b). Even though in all cases the surface becomes more hydrophilic with time, the control surface stays hydrophobic (WCA changes from 137° at the 1st second of application to 108.3° at 7ths), while the average wettability of the treated samples is enhanced up to 60.9-61.8°. The most striking is the case of dH_2O drop application along the grooves created by the laser on the group 2 fs PLA surface, when the WCA drops from 25.34° to 19.8° for the period of application.

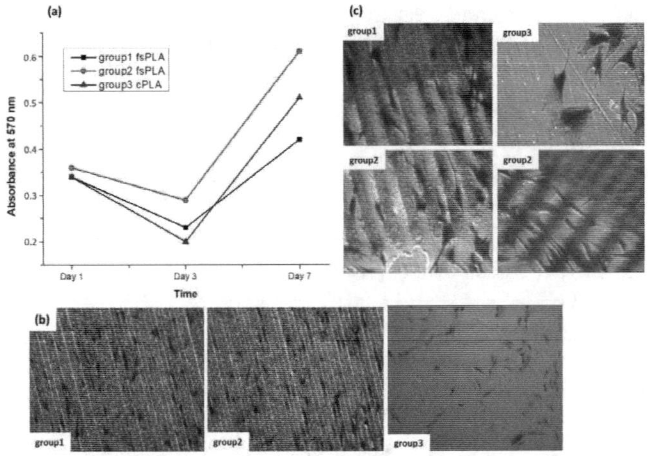

Figure 4. (a) Graph of the fibroblast proliferation; CrystalViolet optical images (nuclei stained in deep purple) of: **(b)** the cell-seeded PLA matrices on day 7 observed at 4× magnification; **(c)** the fibroblast cells "entering" and aligning along the fs laser created grooves on group 1 and 2 fsPLA, compared to group 3 cPLA observed at 10× magnification.

The preliminary cell seeding results demonstrate a notable orientation of the fibroblast cells along the grooves formed by the laser in both groups 1 and 2, as compared to the chaotic spreading of cells cultured on the control PLA group 3 – figure 4 (b-c). In all cases, a minimum in cell proliferation is seen on day 3, while a slightly enhanced fibroblast proliferation is observed in group 2 on day 7 compared to the control PLA samples (figure 4(a)) – the results do not show notable differences between all three PLA groups with respect to the cells viability. The enhanced hydrophilicity of the PLA scaffolds, in combination with the surface micro-roughness and porosity achieved in PLA structures, has been proved to be a key parameter for cells adhesion and proliferation [1].

4. Conclusions

The fs laser pulse surface micro-patterning could enhance the biocompatibility of the studied 2D PLA matrices as it tunes the biological properties of the PLA with respect to surface structuring – roughness, porosity, and wettability with no indication of chemical structure alterations. The data obtained confirm that fs laser processing is a technique suitable for biopolymer surface functionalization as it could direct cells matrix ingrowth; a further precise selection of the applied laser parameters could significantly improve the biomedical applications of PLA-based temporary implants, as optimization of the cell seeding conditions could be achieved, which could make PLA bone tissue engineering interbody future application more successful and adaptable to personal recipient needs.

Acknowledgments
This research was funded by the European Union's H2020 research and innovation program under the Marie Sklodowska-Curie Grant Agreement No. 861138 AIMed and H2020 FET Open METAFAST Grant Agreement No. 899673, and the Bulgarian National Science Fund under grant number No. KP-06-H38/5 (2019).

References

[1] Scheinpflug J, Pfeiffenberger M, Damerau A, Schwarz F, Textor M, Lang A and Schulze F 2018 *Genes (Basel).* **9(5)** 247
[2] Bose S, Roy M and Bandyopadhyay A 2012 *Trends Biotechnol.* **30** 546–54
[3] Hutmacher D W, Schantz, J T, Lam C X F, Tan K C and Lim T C 2007 *J Tissue Eng Regener Med.* **1** 245–60
[4] Rasal R M and Hirt D E 2010 *Macromol Mater Eng.* **295** 204–9

[5] Gomez S, Vlad M D, Lopez J and Fernandez E 2016 *Acta Biomater.* **42** 341–350
[6] Hutmacher DW, Sittinger M and Risbud M V 2004 *Trends Biotechnology.* **22** 354–62
[7] Schagemann J C, Chung H W, Mrosek E H, Stone J J, Fitzsimmons J S, O'Driscoll S W and Reinholz G G 2010 *J Biomed Mater Res* A **93** 454–463
[8] Holloway J L, Lowman A M and Palmese G R 2010 *Acta Biomater.* **6** 4716–4724
[9] Tyler B, Gullotti D, Mangraviti A, Utsuki T and Brem H 2016 *Adv Drug Deliv Rev.* **107** 163–175
[10] Serra T, Mateos-Timoneda M A, Planell J A and Navarro M 2013 *Organogenesis.* **9(4)** 239–44
[11] Giordano R A, Wu B M, Borland S W, Cima L G, Sachs E M and Cima M J 1996 *J Biomater Sci Polym Ed.* **8(1)** 63–75
[12] Ronca A et al 2014 *J Biomater Appl.* **29(5)** 715–27
[13] Hamad K 2015 *Express Polym Lett.* **9(5)** 435–55
[14] Terakawa M 2018 *Appl. Sci.* **8** 1123
[15] Li H, Wen F, Wong Y S, Boey F Y, Subbu V S, Leong D T, Ng K W, Ng G K and Tan L P 2012 *Acta Biomaterialia.* **8** 531–539
[16] Lee B, Jeon H, Wang A, Yan Z, Yu J, Grigoropoulos C and Li S 2012 *Acta Biomaterialia.* **8** 2648–2658

Journal of Physics: Conference Series **2240** (2022) 012043 doi:10.1088/1742-6596/2240/1/012043

Duplex surface modification of tool steel – deposition of a WN film and subsequent electron-beam modification

S Valkov[1,2,4], M Ormanova[1], N Ivanov[1], D Dechev[1], R Bezdushnyi[3] and P Petrov[1]

[1]Acad. Emil Djakov Institute of Electronics, Bulgarian Academy of Sciences, 72 Tsarigradsko Chaussee, 1784 Sofia, Bulgaria

[2]Technical University of Gabrovo, 4 H. Dimitar Str., 5300 Gabrovo, Bulgaria

[3]Faculty of Physics, St. Kliment Ohridski University of Sofia, 5 James Bourchier Blvd., 1164 Sofia, Bulgaria

E-mail: stsvalkov@gmail.com

Abstract. In this study, we demonstrate a combined treatment technique of R18 high-speed tool steel comprising deposition of a WN coating followed by electron-beam surface treatment (EBT). In order to understand the influence of the deposited WN coating on the resultant structure and properties, an uncoated specimen was directly electron-beam treated under the same EBT technological conditions. The structure of the specimens was studied by X-ray diffraction, scanning electron microscopy, and energy-dispersive X-ray spectroscopy. The microhardness was also investigated. The phase composition consisted of a double phase structure of Fe and M_6C carbides in both cases. The microstructure was significantly finer in the case of an EBT of the steel with WN coating, leading to a much higher microhardness in comparison with the case without WN.

1. Introduction

The new materials used in many industrial branches should satisfy a large number of requirements regarding their mechanical properties, high-temperature performance, etc. Usually, they are exposed to wear, corrosion, thermal and other loads mostly applied to their surface, which requires undertaking appropriate measures against possible damage [1] by using appropriate techniques for surface modification, such as ion implantation [2], thin-films deposition [3], surface treatment by high-energy fluxes [4], etc.

Currently, a widely employed technique for modifying and improving the surface properties of materials is electron-beam treatment (EBT), whereby the treated surface area is irradiated by accelerated electrons. The kinetic energy of the electrons is transferred into heat and a heat distribution from the surface to the bulk is formed. The heating and cooling rates are quite high (about 10^5 K/s in continuous mode) leading to melting of the surface and changes in the phase composition, the microstructure, the chemical composition, and a possible improvement in the surface properties [5].

The authors of [6] studied the effect of electron-beam treatment of 30CrMnSiA steel on the surface hardening and microstructure. The results showed an increase in the microhardness due to the formation of grains of different types and sizes. Similarly, Wei et al. [7] applied electron-beam surface modification of 5CrMoMn steel and concluded that the surface roughness, wear-resistance, and

[4] To whom any correspondence should be addressed.

Content from this work may be used under the terms of the Creative Commons Attribution 3.0 licence. Any further distribution of this work must maintain attribution to the author(s) and the title of the work, journal citation and DOI.
Published under licence by IOP Publishing Ltd

microhardness were significantly improved. The authors of [8] demonstrated that the corrosion resistance of 40CrNiMo7 steel was enhanced after surface modification by an electron beam.

Currently, combinations of several methods for surface treatment, including electron-beam modification, thin films deposition, plasma nitriding etc. are being used by many researchers, since one could thus combine the advantages of several techniques. The authors of [9] studied the effect of combining plasma nitriding and electron-beam surface treatment on 40X steel. The results showed an increase in the microhardness and wear resistance. Ormanova et al. [10] demonstrated a combined technique for surface modification, which includes electron-beam treatment and subsequent plasma nitriding, and electron-beam treatment followed by a plasma nitriding and subsequent electron-beam treatment. The highest microhardness value reported was 760 HV.

The authors of [11] studied the hardening of tool steel by combining deposition of a TiAlN film with subsequent electron-beam treatment. Their results showed that the prior coating deposition leads to the formation of a much harder zone in comparison with the EBT of uncoated steel.

However, data on combining a WN coating deposition with subsequent electron-beam treatment is currently lacking in the scientific literature. Therefore, the aim of this study is investigating the possibilities of a combined surface treatment by deposition of a WN coating and electron-beam treatment in view of improving the structure and microhardness of R18 high-speed tool steel (0.83%C; 0.3%Si; 0.4%Mn; 0.6%Ni; 0.03%S; 0.03%P; 4.4%Cr; 1%Mo; 17.5%W; 1.4%V; 0.5%Co; 0.25%Cu; Fe-balance, in wt %).

2. Experimental part

A WN coating with a thickness of 1.8 μm was deposited by DC magnetron sputtering on the surface of a R18 high-speed tool steel substrate. During deposition, the discharge voltage was 580 V; the discharge current, 1 A; the working pressure, 1.2×10^{-1} Pa; and the deposition time, 60 minutes. The specimen was subsequently electron-beam treated by a scanning electron beam. During the EBT process, the accelerating voltage was 50 kV; the electron beam current, 20 mA; the electron beam scanning frequency, 200 Hz; and the sample motion, 5 mm/s. In order to understand the influence of the deposited WN coating on the resultant structure and properties, another specimen was directly electron-beam treated under the same technological conditions without a prior WN coating deposition.

The phase composition of the specimens was studied by X-ray diffraction. The experiments were realized in a symmetrical Bragg-Brentano (B-B) geometry using CuKα characteristic radiation (1.54 Å). The patterns were obtained within the range of 30 – 80° at a 2θ scale with a step of 0.1° and a counting time of 10 s per step.

The microstructure and chemical composition were studied by a LYRA XMU (Tescan) scanning electron microscope equipped with an EDX detector. During the experiments, secondary electrons were used.

The microhardness was measured by an EMCO Test DuraScan 20 G5 device with a load of 25 g, time 10 s and gradient 10 g/s.

3. Results and discussion

The X-ray diffraction patterns of the specimens considered are shown in figure 1. The phase composition in both cases is in the form of a double phase structure of Fe, which is the main element in the steel, and M_6C carbides. It is seen that the application of a WN coating with a thickness of 1.8 μm does not influence the phase composition of the specimens under the chosen technological conditions of the electron-beam treatment process. Normally, in high-speed steels three types of carbides exist, namely M_6C, M_2C, and MC. According to the authors of [12], the main carbide-based phase in high-speed steels is M_2C. However, this compound is metastable and can be easily decomposed and transformed to M_6C or MC after heat treatment and annealing. These statements are in agreement with the results presented here, where the carbides identified on the surface of the treated materials are M_6C. It is known that the temperature of the treated materials during electron-beam surface modification reaches quite high values that are sufficient for decomposition and transformation of M_2C to M_6C.

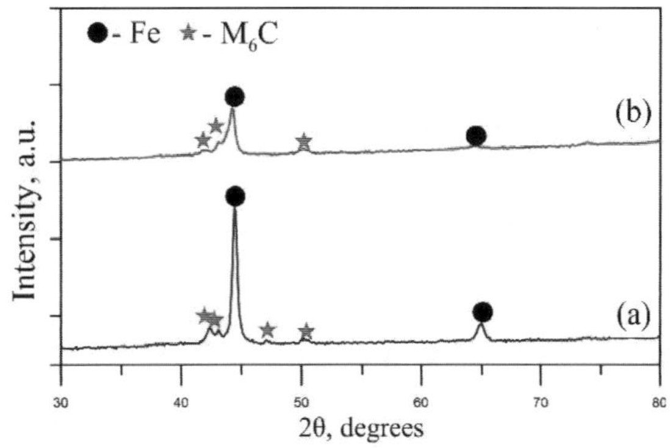

Figure 1. XRD patterns of electron-beam surface modified high-speed steel (a) with a WN coating; (b) without a WN coating.

Cross-sectional SEM images of the electron-beam treated steels with and without a WN coating are shown in figures 2 (a) and (b), respectively. In both cases, the treated zone is marked as zone A, while the base material (i.e., the high-speed tool steel substrate) is indicated as B. It is well visible that the thickness of the treated zone is significantly larger in the case of electron-beam treatment of the uncoated sample when it is about 75 μm, while in the electron-beam modified specimen with a deposited tungsten nitride film, it is about 40 μm. Therefore, it could be concluded that applying the discussed coating on the modified steel surface has a significant influence on the thickness of the treated zone. According to the authors of [5], the kinetic energy of the electrons is transferred into heat, and the technological conditions of the electron-beam treatment process can be optimized to melt the surface, produce a finer microstructure, provoke phase transformations, etc. The formation of a significantly thinner modified zone in the case of deposition of WN coating is attributed to part of the electron energy being absorbed in the film. This leads to a reduction in the electrons penetration level and, therefore, in the thickness of the treated area.

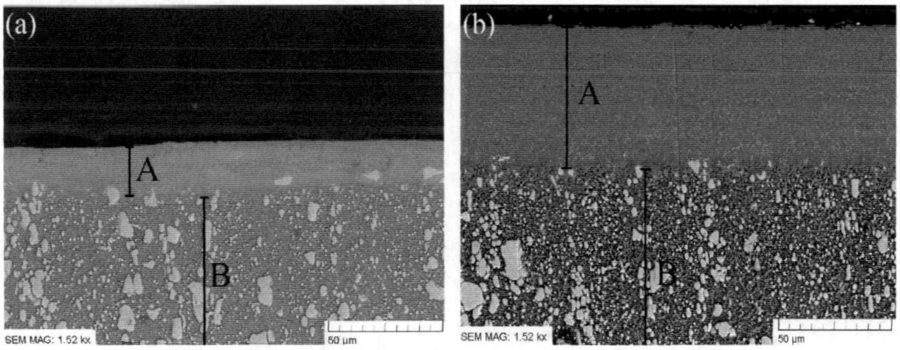

Figure 2. Cross-sectional SEM micrographs of electron-beam surface modified high-speed steel (a) with a WN coating; (b) without a WN coating.

Figures 3 (a) and (b) present higher-magnification SEM micrographs of the electron-beam treated high-speed tool steel with and without a WN coating, respectively. The chemical composition of the modified zones of both samples was studied by energy-dispersive X-ray spectroscopy (EDX) and the results are summarized in Table 1. The carbide structures that are visibly brighter are well observable on the SEM micrographs. Also, the microstructure is significantly finer in the case of an EBT of the steel with a WN coating in comparison with the case without WN. This could be attributed to the higher cooling rate during the electron-beam surface modification in the case with a deposited coating. It is known that higher values of the thermal cycle gradient correspond to a finer microstructure [13]. The results obtained for the chemical composition of the treated zones of both samples point to the existence of some amount of nitrogen element in the case of EBT of the steel with WN coating.

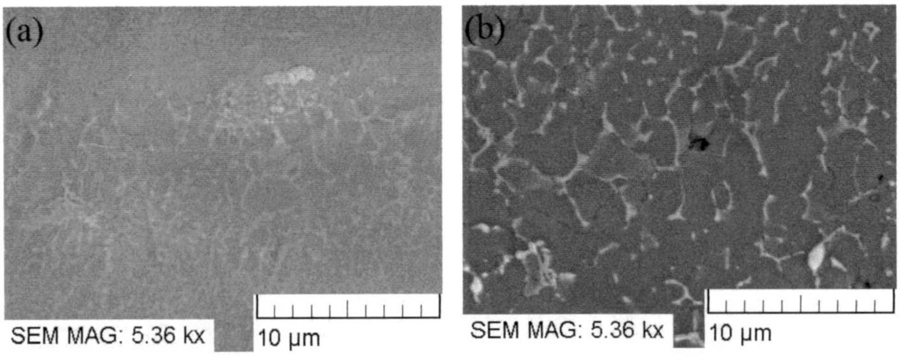

Figure 3. Scanning electron microscopy study of the microstructure of the electron-beam modified high-speed steel (a) with a WN coating; (b) without a WN coating.

Table 1. Chemical composition of the EBT zones with and without a WN coating.

	Fe, wt%	C, wt%	N, wt%	W, wt%	Cr, wt%	V, wt%
with WN	75.79	2.31	1.25	14.89	4.67	1.08
without WN	74.42	1.97	-	18.54	4.04	1.02

Figure 4 shows the results obtained for the microhardness of the electron-beam treated surfaces of the steel with and without WN coating and that of the base material. It is obvious that the microhardness is significantly higher in the case of electron-beam treatment of the steel. The microhardness of the base material is about 430 HV, while that of the uncoated electron-beam surface-modified specimen is about 720 HV. The electron-beam modified steel with a WN coating exhibits an even higher microhardness, with the value measured reaching 880 HV (or about twice as that of the base material). This great improvement in the properties of modified zones of both specimens is attributed to the formation of a significantly finer microstructure after the EBT procedure, while, as already mentioned, a finer microstructure corresponds to a higher hardness [13]. Since the microstructure of the EBT specimen with a WN coating exhibits finer grains in comparison with the uncoated sample, this is completely in agreement with the higher hardness of the previously coated electron-beam modified steel.

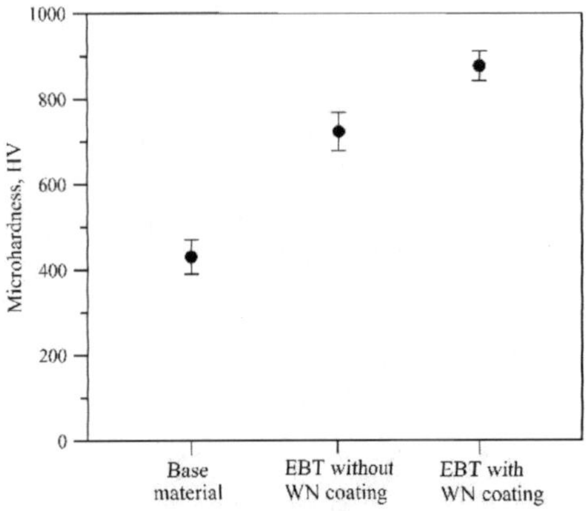

Figure 4. The measured microhardness of the electron-beam treated zones and the base material.

4. Conclusions

The results obtained in the present study demonstrate the possibility of modifying the surface of R18 high-speed tool steel by combining the deposition of a WN coating with a subsequent electron-beam treatment (EBT). In order to analyze the effect of the WN coating on the structure and properties of the treated zone, another specimen was subjected to EBT under the same technological conditions. The phase composition in both cases is a double phase structure of Fe and M_6C carbides. It was found that the combined method (i.e., the deposition of WN and subsequent EBT) leads to the formation of a thinner treated zone (a thickness of about 40 μm) in comparison with the case of EBT with a previously deposited WN coating (about 75 μm). The microstructure is significantly finer in the case of an EBT of the steel with a WN coating, leading to a much higher microhardness than in the case without WN. The microhardness measured in the case of applying the combined method is about 880 HV, while the value of EBT without a previously deposited WN coating is about 720 HV. In both cases, these values are much higher than that of the base high-speed tool steel, whose hardness is 430 HV.

Acknowledgements

This work was supported by the Bulgarian Ministry of Education and Science under the National Research Program "Young scientists and postdoctoral students" approved by DCM #577/17.08.2018.

References

[1] Kessler O 2006 *Surf Coat Technol* **201** 4046–4051
[2] Tsyganov I, Wieser E, Matz W, Mucklich A, Reuther H, Pham M and Richter E 2000 *Thin Solid Films* **376** 188–97
[3] Hampshire J, Kelly P and Teer D 2002 *Thin Solid Films* **420–421** 386–391
[4] Valkov S, Ormanova M and Petrov P 2018 *Advanced Surface Engineering Research* (Intech Open publ. London UK) 69–87
[5] Weglowski M, Blacha S and Phillips A 2016 *Vacuum* **130** 72–92
[6] Fu Y, Hu J, Shen X, Wang Y and Zhao W 2017 *Nucl Instrum Methods Phys Res Section* B **410** 207–214
[7] Wei D, Wang X, Wang R and Cui H 2018 *Vacuum* **149** 118–123
[8] Wang H, Li L, Qiu S, Zhai W, Li Q and Hao S 2020 *Coatings* **10(4)** 311
[9] Dimitrov D, Aprakova M, Valkanov S and Petrov P 1998 *Vacuum* **49** 239–246
[10] Ormanova M, Petrov P and Kovacheva D 2017 *Vacuum* **135** 7–12
[11] Grumbt G, Zenker R, Biermann H, Weigel K, Bewilogua K and Bräuer G 2014 *Adv Eng Mater* **16** 511–516
[12] Pan F, Wang W, Tang A, Wu L, Liu T and Cheng R 2011 *Progress in Natural Sciences: Mater International* **21** 180–186
[13] Petrov P 1997 *Vacuum* **48** 49–50

Laser processing of nitride ceramics

N Nedyalkov[1,3], T Dilova[2], A Dikovska[1], Ru Nikov[1], Ro Nikov[1], M Koleva[1], N Stankova[1], Al Daskalova[1] and L Angelova[1]

[1]Institute of Electronics, Bulgarian Academy of Sciences, 72 Tsarigradsko Chaussee, 1784, Sofia, Bulgaria
[2]Institute of General and Inorganic Chemistry, Bulgarian Academy of Sciences, bld. 11, Acad. Georgi Bonchev Str., 1113, Sofia, Bulgaria

E-mail: nned@ie.bas.bg

Abstract. In this work, the surface modifications are studied of aluminium nitride (AlN) and silicon nitride (Si_3N_4) ceramics irradiated by nanosecond laser pulses. The laser processing is performed by a Nd:YAG laser system at four wavelengths – 266 nm, 355 nm, 532 nm, and 1064 nm. It is found that the average ablation rate per pulse is in the order of few hundreds of nanometers, as the laser treatment leads to the appearance of a variety of micro- and nanostructures on the surface of the material. In the case of AlN, a conductive layer is formed on the surface whose resistance can be modified by varying the processing conditions. Using a model based on the heat-conduction equation, the temperature evolution, ablation depth and temporal dynamics of the ablation process are followed.

1. Introduction

Nitride ceramics are a class of synthesized materials that attract significant interest in different areas of the modern industry. Due to their high hardness, chemical resistance, high temperature stability and good thermal shock resistance related to the low thermal expansion coefficient, they find applications in engine construction; in metallurgy for casting nozzles and crucibles; in materials processing as abrasives and cutting tools [1, 2]. Thanks to their high thermal conductivity (for AlN it is the highest among the nonmetals), these materials are applied to the fabrication of crucial parts of gas turbines and spark plugs. The nitride ceramics, such as AlN and Si_3N_4, are good insulators, which in combination with the high heat conductivity and heat resistance makes them suitable for uses in the high-power electronics [3]. Si_3N_4 is also considered as a material for bio-applications due to its biocompatibility and has already been used in spinal implants and knee joints [4]. However, their high hardness and brittleness make nitride ceramics difficult for machining, as often the process is accompanied by fractures and cracks formation that may significantly alter the basic material's properties [2]. An alternative way that may offer efficient processing is based on laser techniques. At the present state of laser technology, the available systems operate in broad ranges of parameters – wavelength, pulse duration (and continuous mode operation), and intensity. Practically, this offers efficient ways of modifying all known materials. The process of laser machining involves a complex interplay of different mechanisms – laser energy absorption, heating dynamics, chemical reactions, and phase transitions. All these are strongly dependent on the processing conditions – the laser parameters, the

[3] To whom any correspondence should be addressed.

Content from this work may be used under the terms of the Creative Commons Attribution 3.0 licence. Any further distribution of this work must maintain attribution to the author(s) and the title of the work, journal citation and DOI.
Published under licence by IOP Publishing Ltd

material's optical and thermo-physical properties and the role of the ambient medium. Due to this complexity, finding the optimal conditions for highly efficient processing requires detailed studies of each specific case. Such studies devoted to application of lasers as efficient tools in processing nitride ceramics have demonstrated the crucial role of the processing parameters on the ablation efficiency and the reduction of cracks formation [5-8]. Furthermore, ablation in a liquid may reduce significantly the appearance of recast layer and avoid the formation of debris. The efficiency of ceramic ablation, which is basic in drilling, cutting and scribing, has been found to be related to the temperature behavior of these materials. It is known that AlN and Si_3N_4 decompose at high temperature (2000 – 2700 K) and form Al or Si layers on the surface. The dynamics of this process also includes different chemical reactions, which depend on the environment. Thus, the ablation process should take into account heating, decomposition, re-nitration, oxidation [9, 10].

In this work, we present results on laser ablation of AlN and Si_3N_4 ceramics by nanosecond laser pulses at four different wavelengths. The ablation rate and the structures of the ablation zone are compared and discussed. The results represent a further step in the development of laser processing as an efficient tool in ceramic machining.

2. Experimental

Two types of nitride ceramics are used – hot pressed AlN and gas pressure sintered Si_3N_4. The samples are irradiated by a Nd:YAG laser oscillating at its fundamental wavelength of 1064 nm, and its second, 532 nm, third, 355 nm, and fourth, 266 nm harmonics. The laser repetition rate is 10 Hz, with a pulse duration of 15 ns. The laser radiation is focused on the ceramic surface to a spot with a size of about 2 mm. The experiments are performed in air at room temperature. The laser fluence is fixed at about five times the ablation threshold for a given wavelength in order to estimate the role of the latter on the ablation process. The ablation area is analyzed by SEM (Inspect F50, FEI) for surface morphology and the ablation depth is evaluated by means of an optical microscope (OPTICA B-150).

3. Theoretical model

The ablation process induced by nanosecond pulses in a ceramic is thermal in nature. The heat diffusion equation is applied to study the temperature evolution during and after laser irradiation:

$$C\rho\frac{\partial T}{\partial t} = K\frac{\partial^2 T}{\partial^2 z} + A \quad (1)$$

where C, ρ, and K are the heat capacity, density and heat conductivity of the material; A is defined as $A = (1\text{-}R)I\alpha$, R being the reflectivity, α, the absorption coefficient, and I, the power density of the laser pulse. The model accounts for a decomposition reaction, as it is considered that at a certain temperature an Al or Si layer is formed on the AlN or the Si_3N_4 ceramic surface, respectively. Ablation occurs when the temperature of the material exceeds the corresponding boiling point of Al or Si. Since these are higher than the decomposition one, a layer of these elements is formed. The material parameters are taken from Ref. 11-13 for AlN and Al. For Si_3N_4 and Si the parameters are taken from [14,16].

4. Results and discussion

The study of the ablation process in the ceramics includes defining the ablation threshold fluence, F_{tr}. Using the dependence of the ablation depth on the laser fluence applied, the threshold values for the different wavelengths are estimated. The ablation experiments are then performed at fluences that are about five times higher than the corresponding threshold value for the given wavelength. The fluences obtained are presented in table 1.

4.1. AlN

Figure 1 presents the measured ablation depth for the AlN ceramic at the four wavelengths used and ablation by 1 and 100 pulses. The ablation at 1064 nm results in the highest ablation depth. The average ablation rate, defined as the ablation depth per pulse for all wavelengths, is less than

1 μm/pulse. Note, however, that the ablation by a single pulse results in a depth exceeding 1 μm/pulse – at 1064 nm it is about 14 μm. This effect can be explained by the low absorption of the material in the IR and VIS spectral ranges. This ensures a higher penetration depth of the laser radiation and an altered (different from the exponential) depth distribution due to multiple scattering. As a result, above a certain fluence, a big portion of material can be ejected.

Table 1. Laser fluences used in the ceramics ablation experiments.

Ceramic	Wavelength (nm)	F (5×F_{tr}), J/cm^2
AlN		
	266 nm	1
	355 nm	2.7
	532 nm	3.3
	1064 nm	18
Si_3N_4		
	266 nm	0.7
	355 nm	1.5
	532 nm	2
	1064 nm	10

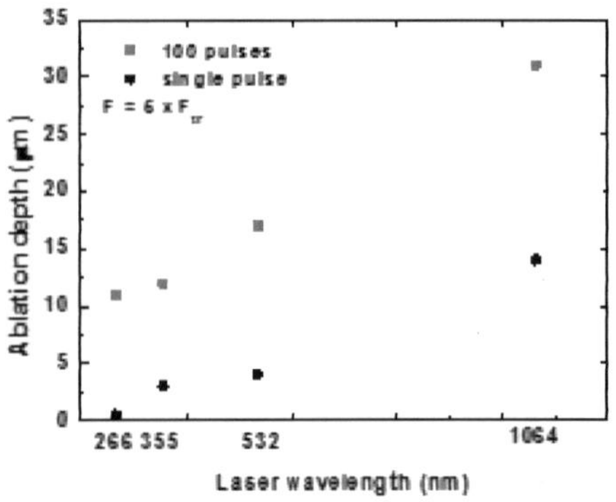

Figure 1. Ablation depth for AlN ceramic for the four wavelengths used and number of applied pulses of 1 and 100. The corresponding fluences are quoted in table 1.

The morphology of the ablated areas for AlN are presented in figure 2 for the same fluences as in figure 1 and for application of 100 laser pulses. A comparison with the native ceramic surface indicates that the ablation in all cases leads to a smoothening of the surface as the sharp edges of the crystallites that form the material are not clearly present. Ablation at 1064 nm forms a ripple structure on the surface. All surfaces become conductive after the laser treatment, while the native ceramic is a good insulator. The reason for this is the decomposition reaction that takes places due to the laser heating – nitrogen is released and an Al layer is formed on the surface. In a previous work [17], we confirmed the existence of metallic Al in the irradiated zone. The resistance of the areas is in order of few tens of ohms for all wavelengths. However, it can be modified efficiently by varying the processing parameters.

Figure 3 shows the dependence of the ablated surface resistance measured by the two-probe method for ablation at 1064 nm and different number of pulses. It is. It is seen that it drops rapidly with the number of applied laser pulses, reaching a saturation value of about 20 Ω, which can be explained by an increase of the conductive layer thickness with the number of pulses applied.

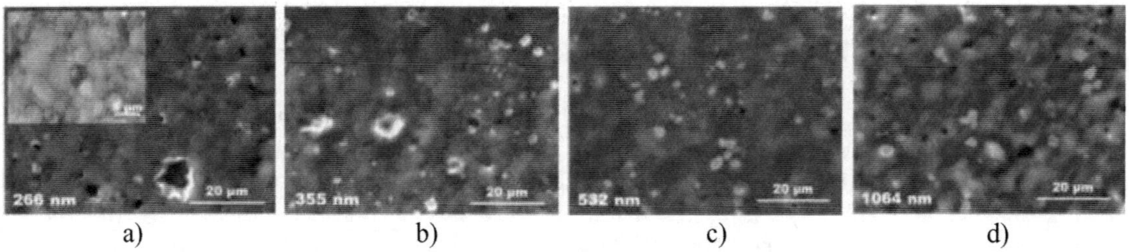

Figure 2. SEM images of AlN ceramic ablated at different wavelengths – 266 nm in a), 355 nm in b), 532 nm in c), and 1064 nm in d). The applied laser fluences are the same as in figure 1. The number pulse of is 100. The inset in c) shows the non-processed ceramic surface.

In order to trace the dynamics of the ablation process, the numerical model based on equation (1) is applied. The ceramic surface temperature evolution after a single laser pulse of fluence of 18 J/cm^2 at 1064 nm is presented in figure 4. It is found that the ablation process starts at about 5 ns after the laser pulse onset. The thickness of the formed layer is about 200 nm and the ablation depth obtained is 13 µm. The latter is in agreement with the experimentally obtained value. The simulation also reveals that the cooling of the material takes several microseconds.

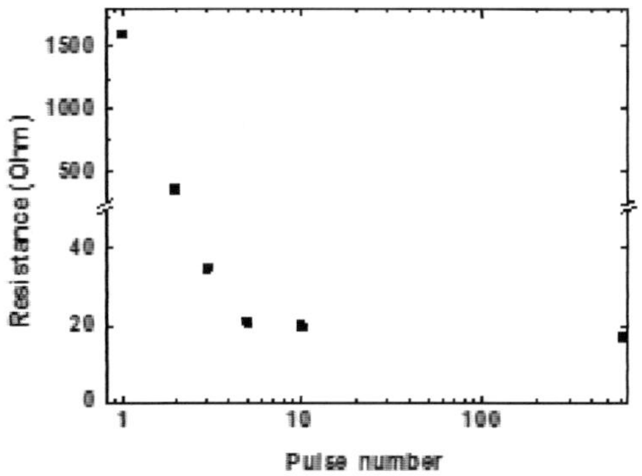

Figure 3. Ablated zone resistance in AlN ceramic ablation at 1064 nm as a function of the number of pulses. The laser fluence is 18 J/cm^2.

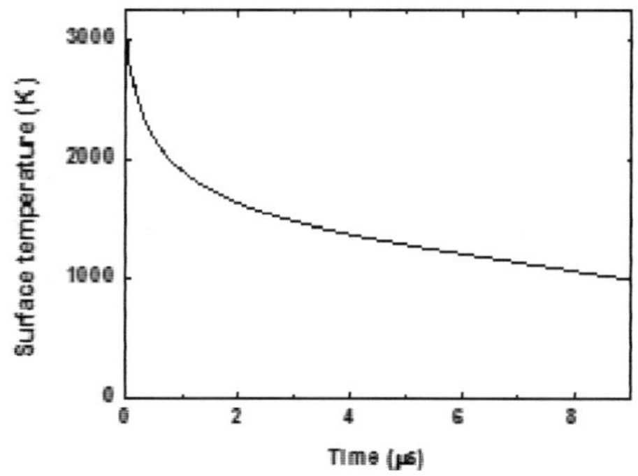

Figure 4. Surface temperature time evolution of AlN ceramic after applying a single laser pulse at 1064 nm and a fluence of 18 J/cm^2.

4.2. Si_3N_4

The ablation thresholds for the Si_3N_4 ceramic at the four wavelengths used are lower compared to the case of AlN. The ablation depths for these wavelengths at fluences five times higher than the ablation threshold are presented in figure 5 in the cases of 1 and 100 pulses.

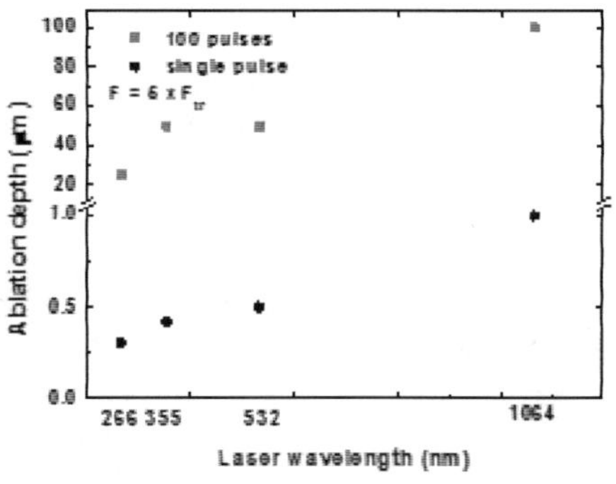

Figure 5. Ablation depth for Si_3N_4 ceramic for the four wavelengths used and number of applied pulses of 1 and 100.

In the case of Si_3N_4 ceramic, the ablation depth per pulse is similar for both single and multipulse ablation for all wavelengths. The ablation process is most efficient at 1064 nm, where the ablation rate is about 1 μm/pulse at the fluence used. The ablation also induces a modification of surface morphology. As seen in figure 6, which displays SEM images of the treated surfaces, the initially polished surface changes after laser irradiation to a rougher surface. The inset in the image for 266 nm presents the polished surface of the native ceramic. The process is related to decomposition of the ceramic and formation of a molten layer that solidifies and leaves a surface that becomes smoother as the laser wavelength is decreased. The increased absorption coefficient at the shorter wavelengths results in a more precise surface modification, whereby the smaller ablation depth per pulse ensures accumulation of the molten material.

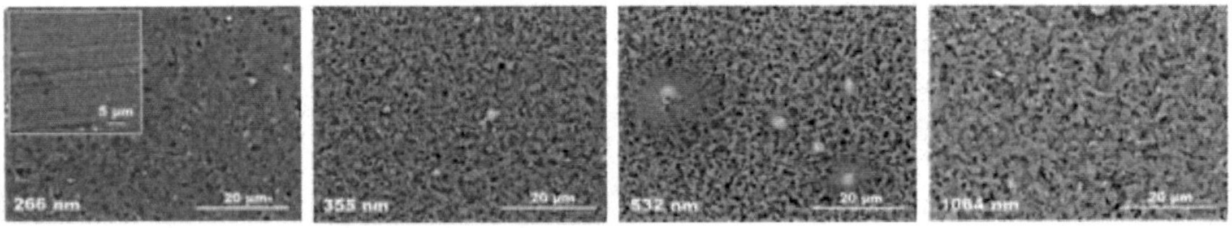

a) b) c) d)

Figure 6. SEM images of Si_3N_4 ceramic ablated at different wavelengths – 266 nm in a), 355 nm in b), 532 nm in c, and 1064 nm in d). The laser fluences are applied given in figure 1. The number of pulses is 100. The inset in a) presents the non-processed ceramic surface.

The simulation of the ablation process in Si_3N_4 using equation (1) for 1064 nm and a laser fluence of 10 J/cm^2 indicates that the ablation process starts about 1 ns after the laser pulse due to the high absorption of the material ($\alpha \approx 0^4$ cm^{-1} (compared with $\alpha \approx 10^2$ cm^{-1} for AlN)). The calculated ablation depth is 960 nm, which again is in agreement with the experimental data. The evolution of the surface temperature under these conditions is illustrated in figure 7.

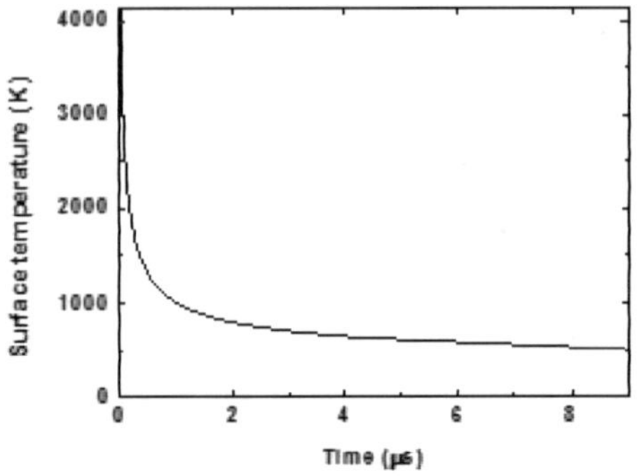

Figure 7. Surface temperature time evolution of Si_3N_4 ceramic after application of a single laser pulse for a fluence of 10 J/cm^2 at 1064 nm.

5. Conclusions

Results are presented on laser ablation of AlN and Si_3N_4 ceramics by nanosecond laser pulses at four wavelengths. The ablation characteristics are obtained at fluences five times higher than the threshold one for given wavelength. The average ablation rate for both ceramics in a multipulse regime is few hundreds of nanometers per pulse. In the case of AlN, the interaction of a single laser pulse with the native surface results in an ablation rate of few microns per pulse. The morphology of the ablated zones shows traces of formation of a layer induced by decomposition of the ceramic. In the case of AlN, this layer is conductive; its resistivity can be varied in the range of kΩ to tens of Ω by varying the processing conditions. The simulation model developed based in the heat conduction equation and taking into account the decomposition process shows that under the conditions discussed the ablation starts in an early stage after the laser pulse onset. The cooling dynamics extends to several microseconds. The results obtained can be used in the development of techniques for ceramics processing.

Acknowledgements

The authors acknowledge the financial support by the Bulgarian National Science Fund under project KP-06-H47/11.

References

[1] Ohji T and Singh M 2016 *Engineered Ceramics: Current Status and Future Prospects* (New Jersy: John Wiley Sons Inc)
[2] Wang H, Lin H, Wang Ch, Zheng L and Hu X 2017 *J. Eur. Ceram. Soc.* **37** 1157
[3] Hu F, Xie Z P and Zhang J 2020 *Rare Met.* **39** 463
[4] Bal B S and Rahaman M N 2021 *Acta Biomaterialia* **8** 2889
[5] Wang H, Chen X and Zhang W 2019 *J. Laser Appl.* **31** 042011
[6] Hirayama Y, Yabe H and Obara M 2001 *J. Appl. Phys.* **89** 2943
[7] Samant A N and Dahotre N B 2009 *J. Eur. Ceram. Soc.* **29** 969
[8] Kim S H, Sohn I-B and Jeong S 2009 *Appl. Surf. Sci.* **255** 9717
[9] Lei S, Shin Y C and Incropera F P 2001 *J. Manufact. Sci. Engin.* **123** 639
[10] Siqi C, Pedraza A J and Allard L F 1995 *J. Mater. Res.* **10** 54
[11] Sainz de Baranda P, Knudsen A K, Pruitt M E and Ruh E 1994 *J. Am. Chem. Soc.* **77** 1846
[12] Konov V, Dausinger F, Garnov S V, Klimentov S M, Kononenko T V and Tzarkova O G 1997 *Proc. SPIE* **2991** 151
[13] Leitner M, Leitner Th, Schmon Al, Aziz K and Pottlacher G 2017 *Metallurg. Mater. Trans.* A **48** 3036
[14] Krstic Z and Krstic V D 2012 *J. Mater. Sci.* **47** 535
[15] Mills K C and Courtney L 2000 *ISIJ Internation.* **40** S130
[16] Jellison G E and Lowndes D H 1987 *Appl. Phys. Lett.* **51** 352
[17] Nedyalkov N N, Dikovska A, Nikov R, Atanasova G, Hayashi S and Terakawa M 2021 *Opt. Las. Techn.* **144** 107402

Laser annealing of bimetal porous structures produced by PLD in open air

Ru Nikov[1,3], Ro Nikov[1], N Nedyalkov[1], A Dikovska[1] and K Grochowska[2]

[1]Institute of Electronics, Bulgarian Academy of Sciences, 72 Tsarigradsko Chaussee, Sofia 1784, Bulgaria
[2]Center for Plasma and Laser Engineering, The Szewalski Institute of Fluid-Flow Machinery, Polish Academy of Sciences, 14 Fiszera Str., 80-231 Gdańsk, Poland

E-mail: rumen_nikov24@abv.bg

Abstract. This work presents experimental results on laser annealing of bimetal structures produced by pulsed laser deposition in air at atmosphere pressure. The initial structures are deposited in open air by laser ablation of a rotating target consisting of two sections – Au and Ag. The ablation process is carried out by nanosecond pulses delivered by a Nd:YAG laser system emitting the wavelength of 355 nm. In open air, the laser ablation results in the formation of nanoparticles (NPs) and NP aggregates; during a prolonged deposition, these grow on the substrate into a 3D porous structure. The as-deposited structures are then annealed by laser pulses using the same laser system. The morphology of the annealed samples is studied in relation to the laser processing by varying the laser fluence and the number of the laser pulses. It is found that under certain conditions, the laser annealing leads to the formation of a 2D array of bimetal NPs on the substrate. The optical response of such structures composed by noble metals (such as Au and Ag) or their alloys is associated with a strong absorption in the visible spectral range known as surface plasmon resonance. Special attention is paid to the influence of the annealing parameters on the optical properties of the samples prepared.

1. Introduction

Advances in nanoscience and nanotechnology over the past two decades have made it possible to fabricate nanostructures with a variety of morphologies and materials. Nanoparticles of noble metals, such as Au and Ag, have been among the most widely studied structures. The great interest in these structures is related to their specific optical properties, which are not observed in their bulk counterparts. The optical response of noble metal nanostructures is expressed in a strong absorption in the UV-visible range of the spectrum [1]. This absorption is due to the excitation of surface plasmons – collective oscillations of the free electrons at the metal-dielectric interface; the maximum absorption position in the optical spectrum is known as surface plasmon resonance (SPR). Furthermore, the intensity of the electromagnetic field is greatly enhanced and confined close to the particle, which allows for light manipulation beyond the diffraction limit of the far-field optics [2]. This enhancement is the basis of many important areas of research in plasmonics, including surface enhanced Raman scattering (SERS) [3], nanoparticle antennas [4], quantum plasmonics [5], nonlinear plasmonics [6]

[3] To whom any correspondence should be addressed.

Content from this work may be used under the terms of the Creative Commons Attribution 3.0 licence. Any further distribution of this work must maintain attribution to the author(s) and the title of the work, journal citation and DOI.
Published under licence by IOP Publishing Ltd

and plasma chemistry [7]. Even more impressive properties can be obtained if the nanostructure is composed of two or more materials, such as composite, bimetallic and alloyed nanomaterials [8]. Such structures can enhance some of the properties associated with their single components and also express novel properties unattainable in single-metal nanostructures [9, 10]. For example, the SPR wavelength of Au-Ag alloyed nanorods can be precisely tuned in a wide spectral range by changing the ratio between the two metals [11].

To date, a variety of methods have been developed for the fabrication of alloyed or composite nanostructures, including laser ablation [12], solvothermal synthesis [13], co-reduction of salts [14] and replacement reactions [15]. However, most of the common fabrication methods are based on chemical processes, which can lead to residual products in the samples thus limiting their application in areas such as biophotonics. Laser annealing of thin films on substrates is a physical technique for fabrication of monolayers of single metal or composite nanoparticles [16, 17]. The method has been developed for a variety of metals on different substrates and allows a precise control of the optical properties of the structure by changing the laser parameters, the thickness of the film and its composition. One of the disadvantages of the method is that the deposition of thin films is typically carried out in a vacuum chamber at ultrahigh vacuum. The need for a specific environment reduces the flexibility of the method which limits its use in industrial applications.

In this paper, we demonstrate the fabrication of composite nanostructures by laser annealing of Au-Ag porous structures. These structures were produced by pulsed laser deposition (PLD) from a two-segment target in air at atmospheric pressure (open air), which eliminates the need of a vacuum system. It was found that under certain conditions the laser annealing can lead to the formation of a monolayer of bimetal nanoparticles on the substrate. The influence of the laser fluence and the number of laser pulses on the morphology and optical properties of the annealed samples was studied and discussed.

2. Experimental

Initially, bimetal porous nanostructures were produced by pulsed laser deposition in air at atmospheric pressure. A rotating target composed of two sections – Au (99.99%) and Ag (99.99%) was used in the PLD experiments. The irradiation area on both target metals was equal. The ablation process was carried out by ns-laser pulses delivered by a Nd:YAG laser system operating in a third harmonic generation regime (λ = 355 nm). The laser parameters used during the ablation experiments were pulse duration of 15 ns, laser fluence of 2 J/cm^2 and number of the laser pulses of 3000. The PLD process was performed in open air and the ablated material was deposited on a glass substrate. The target and substrate were set parallel at a distance of 10 mm from each other.

In the next step of the fabrication process, the as-deposited porous structure was modified by additional laser processing. This process, known as laser annealing, was performed at the same wavelength (355 nm) used in the PLD experiments. The laser annealing was implemented by one or ten laser pulses for different laser fluences – F = 240 mJ/cm^2 and F =300 mJ/cm^2 in order to follow the processing parameters effect on the structural modifications taking place. The samples' morphology after the annealing process was studied by means of scanning electron microscopy (SEM) (Quanta FEG 250 FEI, operating voltage 10 kV). The optical properties of the annealed samples were analyzed on the basis of their transmission spectra taken by an HR 4000 spectrometer (Ocean Optics).

3. Results and discussion

The PLD process from a Au-Ag target in open air resulted in the formation of a dark gray area on the transparent substrate, as seen in the photograph inserted in figure 1 a). On a micron scale, a SEM image in the deposition area reveals the presence of a porous structure composed of aggregated Au and Ag nanoparticles (figure 1 a). It should be noted that these nanoparticles are formed close to the target by rapid condensation of the ablated material due to the strong confinement of the plasma plume in the high-density air environment [18]. Figure 1 b-e) shows SEM images of bimetal structures after laser annealing with different laser parameters: b) 1 pulse at F = 240 mJ/cm^2, c) 1 pulse at

$F = 300$ mJ/cm^2, d) 10 pulses at $F = 240$ mJ/cm^2 and e) 10 pulses at $F = 300$ mJ/cm^2. The photographs inserted in the figure show the processed area in each sample. As can be seen, the color of the annealed spots varies from pale violet to pink depending on the laser processing. The different coloration is related to different optical properties of the samples, which is due to differences in the characteristics of their structure. The SEM image of the area that underwent the weakest laser processing (figure 1 b) shows slight morphological changes compared to the initial structure (figure 1 a). In this case, the as-deposited porous structure is transformed into a two-dimensional (2D) structure consisting of randomly distributed islands with dimensions of a few nanometers (figure 1 b). As in figure 1 a), big nanoparticle aggregates are also found. It should be mentioned that irradiations with a single pulse at a laser fluence bellow ~ 240 mJ/cm^2 did not affect the initial morphology of the samples. Single-pulse laser annealing at the higher fluence (300 mJ/cm^2) resulted in a more significant modification of the structure (figure1 c). The SEM image reveals that some of the nano-islands have merged to form bigger spherical-like particles measuring several tens of nanometers. The process of nanoparticle formation is even more pronounced when the laser annealing is performed by 10 laser pulses, as seen in figure 1 d), e). Multi-pulse irradiation for both laser fluences resulted in a complete modification of the porous structure into a 2D nanoparticle array. The SEM images of these samples reveal that the nanoparticles formed are predominantly spherical in shape. Particle size distributions are also shown in figure 1 for the structures representing a well-defined nanoparticle monolayer. It was found that increasing the laser fluence from 240 mJ/cm^2 to 300 mJ/cm^2 leads to a decrease in the mean particle diameter from 22.1 nm to 20.2 nm. Annealing with the higher fluence resulted also in a reduction in the standard deviation of the distribution.

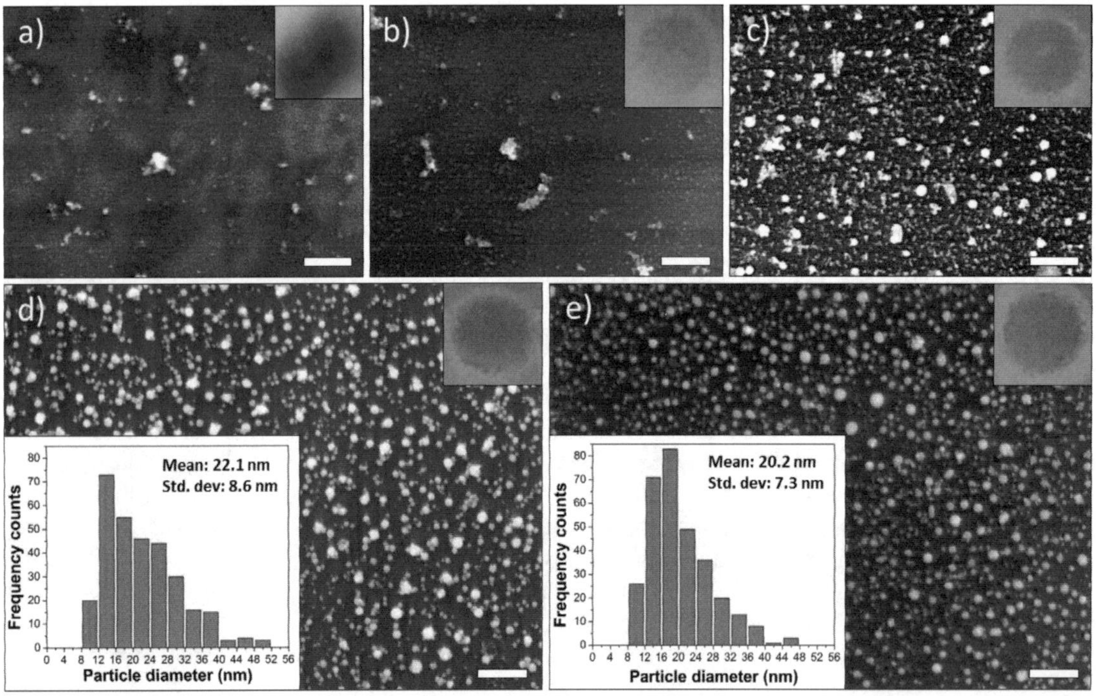

Figure 1. SEM images of Au-Ag structures produced by PLD in open air – a) without annealing, and after laser annealing with b) 1 pulse and $F = 240$ mJ/cm^2, c) 1 pulse and $F = 300$ mJ/cm^2, d) 10 pulses and $F = 240$ mJ/cm^2, and e) 10 pulses and $F = 300$ mJ/cm^2. Photographs (6×6 mm) of the structures are presented in the insets. Particle size distributions of the nanoparticle arrays are presented d) and e). The scale of the SEM images is 200 nm.

The formation of nanoparticle arrays by laser annealing depends on the characteristics of the initial structure and its interaction with the substrate. When the porous nanostructure is irradiated by ns-pulses with a sufficient laser fluence, it melts. Since the morphology with the lowest surface energy is a sphere, the molten metal can be transformed into individual particles with shapes from cap to sphere depending on the wetting between the liquid metal and substrate [16]. It is known that Au and Ag are metals which poorly wet glass substrates; therefore, the as-deposited structures that underwent multi-pulse laser processing are modified into arrays of predominantly spherical nanoparticles. The decrease observed in the mean particle diameter with the increasing laser fluence is probably related to the decomposition of already formed nanoparticles. The use of laser pulses with a fluence higher than 300 mJ/cm^2 even led to ejection of particles from the substrate (not shown).

Figure 2 shows the optical transmission spectra of the Au-Ag structures presented in figure 1. The transmission spectra of pure Au and Ag nanoparticle arrays are also presented for comparison in the inset. Both structures were obtained under the same processing conditions as the sample in figure 1 e). The clear dip observed in the transmission spectra of the annealed samples is related to plasmon excitation in the system. The position of the transmission minimum (SPR) in that range of the spectrum indicates the presence of noble metal nanoparticles in the structure [1]. The absence of two minima in the transmission spectra of the Au-Ag samples which would correspond to single Au and Ag nanoparticles is a strong evidence that the structures produced by laser annealing are composed by bimetallic nanoparticles [17]. It is seen that the laser processing strongly affects the optical properties of the Au-Ag composite structures, as the SPR in their transmission spectra varies in the range from 470 nm to 506 nm. Furthermore, the SPR wavelength for each of the structures is between the SPRs for the same structures of pure Ag and Au – 428 nm and 530 nm, respectively (inset in figure 2). The structures annealed by one laser pulse exhibit a much less expressed SPR effect in the transmission spectrum compared to the structures processed by 10 laser pulses. Single-pulse laser irradiation does not provide enough energy to the initial structure for its complete transformation into a nanoparticle array, which leads to a weak plasmon absorption in the sample. Therefore, clear SPR behavior in the spectrum of the unannealed structure is not observed. Laser processing with 10 pulses and laser fluence of 240 mJ/cm^2 resulted in a pronounced plasmon band with a minimum at 470 nm in the transmission spectrum of the structure. Increasing the laser energy to 300 mJ/cm^2 and applying the same number of laser pulses resulted in a red shift of the transmission minimum to 498 nm; a broadening of the plasmon band in the spectrum of the structure was also observed. It should be noted that the slight dips at ~ 400 nm and ~ 500 nm in the transmission spectrum of the unannealed sample are probably a sign of the presence of single Ag and Au nanoparticles in the structure, respectively. This leads us to the conclusion that the fabrication of Au-Ag composite nanoparticles by PLD in open air can be achieved only after laser annealing of the sample.

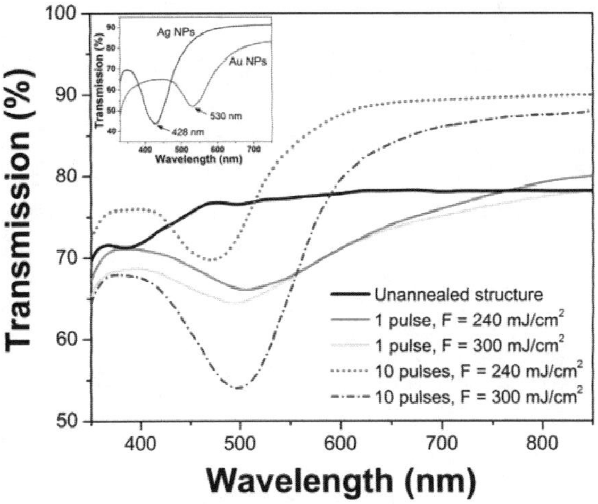

Figure 2. Transmission spectra of the Au-Ag structures obtained under different laser processing (shown in figure 1). The inset shows the transmission spectra of pure Au and Ag structures annealed by 10 laser pulses and laser fluence of 300 mJ/cm^2. Each of the samples was deposited by 3000 laser pulses at laser fluence of 2 J/cm^2.

4. Conclusions

New experimental results on fabrication of Au-Ag composite nanostructures were presented. The structures were produced by an advanced nanofabrication process that involves laser annealing of porous structures instead of classical thin films. The porous structures were obtained by laser deposition of a two-segment Au-Ag target in open air; thus, the use of a vacuum system was not necessary. It was found that the morphology of the annealed structures strongly depends on the laser parameters. The use of multi-pulse laser processing resulted in the modification of the porous structure into a well-defined nanoparticle monolayer, while when a single laser pulse was used, the annealing process was incomplete and the structure was only partially modified. In the structures annealed by 10 laser pulses, where a clear monolayer of nanoparticles was formed, increasing the laser fluence from 240 mJ/cm^2 to 300 mJ/cm^2 led to a decrease in the mean particle diameter in the distribution from 22.1 nm to 20.2 nm. The optical properties of the annealed structures strongly depend on their morphology, so that the SPR in their transmission spectrum can be tuned in the range from 470 nm to 506 nm by changing the laser processing parameters. The presence of only one minimum in the transmission spectrum of the fabricated structures is a proof that they are composed of bimetallic nanoparticles.

Acknowledgements

The authors acknowledge the financial support of project KP-06-N37/20 "Formation and physical properties of composite nanostructures of metal oxides and noble metals" under the "Competition for financial support of basic research projects – 2019" Program of the Bulgarian National Science Fund.

References

[1] Noguez C 2007 *J. Phys. Chem. C* **111** 3806
[2] Gramotnev D K and Bozhevolnyi S I 2010 *Nat. Photon.* **4** 83
[3] Wei H and Xu H X 2013 *Nanoscale* **5** 10794
[4] Li Z, Shegai T, Haran G and Xu H 2009 *ACS Nano* **3** 637
[5] Savage K J, Hawkeye M M, Esteban R, Borisov A G, Aizpurua J and Baumberg J J 2012 *Nature* **491** 574
[6] Kim S, Jin J, Kim Y J, Park I Y, Kim Y and Kim S W 2008 *Nature* **453** 757
[7] Sun M, Zhang Z, Zheng H and Xu H 2012 *Sci. Rep.* **2** 647
[8] Duan S and Wang R 2013 *Prog. Nat. Sci.: Mater. Inter.* **23** 113
[9] Mu R T, Fu Q A, Xu H, Zhang H I, Huang Y Y, Jiang Z, Zhang S O, Tan D L and Bao X H 2011 *J. Am. Chem. Soc.* **133** 1978
[10] Bochicchio D and Ferrando R 2010 *Nano Lett.* **10** 4211
[11] Ni Y, Kan C, He L, Zhu X, Jiang M and Shi D 2019 *Photon. Res.* **7** 558
[12] Lee I, Han S W and Kim K 2001 *Chem. Commun.* **2001** 1782
[13] Wang C, Peng S, Chan R and Sun S 2009 *Small* **5** 567
[14] Rajendra R, Bhatia P, Justin A, Sharma S and Ballav N 2015 *J. Phys. Chem. C* **119** 5604
[15] Sun Y G and Xia Y N 2003 *Nano Lett.* **3** 1569
[16] Krishna H, Favazza C, Gangopadhyay A K and Kalyanaraman R 2008 *JOM* **60** 37
[17] Beliatis M J, Henley S J and Silva S R P 2011 *Opt. Lett.* **36** 1362
[18] Nikov Ru G, Dikovska A Og, Nedyalkov N N and Atanasov P 2017 *Proc. SPIE* **10226** 102260F

Electrical characterization of memory capacitors for non-volatile memory applications based on nanolaminated HfO_2/Al_2O_3 and Al-doped HfO_2 stacks

D Spassov, A Paskaleva, T Stanchev[1] and Tz Ivanov

Institute of Solid State Physics, Bulgarian Academy of Sciences, 72 Tsarigradsko Chaussee, Sofia 1784, Bulgaria

E-mail: stanchev@issp.bas.bg

Abstract. Memory capacitors with atomic-layer-deposited HfO_2/Al_2O_3 nanolaminated layers and Al-doped HfO_2 charge trapping layers were investigated through capacitance-voltage (*C-V*) and current-voltage (*I-V*) measurements. The dielectric constant of the multi-dielectric stack comprising 20-nm Al_2O_3 blocking oxide, a HfO_2-based layer and 2.4-nm tunnel SiO_2 does not depend on the manner of Al-introduction in HfO_2.The stacks exhibit a negative oxide charge of about -5.1×10^{11} cm^{-2} and -2.5×10^{11} cm^{-2} for the structures with nanolaminated and doped layers, respectively. The Al-doping of HfO_2 is found to produce lower leakage currents. A sublinear behavior of the current-voltage curves is observed in the range of $-20 \div +10$ V for both HfO_2-based stacks. Memory windows of $\sim$ 1 V when charging with ±27-V voltage pulses are obtained; the data suggests that electron trapping is better pronounced in the HfO_2/Al_2O_3 nanolaminate, while positive charge accumulation prevails in the Al-doped HfO_2 layers.

1. Introduction

The current trends in the design of non-volatile semiconductor memory devices assume switching from the widely used in the mass production "floating gate" technology to the charge-trapping memory (CTM) concept. The CTM is considered as more suitable for the need of continuously shrinking the memory storage cells to increase the volume of stored information as dictated by the non-volatile memory market requirements [1]. The first CTMs were built of Si_3N_4 using a SiO_2/Si_3N_4 or a $SiO_2/Si_3N_4/SiO_2$ three-layer stack as a gate dielectric of a field-effect transistor (MOSFET) [2]. Since the modern MOSFET technology for ultra-high-density circuits requires the employment of high-*k* dielectrics in the gate stacks, it is quite tempting to use these materials as charge storage media; moreover, the high-*k* insulators turned out to be trap-rich materials. Among the various high-*k* dielectrics, HfO_2-based dielectric stacks have been recently pointed out as very promising candidates for replacing Si_3N_4 in the future CTM generations [1, 3-6]. The attractiveness of employing HfO_2 as a charge-trapping material is additionally reinforced by the fact that the HfO_2 MOSFET technology has already reached a state of maturity. Previously, we have reported some very promising from a CTM point of view charge trapping and storage results for HfO_2/Al_2O_3 nanolaminated stacks [3-5] considering the influence of the annealing ambient and the stack structure on their ability to accumulate and store electrical charges.

[1] To whom any correspondence should be addressed.

Content from this work may be used under the terms of the Creative Commons Attribution 3.0 licence. Any further distribution of this work must maintain attribution to the author(s) and the title of the work, journal citation and DOI.
Published under licence by IOP Publishing Ltd

One of the issues which are still not clarified is the relation of Al with the charge trapping sites in the layer and their spatial distribution, as well as its impact on the stack's electrical characteristics. In the present work, we report on the electrical properties of HfO_2-Al_2O_3 based multidielectric stacks with a structure close to the real memory capacitors in an effort to elucidate the effect of the manner of introducing Al.

2. Experimental details

P-type Si (100) wafers with resistivity of 4 Ω cm were used as substrates for the metal-insulator-silicon (MIS) structures. After standard RCA (Radio Corporation of America) chemical cleaning, an ultrathin (2.4 nm) SiO_2 film was grown on Si by thermal oxidation. Here on, this SiO_2 layer is referred to as TO (tunnel oxide). HfO_2-based charge trapping (CT) stacks were deposited on top of the TO layer by atomic layer deposition (ALD) at 135 °C. The stacks structures were composed of bilayer building blocks consisting of HfO_2 and Al_2O_3 sublayers. The precursors employed were tetrakis(dimethylamido)hafnium (TDMAH) and trimethylaluminium (TMA) for HfO_2 and Al_2O_3 sublayers, respectively. The oxidant was H_2O vapor. The bilayer blocks of the nanolaminated HfO_2/Al_2O_3 stack were deposited by 5 ALD cycles (cy) for Al_2O_3 sublayer followed by 20 cy HfO_2 deposition; the bi-layer block was repeated 5 times. To achieve the effect of Al-doping of HfO_2, the number of HfO_2 and Al_2O_3 cycles were reduced to 4 and 1 cy, respectively; the bilayer block was repeated 25 times to obtain the same stack thickness as that of the nanolaminated film. The overall thickness of both HfO_2-based stacks is ~16.5 nm, estimated by the ellipsometrically determined deposition rates. Finally, a 20-nm-thick Al_2O_3 layer, referred to as blocking oxide (BO), was deposited on top of the CT layer under the same conditions as the Al_2O_3 sublayers. MIS capacitors were formed by evaporation of Al gate (top) and backside electrodes, followed by photolithographical patterning of the gate electrode (area of 2.25×10^{-4} cm^2), figure 1. The electrical characterization was performed by capacitance-voltage (*C-V*) and current-voltage (*I-V*) measurements by Agilent E4980A LCR and Keithley 4200 CSC, respectively. (The methods and basic interpretation concepts used have been thoroughly elucidated in [7]). All measurements were conducted in a dark chamber at room temperature with *V* applied with respect to the top electrode.

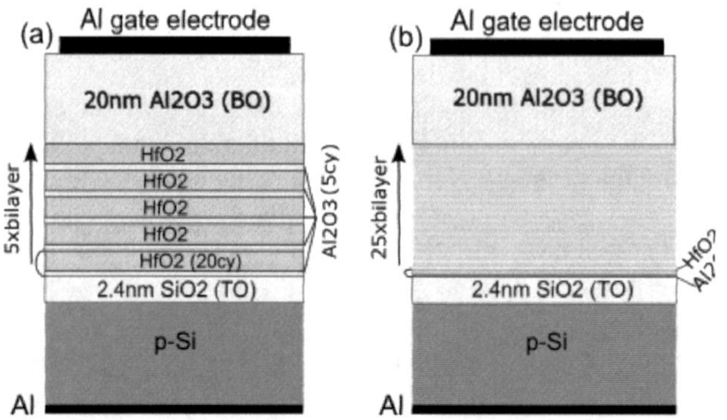

Figure 1. Schematics of the MIS capacitors: (a) nanolaminated layer; (b) doped one.

3. Results and discussions

Figure 2 depicts representative *C-V* curves of the MIS capacitors measured at three different frequencies. The frequency variations of the accumulation capacitance are about 2 % for both HfO_2 stacks. Generally, the frequency dependence of the *C-V* characteristics is related to the dispersion of the permittivity, to the impact of the parasitic contact impedance or to the influence of the interface states at the dielectric/Si interface. The weak frequency dependence of the accumulation capacitance suggests that in 10 kHz÷1 MHz range the dielectric constant of the stacks is independent of *f*. We suggest that the small dispersion of *C-V* is due to the influence of the fast interface states at the Si interface; this suggestion is supported by the shift observed of the curves towards $+V$ as f decreases (figure 1). The dielectric constant of the whole multilayered gate stacks was obtained from the C-V curves at 1 MHz employing Kar's method for evaluating the accumulation capacitance [8]. The dielectric constant values, ε_{eff}, are almost the same – 8.6±0.3 and 8.5±0.1 for the structures with nanolaminated and with doped CT films, respectively. The ε_{eff} values could be reasonably well fitted by a simple model assuming that the overall capacitance is defined by the serially connected capacitances of BO, TO and HfO_2 and Al_2O_3 sublayers with permittivities of: $\varepsilon_{SiO2} = 3.9$, $\varepsilon_{HfO2} = 15$ and $\varepsilon_{Al2O3} = 8$. Note that, apart from the well-established permittivity of thermal SiO_2, ε_{HfO2} and ε_{Al2O3} are not known and usually vary with the deposition conditions. The values here are in a good agreement with the data for layers fabricated under similar conditions [9, 10]. The results imply that the manner of Al introduction in HfO_2 does not affect significantly the dielectric constant of the HfO_2-Al_2O_3 CT stacks as evaluated by C-V measurements. The Maxwell-Wagner polarization effect leading to a giant gate stack permittivity as in [11] is not observed, and the overall dielectric constant of the gate stack is expectedly dominated by the BO.

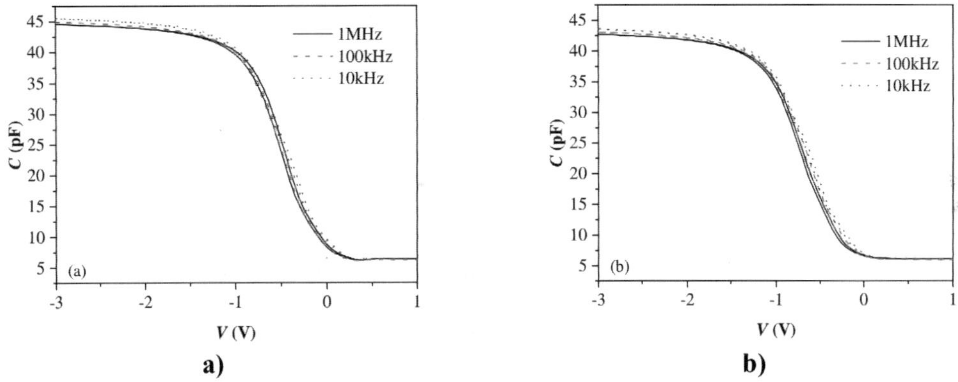

Figure 2. C-V characteristics measured at 1 MHz, 100 kHz and 10 kHz: **a)** nanolaminated stack; **b)** Al-doped HfO_2 charge trapping layer.

The electrical charges density in the gate dielectric, Q_{ox}, was estimated from the 1-MHz C-V curves through the values of the flat-band voltage, V_{fb}, according to the conventional relation:

$$V_{fb} = \varphi_{ms} - Q_{ox}/C_0 \quad (1)$$

where φ_{ms} is the Al gate-Si work-function difference and C_0, the gate stack capacitance. Equation 1, in fact, gives the values of Q_{ox} projected onto the Si interface; i.e., the oxide charges are assumed to be located at the very Si interface. The measured V_{fb} is −0.274±0.065 V for the structures with HfO_2/Al_2O_3 nanolaminated films and −0.491±0.079 V for the capacitors with an Al-doped HfO_2 layer. Both V_{fb} suggest the presence of negative oxide charges in the high-k stack dielectric. The capacitors with nanolaminated layers exhibit ~ 2× higher negative Q_{ox} ($-5.1\times10^{11}\pm1.2\times10^{11}$ cm^{-2}) than the structures with Al-doped CT films ($-2.5\times10^{11}\pm9\times10^{10}$ cm^{-2}). The measurements conducted on MIS capacitors with only a 2.4-nm SiO_2 layer reveal that the charge in TO films is positive ~ 1.9×10^{12} cm^{-2}. Hence, the Al_2O_3(BO)/HfO_2-based (CT) structure causes V_{fb} shifts corresponding to a negative oxide charge. At present, the origin of the negative Q_{ox} is not clear. Both HfO_2 and Al_2O_3 ALD films have been reported to show negative oxide charges [10,12], although for HfO_2 a more frequently positive Q_{ox} has been found [10]. Previously [6], positive Q_{ox} was observed for HfO_2/Al_2O_3 nanolaminated layers in test capacitor structures without BO and TO. A dipole formation between the first Al_2O_3 sublayer and the SiO_2 TO could partially explain the V_{fb} values leading to a negative Q_{ox} (a V_{fb} shift due to an Al_2O_3-SiO_2 dipole ~ 0.55 eV was found in [13]). The possibility that the Al_2O_3 BO layer introduces negative charge has to be also considered. A simulation with the Band diagram tool [14] reveals that an oxide charge of about -6×10^{12} cm^{-2} with a centroid situated in the middle of the BO layer, combined with the dipole potential, could explain the observed V_{fb}. Note that a negative Q_{ox} (-6.4×10^{12} cm^{-2}) was detected earlier [5] for HfO_2/Al_2O_3 nanolaminates with thicker (~3 nm) Al_2O_3 sublayers. Hence, it could be suggested that the Al_2O_3 sublayers contribute a negative charge that is thickness-dependent; for certain thicknesses it might prevail over the positive charge associated with HfO_2 and SiO_2, thus determining the net negative Q_{ox}.

The *C-V* curves exhibit a small ~ −23 mV (as measured at 1 MHz) clockwise hysteresis upon cycling the applied voltage from −3 V to 1 V and backwards. The hysteresis is the same for both capacitor types.

The test capacitors with a SiO_2 TO only also show about −20 mV clockwise hysteresis, suggesting that it is an attribute of SiO_2. The direction of the hysteresis can be due either to a capture of negative charges in SiO_2 border traps under $-V$ and their release at $+V$, or to a movement of positive mobile charges (holes or ions, e.g. H^+) near the Si interface. The estimated density of states/mobile charges is ~ 2.7×10^{10} cm^{-2}.

The *J-V* characteristics (figure 3) of the capacitors with Al-doped films demonstrate lower leakage currents. The difference between J of nanolaminated and doped structures is the most prominent in the high electric field region ($V >$ ~ +10 V and $V <$ −20 V), where a change of the conduction mechanism occurs. The current reduction in Al-doped layers can be attributed to the more effective Al compensation of the oxygen vacancy donor states in HfO_2 [15], or alternatively to the influence of the higher negative Q_{ox}, which enhances the electric field at the Si interface under a positive applied V. The observed asymmetry of the *J-V* curves with respect to 0 V could be related to the influence of BO and TO. The conduction through the SiO_2 TO is direct tunneling, while most likely some bulk-limited mechanism operates in the Al_2O_3 BO limiting the charge injection through the gate electrode. Figure 3 b) presents a simple *log-log* plot of the *J-V* data. The *J-V* dependence in the range of −20÷8 V is sublinear for both structures and the corresponding slopes are almost the same. The sublinear *J-V* segments are most likely due to the charge trapping in the stacks and the accompanying change of the electric field in the dielectric. The high-field conduction will be a subject of a further more extended study.

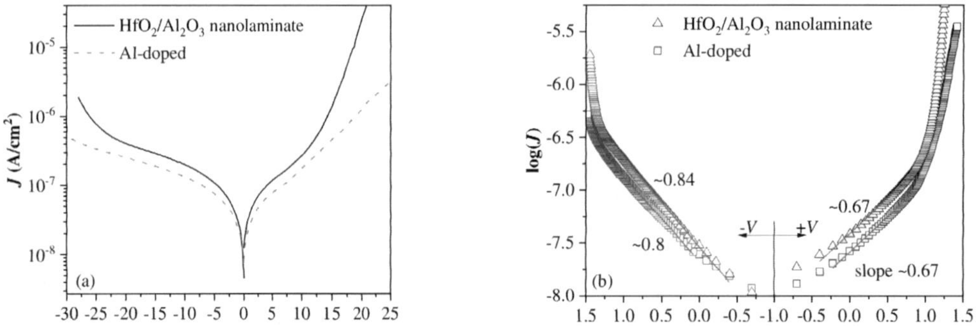

a) **b)**

Figure 3. Leakage currents **a)** of the investigated capacitors and corresponding *log-log* plot **b)**.

The ability of the investigated stacked capacitors to accumulate electrical charges and hence to realize a memory effect is illustrated in figure 4. The electric charges are introduced into the stacks by applying a voltage pulse with an amplitude $V_p = \pm27$ V and a duration of 1 s and consecutive *C-V* measurement [8, 9]. The value of V_p is chosen to be sufficiently high to ensure electron and hole injection from the Si substrate for a positive and a negative pulse polarity, respectively. As demonstrated in figure 4 a), after a $+V_p$ pulse, the *C-V* curves are shifted to positive V values as a result of negative charge trapping, while negative pulses shift the curves to the $-V$ direction due to the hole trapping. (The voltage difference between the *C-V* curves after positive and negative pulses forms the so called "memory window"). At the chosen V_p, the electron trapping in nanolaminated stacks is more

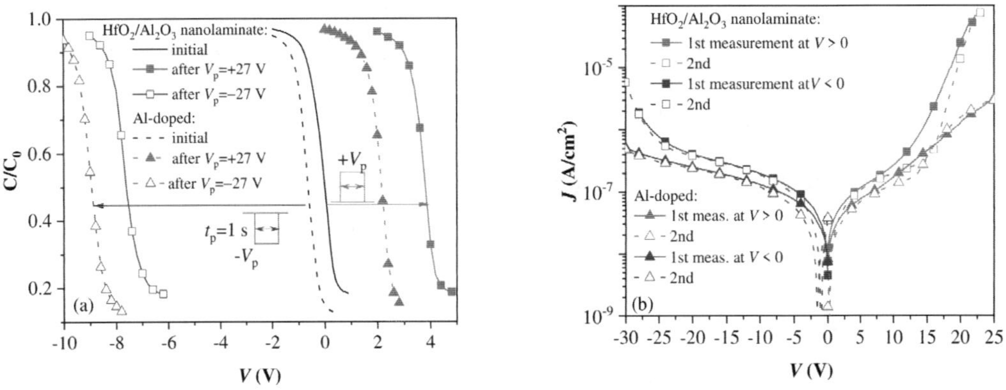

a) **b)**

Figure 4. Effect of the charge trapping on *C-V* **a)** and *J-V* **b)** curves.

intensive, while the hole trapping is stronger in doped films. This result could also reflect the higher current through the nanolaminated stacks. The charge trapping effect on the *J-V* curves is demonstrated in figure 4 b). The second measurement of *J* at +*V* shifts the *J-V* curve to a more positive *V*, because the electric field induced by the trapped electrons during the first trace is in opposite direction to the applied one. As in the case of the *C-V* measurement, the shift of the 2nd *J-V* curve for doped stacks is smaller. Except of a small current shift in the region of −5÷0 V, the 2nd *J-V* curve at negative bias for both stack types almost coincides with the 1st one, demonstrating the negligible electron injection from the gate electrode.

4. Conclusions

The electrical properties of memory MIS capacitors with nanolaminated HfO_2/Al_2O_3 layers and Al-doped HfO_2 layers comprising dedicated SiO_2 tunneling and Al_2O_3 blocking oxides were investigated through *C-V* and *I-V* measurements. It is demonstrated that the manner of Al introduction into HfO_2 matrix does not affect significantly the overall dielectric constant and the initial oxide charge, both of which are dominated by the Al_2O_3 top blocking oxide. The Al-doped HfO_2 layers, however, exhibit lower leakage currents, suggesting that the shallow centers primarily responsible for low electric field conduction at room temperature are more effectively subdued than in the case of nanolamination. Both capacitor types show very promising memory characteristics by realizing memory windows of about 10 V.

Acknowledgements
The work is supported by the Bulgarian National Science Fund under contract KP-06 H37/32

References

[1] Zhao C, Zhao C Z, Taylor S and Chalker P R 2014 *Materials* **7** 5117
[2] Chen P C Y 1977 *IEEE Transactions on Electron Devices* **24** 584
[3] Zhu C, Huo Z, Xu Z, Zhang M, Wang Q, Liu J, Long S and Liu M 2010 *Appl. Phys. Lett.* **97** 253503
[4] Paskaleva A, Rommel M, Hutzler A, Spassov D and Bauer A J 2015 *ACS Appl. Mater. Interfaces* **7** 17032
[5] Spassov D, Paskaleva A, Krajewski T A, Guziewicz E and Luka G 2018 *Nanotechnology* **29** 505206
[6] Spassov D, Paskaleva A, Krajewski T A, Guziewicz E, Luka G and Ivanov T 2018 *Phys. Status Solidi* A **215** 1700854
[7] Nicollian E H and Brews J R 2002 *MOS (Metal Oxide Semiconductor) Physics and Technology* (New York: Wiley)

[8] Kar S, Rawat S, Rakheja S and Reddy D 2005 *IEEE Trans. Electron Devices* **52** 1187
[9] Taube A, Gierałtowska S, Gutt T, Małachowski T, Pasternak I, Wojciechowski T, Rzodkiewicz W, Sawickicand M and Piotrowska A 2010 *Acta Physica Polonica A* **119** 696
[10] Simon D K, Jordan P M, Mikolajick T and Dirnstorfer I 2015 *ACS Appl Mater Interfaces* **7** 28215
[11] Li W, Auciello O, Premnath R N and Kabius B 2010 *Appl. Phys. Lett.* **96** 162907
[12] Wilk G D and Muller D A 2003 *Appl. Phys. Lett.* **83** 3984
[13] Iwamoto K, Kamimuta Y, Arito Ogawa I, Watanabe Y, Migita S, Mizubayashi W, Morita Y, Takahashi M, Ota H, Nabatame T and Torium A 2008 *Appl. Phys. Lett.* **92** 132907
[14] Southwick III R G, Sup A, Jain A and Knowlton W B 2011 *IEEE Trans. Device Mater. Reliab.* **11** 236
[15] Li Q, Koo K M, Lau W M, Lee P F, Dai J Y, Hou Z F and Gong X G 2006 *Appl. Phys. Lett.* **88** 182903

Effect of growth temperature on nitrogen incorporation into GaAsN during liquid-phase epitaxy

M Milanova[1], V Donchev[2,4], S Georgiev[2], K Kirilov[2] and P Terziyska[3]

[1]Central Laboratory of Applied Physics, Bulgarian Academy of Sciences,61 St. Petersburg Blvd., 4000 Plovdiv, Bulgaria
[2]Faculty of Physics, St. Kliment Ohridski University of Sofia, 5 James Bourchier Blvd. 1164 Sofia, Bulgaria
[3]Institute of Solid-State Physics, Bulgarian Academy of Sciences,72 Tsarigradsko Chaussee Blvd.,1784 Sofia, Bulgaria

E-mail: vtd@phys.uni-sofia.bg

Abstract. We studied the growth temperature effect on nitrogen incorporation into GaAsN layers grown by liquid-phase epitaxy (LPE). The growth of dilute nitrides at nearly equilibrium conditions during LPE presents several major challenges: low solubility of nitrogen in Ga-melt, small incorporation efficiency in the solid and high volatility of nitrogen. GaAsN layers are grown from three different initial epitaxy temperatures: 580 °C, 680 °C and 780° C using GaN powder in the melt as a source of nitrogen. The N content in GaAsN is estimated from X-ray diffraction curves applying Vegard's law. The local microstructure and surface morphology of the grown layers are studied by Fourier-transform infrared spectroscopy and atomic force microscopy measurements, respectively. The absorption edge and the composition dependence of the effective band gap are determined by employing optical transmission and surface photovoltage spectroscopies.

1. Introduction

Dilute nitrides, such as GaAsN, InGaAsN, are a novel class of compound semiconductors, whose fundamental properties are dramatically modified through the substitution of a relatively small fraction of host atoms with nitrogen. Their unusual properties are due to the large electronegativity and the small atomic radius of nitrogen. Currently, their study is of considerable interest for applications in modern optoelectronics [1–4]. The substitution of a few percent of As atoms in GaAs by N atoms leads to substantial changes in the optical absorption because of a strong reduction in the bandgap energy [5]. However, the small quantity of nitrogen incorporated into GaAs deteriorates the crystalline and optoelectronic properties of the dilute nitride materials. Ideally, the N atoms are included primarily in the As-sublattice, replacing As atoms. However, it is known that there are some other N configurations, such as an isolated N interstitial; an N-As split interstitial; an N-N split interstitial, etc. [6]. The existence of high-density N-related point defects in the grown layers and structures significantly affects the device performance, since they introduce band gap levels affecting the carrier collection and recombination in the transport process. Technologically, the incorporation probability of nitrogen in GaAs is very small and strongly depends on the growth conditions. GaAsN-based

[4] To whom any correspondence should be addressed.

Content from this work may be used under the terms of the Creative Commons Attribution 3.0 licence. Any further distribution of this work must maintain attribution to the author(s) and the title of the work, journal citation and DOI.
Published under licence by IOP Publishing Ltd

compounds are primarily grown by kinetic growth methods: metalorganic vapor-phase epitaxy [1] molecular-beam epitaxy [4], and chemical-beam epitaxy [7], but (up to now) the material quality has been inferior to that of GaAs. In contrast to other epitaxial methods, the crystallization during the liquid phase epitaxy (LPE) growth is carried out under near-equilibrium conditions, which provide a higher-quality material in terms of lifetime, mobility and absence of defects in the epitaxial layers. However, the growth of dilute nitrides by LPE presents several major difficulties, such as low N solubility in the Ga melt, low N incorporation efficiency into the GaAs crystal lattice, and high volatility of N [8]. Therefore, it remains a challenge to grow GaAsN with low nitrogen content with controlled composition and without phase separation.

In this work, we study the effect of the growth temperature on the nitrogen incorporation into the As-sublattice of GaAs. First, we investigate the structural quality, composition, local microstructure and surface morphology of the grown layers by high-resolution X-ray diffraction (XRD), Fourier transform infrared spectroscopy (FTIR) and atomic force microscopy (AFM). Second, we compare the compositionally dependent absorption edge of the layers grown at different temperatures measured by optical transmission and surface photovoltage (SPV) spectroscopy. Last, we demonstrate a significant red shift of the photosensitivity spectral range achieved by the incorporation of indium atoms in the crystal lattice of the compound.

2. Experiment

A series of GaAsN epitaxial layers were grown by the horizontal graphite slide-boat technique for LPE on (100) n-type GaAs:Si substrates. The starting materials consisted of 6N pure Ga metal as a solvent, and polycrystalline GaAs and GaN as sources for As and N, respectively. The nitrogen content in the melt was 0.5 at % for all grown samples. The charged boat was heated at about 50 °C higher than the initial epitaxy temperature (T_{ie}) for 1 h in a purified H_2 gas flow to dissolve the source materials and decrease the contaminations in the melt. Epitaxial layers with thicknesses of 1 – 2 µm were grown from 3 different initial epitaxy temperatures: 580 °C, 680 °C and 780 °C at a cooling rate of 0.8 °C/min from melt supercooled of 10 °C. Additionally, GaInAsN layers containing a few percent of In were grown at low-temperatures (580 – 577 °C) from In-rich melt to study the influence of In on the incorporation efficiency of nitrogen into the crystal lattice.

The XRD method was used to determine the lattice parameter, the lattice mismatch and the crystalline quality of the layers. The measurements were performed on a PANalitical X'Pert PRO MRD diffractometer with a copper anode. The surface morphology was characterized by AFM in tapping mode employing an MFP-3D Origin apparatus. The local bonding of nitrogen in the material was studied by room temperature FTIR spectroscopy measurements with a spectral resolution of 2 cm^{-1} on a Shimadzu FTIR spectrophotometer. Optical transmission and SPV spectroscopies were used to determine the absorption edge of the GaAsN grown layers. For the transmission measurements, the samples were thinned down to about 100 µm by etching the substrate by $H_2O_2/NH_3OH/H_2O$. The transmission spectra were measured in the spectral range 850 – 2000 nm. The SPV spectra were measured at room temperature in a chopped light configuration (94 Hz) applying the set-up and measurement procedure described elsewhere [9].

3. Results and discussion

3.1. Structural characterization

Figure 1 shows X-ray diffraction (004) curves for two samples grown from different initial epitaxy temperatures T_{ie} of 580 °C and 780 °C. The two small peaks on the right side of the (004) GaAs substrate peak are associated with the GaAsN layers. The lattice mismatch $\Delta a/a_0$ for the samples determined from the XRD data is about 0.02% for the layer grown from 580 °C and 0.04% for that grown from 780 °C. Using Vegard's law, the content of nitrogen in the As sublattice was found to be 0.1% and 0.2% in the GaAsN layers grown in the low and high-temperature ranges, respectively.

An AFM topography image of one of the grown GaAsN layers is plotted in figure 2. A weakly developed terraced surface is observed that is typical for LPE grown layers. The root-mean-square (RMS) surface roughness measured in different samples is in the range 0.2 – 0.3 nm. These low values are due to the small lattice mismatch between layer and substrate.

Figure 3 presents the FTIR absorption spectrum of a GaAsN epitaxial layer. It reveals a well-defined peak at 472 cm^{-1} attributed to a local vibrational mode of nitrogen at arsenic sites in GaAs in accordance with [10]. In ternary GaAsN alloys, the macroscopic composition uniquely defines the number of bonds and local structure of the alloy: the nitrogen atom is bonded to four Ga atoms (Ga4N) in the GaAs lattice.

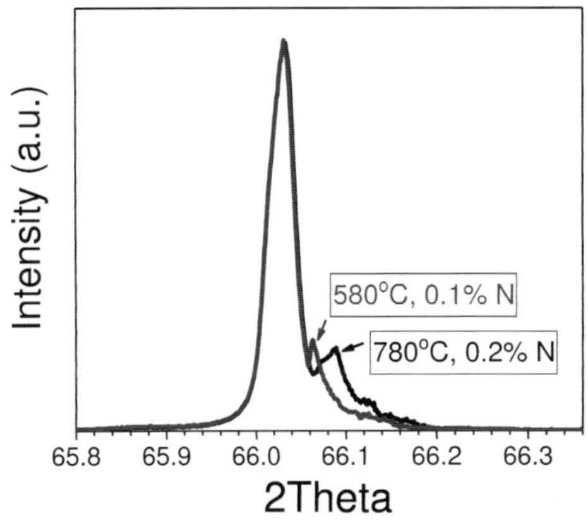

Figure 1. X-ray diffraction (004) curves of two GaAsN samples grown from different initial temperatures and with different content of nitrogen.

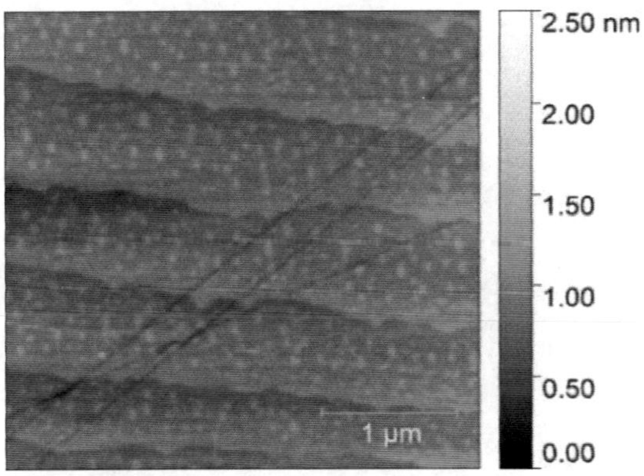

Figure 2. AFM surface topography image of GaAsN grown at T_{ie} = 780 °C. RMS is 0.2 nm.

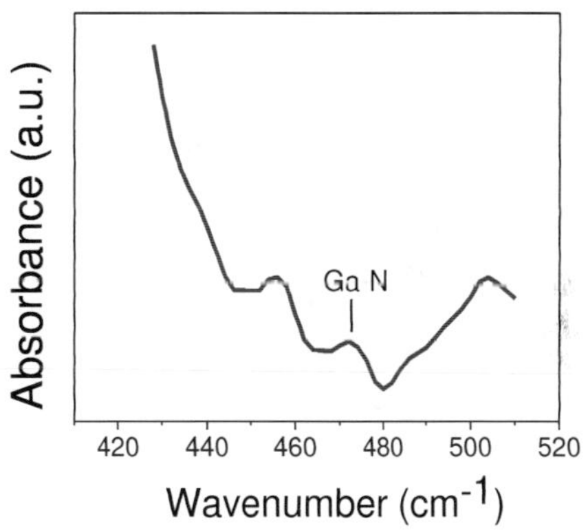

Figure 3. FTIR spectrum of GaAsN sample grown at T_{ie} = 580 °C.

3.2. Optical characterization

Figure 4 shows the normalized transmission spectra of three typical $GaAs_{1-x}N_x$ layers grown from different initial epitaxy temperature T_{ie}. The absorption edge is redshifted with respect to that of GaAs. The largest red shift is observed for the sample grown from the highest T_{ie} of 780 °C and containing 0.2% nitrogen in the layer. Interestingly, the sample grown from the intermediate T_{ie} of 680 °C exhibits a negligible red shift.

The same non-monotonic dependence of the absorption edge values on the growth temperature is observed in the SPV spectra of the same samples. Figure 5 represents the normalized SPV amplitude spectra of the three $GaAs_{1-x}N_x$ samples compared with that of the GaAs substrate. The SPV amplitude is considered as emulating the optical absorption. The absorption edge energy positions determined from the Tauc plot are 1.407 eV, 1.392 eV and 1.371 eV for the samples grown from T_{ie} of 680 °C,

580 °C and 780 °C, respectively. The N content determined from the band gap edge values using the band anticrossing model [11] is $x = 0.05$ at %, 0.1 at % and 0.2 at % for T_{ie} of 680 °C, 580 °C and 780 °C, respectively.

Because of the high volatility of nitrogen, low-temperature epitaxial growth (T < 580 °C) is preferred. At temperatures higher by 100 °C (T_{ie} = 680 °C), the volatility increase leads to a very small N-incorporation efficiency. Since the LPE growth is diffusion limited, two main processes define the N-incorporation in the compound: the diffusion of nitrogen into the As-sublattice and the out-diffusion of substitutional N atoms. Thermodynamic calculations of the diffusion coefficient of nitrogen into GaAs and the escape velocity in the temperature range 650 – 900 °C show that at equilibrium, the N-diffusion into the GaAsN material dominates slightly over the N-diffusion out of the alloy [12]. The increase of nitrogen incorporation efficiency at $T_{ie} \sim 780$ °C could be explained by the higher solubility of nitrogen in the Ga-melt and kick-out of arsenic atoms at these temperatures.

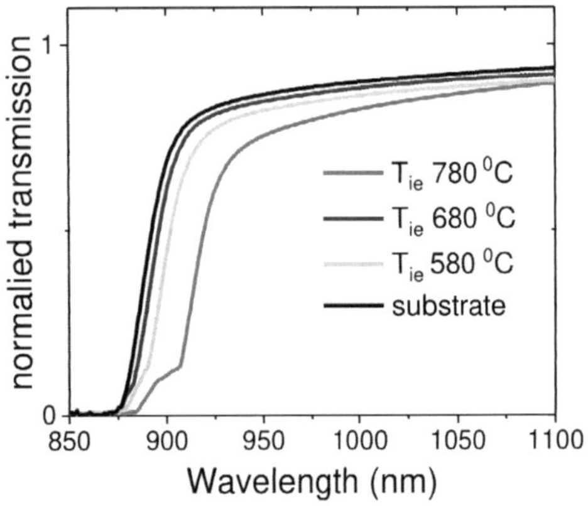

Figure 4. Normalized transmission spectra of GaAsN samples grown from different initial epitaxy temperatures.

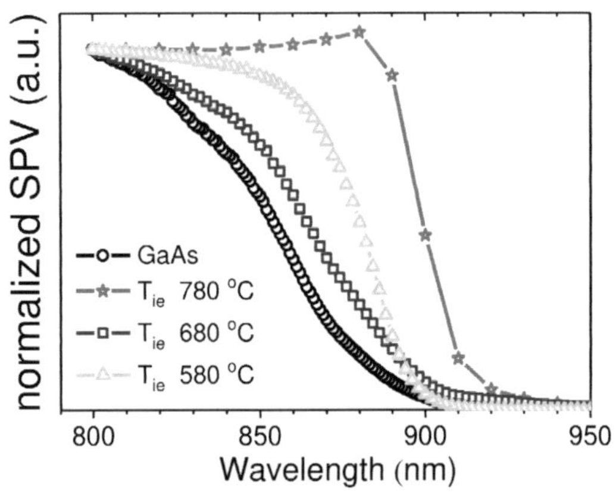

Figure 5. Normalized SPV spectra of GaAsN samples grown from different epitaxy temperatures.

However, it is difficult to obtain uniform epitaxial layers using conventional LPE at high temperatures, since both group V elements in GaAsN exhibit high volatility. On the other hand, crystallization at lower temperatures facilitates the control of the layer composition and makes it possible to avoid the formation of structural defects with high-density.

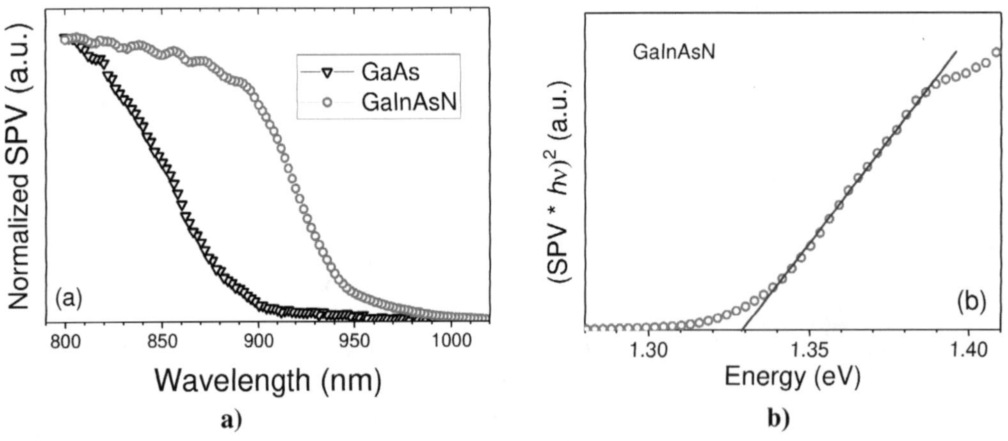

Figure 6. a) Normalized SPV spectra for GaAs and GaInAsN; b) Tauc plot of GaInAsN.

The nitrogen incorporation efficiency at low temperatures could be increased by introducing several percent of In atoms in the crystal lattice [8]. Figure 6 (a) presents the SPV spectrum of a GaInAsN layer grown by low-temperature LPE (T_{ie} < 580 °C). Using the Tauc plot (figure 6 (b)), a bandgap energy of 1.329 eV is found, which is 94 meV lower than the bandgap of GaAs. The In content in the sample determined by EDX measurements is 3%, while the nitrogen content determined from HRXRD and using Vegard's law is 0.3%. The increase of the nitrogen content in the crystal lattice of GaInAsN compensates for the compressive strain caused by the large indium atoms under lattice-matched LPE growth conditions. Thus, GaInAsN with a lower bandgap than that of GaAs can be used in solar cells or IR detectors applications.

4. Conclusions

Successful LPE growth of GaAsN layers on GaAs substrates is demonstrated. The growth temperature has a significant effect on the incorporation efficiency of nitrogen in the crystal lattice. GaAsN layers with a higher nitrogen content of 0.2% are obtained for growth from 780 °C, while the lowest nitrogen incorporation is measured for layers grown from an intermediate initial epitaxy temperature of 680 °C. The incorporation of indium in the crystal lattice favors the nitrogen incorporation. This way, nitrogen content up to 0.3% is achieved at low temperature (T < 580 °C) LPE growth. As a result, the IR photosensitivity of low-temperature grown GaInAsN layers is extended to ~935 nm. Therefore, this dilute nitride material can be used in solar cells or IR detectors applications.

Acknowledgements

This work was supported by the European Regional Development Fund within the Operational Program "Science and Education for Smart Growth 2014 - 2020" under the Project CoE "National center of mechatronics and clean technologies "BG05M2OP001-1.001-0008 and by the Bulgarian Ministry of Education and Science under the National Research Program E+: Low Carbon Energy for the Transport and Households, grant agreement D01-214/2018.

References

[1] Dawidowski W et al 2021 *Energies* **14** 4651
[2] Joshya R S, Carrère H, Ibarra-Sierra V G, Sandoval-Santana J C, Kalevich V K, Ivchenko E L, Marie X, Amand T, Kunold A and Balocchi A 2021 *Adv. Funct. Mater.* **31** 2102003
[3] Boumesjed A, Mazari H, Ameur K, Benseddik N, Kadid S, Benyahya N and Benamara Z 2019 *Int. J. Ambient Energy* 1–6
[4] Guina M 2019 Progress in dilute nitride lasers for high-power and photonic-integration applications *SPIE 10939, Novel In-Plane Semiconductor Lasers XVIII* vol 10939 (SPIE) p 1093902
[5] Kawano M, Minematsu R, Haraguchi T, Fukuyama A and Suzuki H 2020 *Jpn. J. Appl. Phys.* **59** SGGF10
[6] Jen T, Vardar G, Wang Y Q and Goldman R S *Appl. Phys. Lett.* **107** 221904
[7] Wang L, Elleuch O, Shirahata Y, Kojima N, Ohshita Y and Yamaguchi M 2016 *J. Cryst. Growth* **437** 6–9
[8] Donchev V, Milanova M, Lemieux J, Shtinkov N and Ivanov I G 2016 *J. Phys. Conf. Ser.* **700** 012028
[9] Donchev V 2019 *Mater. Res. Express* **6** 103001
[10] Alt H C, Gomeniuk Y, Ebbinghaus G, Ramakrishnan A and Riechert H 2003 *Semicond. Sci. Technol.* **18** 303–6
[11] Shan W, Walukiewicz W, Ager J W, Haller E E, Geisz J F, Friedman D J, Olson J M and Kurtz S R 1999 *Phys. Rev. Lett.* **82** 1221–4
[12] Stolwijk N A, Bösker G, Andersson T G and Södervall U 2003 *Phys. B Condens. Matter* **340–342** 367–70

Journal of Physics: Conference Series **2240** (2022) 012048 doi:10.1088/1742-6596/2240/1/012048

Nanosecond laser ablation of nitride ceramics in liquid

Ro Nikov[1,5], N Nedyalkov[1], T Nikova[2,3] and L Aleksandrov[4]

[1]E. Djakov Institute of Electronics, Bulgarian Academy of Sciences, 72 Tsarigradsko Chaussee, 1784 Sofia, Bulgaria

[2]Institute of Optical Materials and Technologies, Bulgarian Academy of Sciences, G. Bonchev Street, bl. 109, Sofia 1113, Bulgaria

[3]University of Architecture, Civil Engineering and Geodesy, 1 Hr. Smirnenski Blvd., 1046 Sofia, Bulgaria

[4]Institute of General and Inorganic Chemistry, Bulgarian Academy of Sciences, Acad. G. Bonchev Str., Bl. 11, 1113 Sofia, Bulgaria

E-mail: rosen_nikov@abv.bg

Abstract. The work presents results on nanosecond laser ablation of nitride ceramics immersed in a liquid medium. AlN and Si_3N_4 ceramic plates were used targets, and double distilled water was used as a liquid medium. Radiation of a nanosecond Nd:YAG laser system was used in the ablation process with the purpose of studying the influence of the laser fluence and number of pulses on the target surface morphology. The morphology and phase composition of the structured surfaces were explored by scanning electron microscopy and X-ray diffraction measurements, respectively. The optical properties of the obtained colloids were investigated by transmission measurements in the UV-Vis spectral ranges.

1. Introduction

Following the advent of lasers in the 1960s, laser processing has become an attractive tool for researchers and technologists as allowing a controlled and precise modification of the properties of almost any type of material [1]. Such a modification arises from the complex nature of laser-matter interaction, in which the electromagnetic energy delivered by the laser beam to the material is converted first into electronic excitation, then into thermal, chemical and mechanical energy. As a result, the laser treatment can lead to both morphological and chemical changes in the irradiated material. Laser processing of materials almost always involves the phenomenon of laser ablation, which is a thermal or a non-thermal process of removing atoms, atomic clusters, or droplets from a solid surface by irradiating it with an intense and in most cases pulsed laser beam [2]. As one of the most popular and important techniques for material processing, laser ablation can be used for drilling extremely small deep holes through very hard materials, such as metals or ceramics [3, 4], producing thin films [5] and synthesizing nanoparticles [6]. In particular, laser ablation is widely used in processing ceramic materials, allowing precise cutting, scribing and drilling of micron-sized holes for use in microelectronic devices [7]. However, when laser ablation is performed in air, a heat-affected zone is formed near the drilled holes, together with micro-cracks extending away from the holes [8].

[5] To whom any correspondence should be addressed.

Content from this work may be used under the terms of the Creative Commons Attribution 3.0 licence. Any further distribution of this work must maintain attribution to the author(s) and the title of the work, journal citation and DOI.
Published under licence by IOP Publishing Ltd

Ogura et al. [9] have reported that laser drilling in air results in the accumulation of molten material at the edges of the hole. In contrast, laser processing in a liquid environment can minimize the generation of such defects and pileups around the drilled hole [10]. Moreover, the cooling effect of the liquid can reduce the size of the heat-affected zone, which significantly improves the quality of the produced holes.

In this paper, we present results from a study on nanosecond laser ablation of AlN and Si_3N_4 ceramics immersed in double distilled water. The focus is on studying the morphological and chemical changes that occur in the ceramic samples as a result of laser irradiation. The study also reveals the impact of process parameters, such as the fluence and number of laser pulses on the ablation depth in both ceramics.

2. Experimental

The base materials used were hot-pressed AlN and gas-pressure-sintered Si_3N_4 ceramics (Unipretec). The ceramic samples were irradiated with pulses with a duration of 15 ns emitted by a Nd:YAG laser system at the fundamental wavelength of 1064 nm. The experimental setup used for the laser ablation of both ceramics in double distilled water is similar to that used in [11]. Briefly, the laser radiation was focused by a lens with a focal length of 300 mm on the sample surface. The thickness of the water layer over the sample surface was 5 mm. Different laser fluences above the ablation thresholds for the two ceramics were used varying in the range from 3 J/cm^{2}to 31 J/cm^2. The changes in the surface morphology of the ceramics after irradiation with a different number of laser pulses were traced by a scanning electron microscope (SEM) (JEOL 5510). X-ray diffraction (XRD) analysis (Bruker D8 Advance diffractometer, CuKα radiation) was also performed to investigate the structural changes in the ceramics. The ablation depth of the laser-processed ceramics was evaluated by an optical microscope. The transmission spectra of the colloids produced by laser irradiation of the ceramics in water were measured by an Ocean Optics HR 4000 spectrometer (spectral range 200 – 900 nm) equipped with a white light source.

3. Results and discussion

Figure 1 presents XRD patterns of the laser-processed surface of the AlN and Si_3N_4 ceramics in water environment.

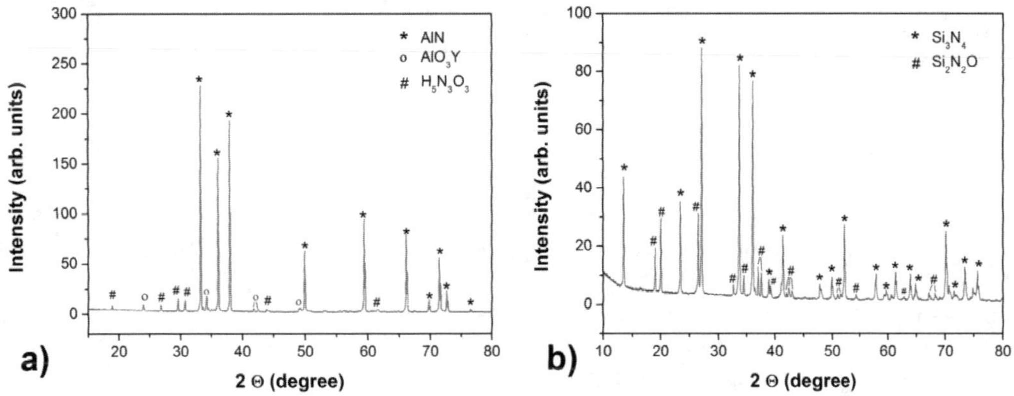

Figure 1. XRD spectra of the AlN (a) and Si_3N_4 (b) samples surface after laser irradiation in an aqueous environment.

The fundamental laser wavelength (λ = 1064 nm) was used to ablate the AlN and Si_3N_4 samples at a laser fluence of 8 J/cm^2 and 11 J/cm^2, respectively. Three different phases, namely AlN (ICSD 98-060-8627), AlO_3Y (ICSD 98-008-8261) and $H_5N_5O_3$ (ICSD 98-016-2750) can be identified in the XRD spectrum of the AlN surface after laser treatment (figure 1 a)). The presence of the AlO_3Y phase in the crystal structure of the ceramics is probably due to the addition of Y_2O_3 (less than 5%) during the ceramics preparation procedure in order to stabilize the ceramics. In the case of laser ablation of the Si_3N_4 target in water, two phases, namely Si_3N_4 (ICSD 98-003-4023) and Si_2N_2O (ICSD 98-010-0775), were observed in the modified region as shown in figure 1 b).

The effect of varying the number of laser pulses on the morphology of AlN samples in laser irradiation in water environment is shown in figure 2. The laser irradiation was performed at the laser wavelength of 1064 nm and a fluence of 8 J/cm^2. As seen, after the first 10 pulses, some of the material is removed, so that the initially polished surface of the ceramic is destroyed and a subsurface material appears that is typical for ceramics, namely a dense material composed of micro-sized crystallites (figure 1 a). The roughness of the native AlN ceramics surface is visualized in the upper part of the image. As the number of applied laser pulses is raised (figure 2 b-d), smoothing of the crystallites edges is seen indicating melting of the ceramics in the ablated region. After irradiation with 300 laser pulses, the discrete ceramic structure disappears almost completely, replaced by randomly distributed molten material (figure 2 c-d). Rounded cavities are observed in the irradiated surfaces, especially in the cases of 100 and 300 applied laser pulses. These cavities are sites of nucleation of bubbles caused by the water boiling as a result of the ablation process, which eventually escape from the liquid [RD1].

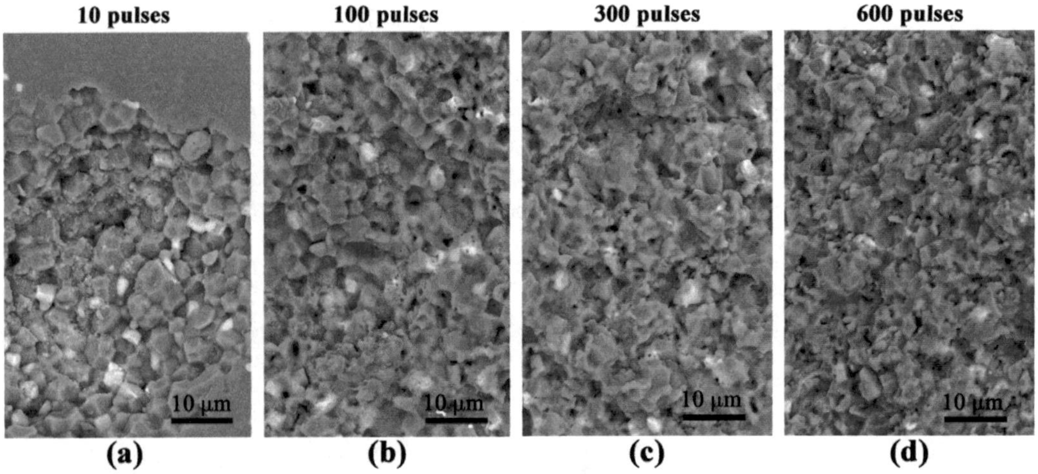

Figure 2. SEM images of AlN samples after laser irradiation in water environment with different number of laser pulses. The wavelength of the laser radiation is 1064 nm and the laser fluence, 8 J/cm^2.

Figure 3 shows SEM images of the Si_3N_4 ceramics surface irradiated with an increasing number of laser pulses. The parameters of the laser treatment are laser wavelength of 1064 nm and fluence of 11 J/cm^2. A SEM image of native Si_3N_4 is inserted in figure 3 (a). Increasing the number of applied laser pulses leads to a melting of the initial discrete structure of the Si_3N_4 ceramics, similarly to the case of AlN ablation in water. However, it is worth noting that in this case the size of the crystallites forming the ceramics is smaller, and a significant melting is observed even after applying a single laser pulse (figure 3 (a)).

Figure 4 a) shows the results of measuring the ablation depth on AlN ceramics surface as a function of the applied number of laser pulses. Increasing the number of laser pulses from 10 to 600 leads to an almost linear increase in the ablation depth, with the depth reaching 80 μm after 600 applied pulses. In contrast, the dependence of the ablation depth on the number of applied laser pulses in the case of laser ablation of Si_3N_4 in water is definitely not linear (figure 4 b)). A rapid increase to 50 μm is then seen in the ablation depth after applying the first 100 laser pulses. In the range of 100 to 600 pulses, the ablation depth increase slows down, reaching 80 μm at 600 pulses, similarly to the

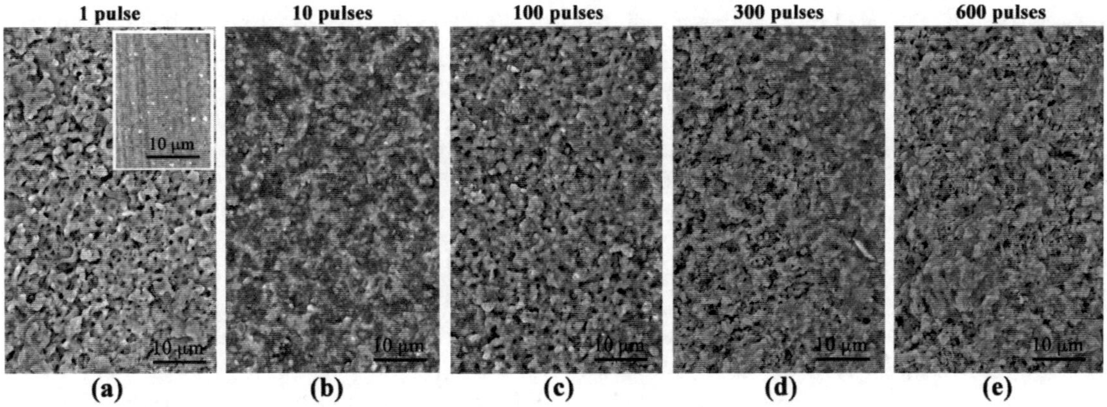

Figure 3. SEM images of Si_3N_4 samples after laser irradiation in water environment with a different number of the laser pulses. The wavelength of the laser radiation is 1064 nm and the laser fluence, 11 J/cm^2. A SEM image illustrating the surface roughness of native Si_3N_4 is inserted in (a).

case of ablation of AlN ceramics. The reasons for the difference in the behavior of the two dependences could be both the difference in the laser fluence used (8 J/cm^2 and 11 J/cm^2) and a different ablation efficiency. Thus, an additional study was performed on the ablation depth variation as a function of the laser fluence applied at a fixed number of laser pulses ($N = 100$). The two dependences for the AlN and Si_3N_4 ceramics are illustrated in figures 4 c) and d), respectively. As can be seen, a significant increase in the ablation depth with the laser fluence is achieved for values of up to 15 J/cm^2 in both ceramics under the process parameters chosen ($\lambda = 1064$ nm, $N = 100$ pulses).

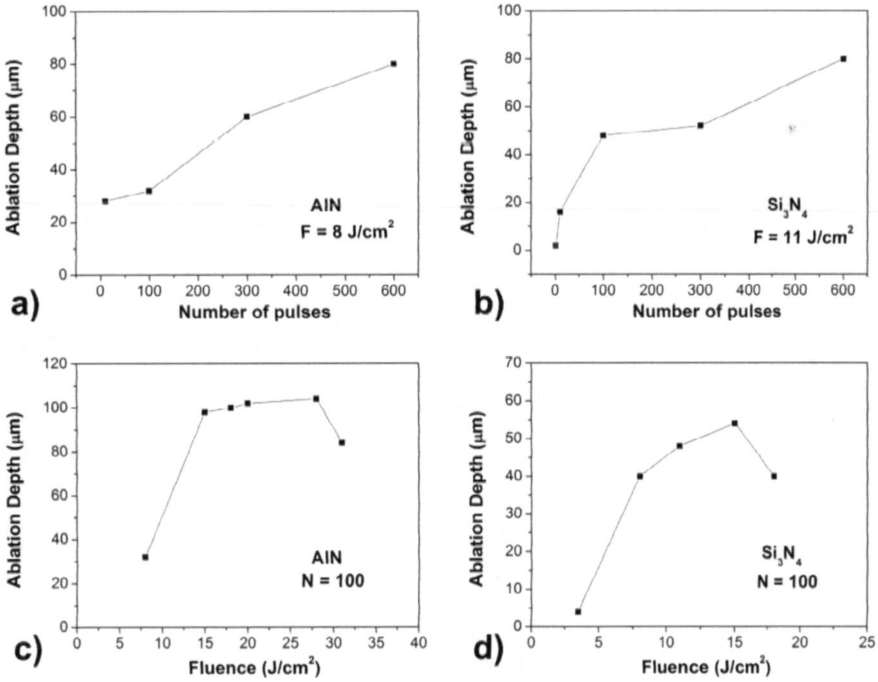

Figure 4. Relationship between the ablation depth and the number of applied laser pulses during laser irradiation of AlN (a) and Si_3N_4 (b) in water. Relationship between the ablation depth and the laser fluence during laser irradiation of AlN (c) and Si_3N_4 (d) in water.

The use of higher fluences (above 15 J/cm^2) leads to a slight increase in the ablation depth in AlN (up to 104 μm) followed by a decrease (figure 4 c)). In the case of Si_3N_4, a peak ablation depth of 54 μm is measured using a fluence of 15 J/cm^2 followed by a decrease in the depth with the further fluence increase (figure 4 d)). The main reason for the ablation efficiency reduction when using higher laser fluences in processing both AlN and Si_3N_4 ceramics is related to the laser radiation being shielded by the material ejected into the liquid. It is known that laser ablation of a target in a liquid medium leads to the ejection from the target surface of debris with sizes varying from a few nanometers to submicron or even micron size [13]. Thus, the use of higher laser fluences leads to a faster saturation with nanoparticles and microparticles of the aqueous layer above the ceramic surface. The nano- and microstructures formed in the liquid, in turn, absorb and scatter the laser radiation, effectively reducing the laser energy deposited on the ceramic surface. Comparing the ablation depths of the two ceramics studied, we found that ablation of AlN results in a depth almost twice as high ($d = 98$ μm) as in Si_3N_4 under the same process parameters ($N = 100$ pulses, $F = 15$ J/cm^2). The explanation t is rooted in the more efficient ablation of the ceramics (lower ablation threshold), which leads to a faster saturation of the water with debris, which in its turn limits the ablation process and hence reduces the ablation depth. To confirm the latter statement, the transmission spectra were measured of the colloids obtained by laser ablation of AlN and Si_3N_4 ceramics in water. The profiles of the two spectra are similar, but it is clear that in the case of ablation of Si_3N_4 the transmission signal is lower by almost 30% than that of AlN (figure 5).

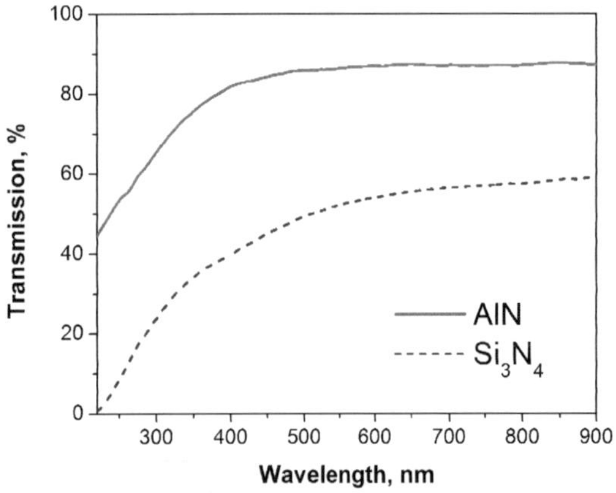

Figure 5. Transmission spectra of colloids produced by laser ablation of AlN and Si_3N_4 ceramic targets in water. The laser radiation wavelength is 1064 nm.

4. Conclusions

The work demonstrates the possibility of modifying the morphology of AlN and Si_3N_4 ceramics in water environment by using nanosecond laser pulses. After laser treatment by 600 laser pulses, a complete melting was observed of the initial crystalline structure of both ceramics studied. In the case of laser ablation of AlN, an almost linear increase was seen in the ablation depth with the applied number of pulses. Laser ablation of Si_3N_4 caused a rapid increase in the ablation depth during the first 100 pulses, followed by a slower growth rate. The dependence of the ablation depth on the laser fluence was studied; it was found that at certain process parameters ($\lambda = 1064$ nm, $N = 100$ pulses) in the case of Si_3N_4, the ablation depth reached a peak value at a fluence of 15 J/cm^2. In laser processing of AlN, a similar dependence of the ablation depth on the laser fluence was observed, with the difference that the peak value of the ablation depth was almost twice as high as that in Si_3N_4. This is related to the more efficient ablation of Si_3N_4 relative to AlN leading to a faster saturation of the water with material ejected from the target, which limits the ablation efficiency.

Acknowledgments
This work was supported by the Bulgarian National Science Fund under project KP-06-N47/11.

Journal of Physics: Conference Series **2240** (2022) 012048 doi:10.1088/1742-6596/2240/1/012048

References

[1] Schaaf P 2010 *Laser processing of materials: fundamentals, applications and developments* (Springer-Verlag Berlin Heidelberg)
[2] Miller J C 1994 *Laser ablation: principles and applications* (Berlin; New York: Springer Verlag)
[3] Chatterjee S, Mahapatra S S, Bharadwaj V, Choubey A, Upadhyay B N and Bindra K S 2019 *Proc IMechE Part B: J Engineering Manufacture* **233(8)** 1872
[4] Wang H, Lin H, Wang C, Zheng L and Hu X 2017 *J. Eur. Ceram. Soc.* **37** 1157
[5] Ogugua S N, Ntwaeaborwa O M and Swart H C 2020 *Coatings* **10** 1078
[6] Amendola V and Meneghetti M 2013 *Phys. Chem. Chem. Phys.* **15** 3027
[7] Rihakova L and Chmelickova H (2017) *Adv. Prod. Eng. Manag.* **12** 412
[8] Sciti D and Bellosi A 2001 *Appl. Surf. Sci.* **180** 92
[9] Ogura H and Yoshida Y 2003 *Jpn. J. Appl. Phys.* **42** 2881
[10] Tsai C-H and Li Ch-Ch 2009 *Journal of Materials Processing Technology* **209** 2838
[11] Nikov R G, Nikolov A S and Atanasov P A 2011 *Proc. of SPIE* **7747** 774708
[12] Wee L, Ling E K, Chi W T and Lim G C 2011 *Int. J. Appl. Ceram. Technol.* **8(6)** 1263
[13] Nikov R G, Nedyalkov N and Karashanova D B 2020 *Appl. Surf. Sci.* **518** 146211

Depth profiling of very thin HfO_2/Al_2O_3 stacks by ellipsometry

Y Karmakov[1,3], A Paskaleva[2] and D Spassov[2]

[1]Department of Solid State Physics and Microelectronics, Faculty of Physics, St. Kliment Ohridski University of Sofia, 5 James Bourchier Blvd., 1164, Sofia, Bulgaria
[2]G. Nadjakov Institute of Solid State Physics, Bulgarian Academy of Sciences, 72 Tsarigradsko Chaussee, 1784, Sofia, Bulgaria

E-mail: karmakov@phys.uni-sofia.bg

Abstract. Ellipsometry (VASE and MAIE) with appropriate algorithms for experimental data interpretation was applied for quantitative characterization of thin Al_2O_3/HfO_2 multilayers formed by atomic layer deposition (ALD) applicable to the fabrication of charge-trapping non-volatile memories. A substantial benefit of the algorithms is the depth profiling of the stacks. In this work, the depth profiles were retrieved of the HfO_2 constituent in very thin (Al_2O_3:HfO_2) stacks embedded in thin Al_2O_3 surrounding layers. The peculiarities in the achieved depth profiles are used to determine the bi-layer blocks and their sub-layer thicknesses. The influence of a rapid thermal annealing in O_2 on the depth profile and the sub-layer thicknesses is studied; substantial changes are thus revealed in the thickness and composition of the stacks. This information could be used in optimizing the dielectric and electrical properties of the stacks and multilayers.

1. Introduction

Nowadays, the ongoing aggressive downscaling of microelectronic devices has stimulated the research on high-*k* dielectrics and, particularly, very thin stacks of metal oxide layers in view of replacing the SiO_2 or SiO_xN_y insulators traditionally used in the metal-oxide-field effect transistor technology. It has been demonstrated that the high-*k* metal oxides are intrinsically trap rich materials, which makes them very attractive for implementation as charge trapping media in the future generations of charge-trapping non-volatile memories. As HfO_2 is already being employed in the production of high-performance logical integrated circuits such as microprocessors, the opportunity of using the HfO_2 technology in non-volatile memory applications is quite attractive. It has also been shown recently that the charge-trapping efficiency and retention characteristics of HfO_2 can be substantially enhanced by incorporation of Al in its matrix, either by creating bi-layered stacks or by Al-doping [1, 2]. Thus, (HfO_2):(Al_2O_3) stacks have been considered as a promising candidate for the next generation of nonvolatile flash memories. Atomic layer deposition (ALD) is the deposition method that made possible the implementation of high-*k* dielectrics in microelectronics. It is capable of producing high-quality thin and ultrathin (few nm) films of a variety of metal oxides and multilayers as (HfO_2):(Al_2O_3) stacks. The noteworthy feature of the ALD growth is the control of the deposited layers on a mono-layer scale, which ensures a constant growth rate for a cycle (GRC) throughout the process. However, in the very beginning of the growth, nucleation inhibition or enhancement is reported which depend

[3] To whom any correspondence should be addressed.

Content from this work may be used under the terms of the Creative Commons Attribution 3.0 licence. Any further distribution of this work must maintain attribution to the author(s) and the title of the work, journal citation and DOI.
Published under licence by IOP Publishing Ltd

strongly on the deposition surface. This results in differences in the properties of very thin layers v/s the thin ones. The properties of the multilayers could be tuned by changing the ratio of deposition cycles (i.e. thicknesses) of sub-layers. However, the characterization of very thin multilayers is challenging due to a lack of reliable experimental tools. The direct experimental methods like HRTEM could provide images with a thickness accuracy of 0.1 nm; the indirect methods are better by several factors, e.g., the outstanding spectroscopic ellipsometry (SE) accuracy reaches down to sub-monolayer thicknesses.

The goal of the present work is to employ SE with appropriate algorithms to determine the physical properties of multilayers and sub-layers, such as (Al_2O_3):(HfO_2), in very thin stacks and blocks of stacks. The physical properties of sub-layers are of paramount importance in achieving better correlation with the electrical measurements of these multilayers and stacks.

2. Ellipsometric data interpretation

In indirect methods, a fitting procedure between the experimental and the modeled data is used with appropriate algorithms. The main equation in ellipsometry is $\rho(\varphi,\lambda) = \tan\psi \, \mathrm{eps}(i\Delta)$, where ψ and Δ are the measured ellipsometric angles and ρ is the Fresnel amplitude reflection coefficients for p- and s-polarized light with incident angle φ and wavelength λ. In our study, experimental data from multi incident angle ellipsometry (MIAE) and variable angles spectroscopic ellipsometry (VASE) measurements of very thin layered samples are interpreted by using the algorithms commented in previous publications [3-5] for characterization of inhomogeneous in depth dielectric layers (figure 1). One substantial benefit of these algorithms is the reduced cross-correlations between the fitting parameters yielding a better fit quality and sensitivity. The depth profiling achieved by these algorithms is used as a complementary tool for precise thickness determination of multilayers with more unknown properties, see Ref. [6]. Briefly, the inverse problem of finding the dielectric function ε of the investigated inhomogeneous in depth sample with thickness d is formulated in a more general way as the problem of solving the non-linear operator equation:

$$\rho^*(\varphi,\lambda) = A[\varepsilon(t),d] \quad (1)$$

where A denotes a non-linear operator representing all mathematical operations involved in the theoretical calculations of $\rho^*(\varphi,\lambda)$ for the constructed model. To solve equation (1), the Newton-Kantorovich type iterative method for non-linear operator equations is used. The final fitting results are estimated by the residual kernel R with a shape of a Dirac delta function. The values of R strongly increase as "spikes" at both ends of the investigated thickness, at the depths where discontinuities exist [4]. Generally, in mixed oxides the dielectric function ε depends on the constituents and their volume fractions. Let the relative volume fraction of HfO_2 in (Al_2O_3):(HfO_2) stacks be denoted as q. If the constituents are depth-dependent, then $\rho^*(\varphi,\lambda) = A[\varepsilon(q(t)),d]$ with $0 < q < 1$ can be solved for the depth profile $q(t)$, as well as the ε depth profile and the refractive index depth profile $N = n + ik$ for a given wavelength λ.

3. Experimental details

Figure 1 and figure 2 show schematically the model and real samples developed for applications in charge trapping non-volatile memories. The top relatively thick Al_2O_3 layer, referred to as blocking oxide (BO), blocks the leakage of the trapped charge to the gate electrode; the middle (Al_2O_3):(HfO_2) stack serves as a charge trapping medium in which the electrical charges injected from the Si substrate are captured in spatially distributed traps; the third thin Al_2O_3 layer, referred as tunnel oxide (TO), insulates the charge trapping layer from the substrate. All oxide layers were deposited by ALD at 135 °C, using tetrakis (dimethylamido) hafnium (TDMAH) and trimethylaluminum (TMA) and H_2O vapor as an oxidant. The growth rate/cycle (GRC) was 0.1 nm/cy for Al_2O_3 and 0.14 nm/cy for HfO_2. These values are used to calculate the nominal thicknesses. After the deposition, some of the samples were annealed (rapid thermal annealing, RTA) in O_2 at 800 °C for 1 min. The SE measurements conducted by a J.A. Woollam M2000D provide the ellipsometric angles ψ and Δ as a function of incident angles (60 ÷ 75) deg and wavelengths (999 ÷ 190) nm. Additionally, SE of a bare Si substrate and of a single 20-cy HfO_2 grown on it was also performed. Many possible dielectric function dispersion data for thin to ultrathin Al_2O_3 and HfO_2 layers reported in the literature [7] were tested as guessed functions in the VASE calculations.

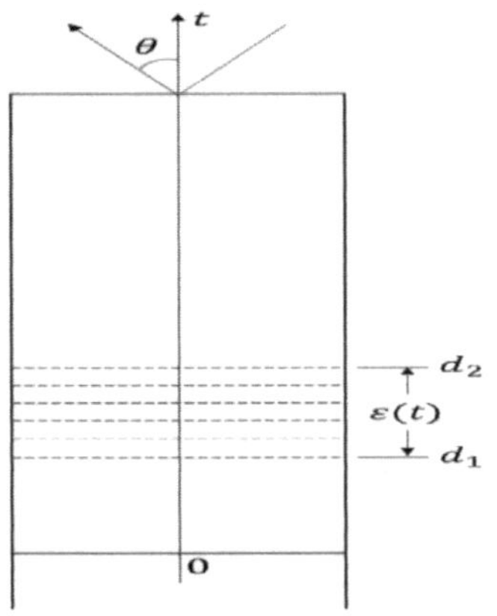

Figure 1. The schematic model used.

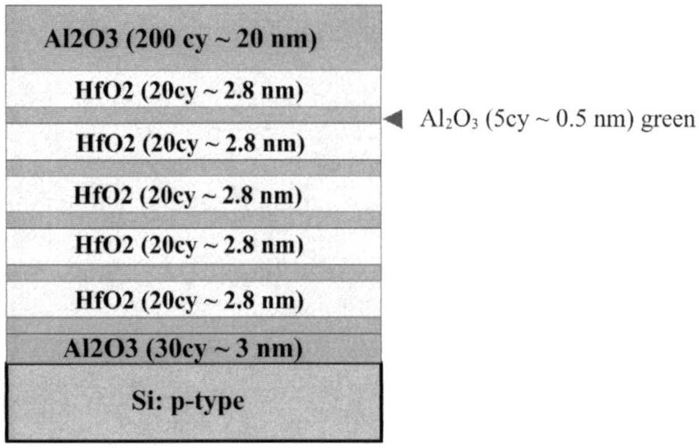

Figure 2. The multilayered structure of the real sample.

The overall dielectric constant *k* of the multilayered dielectric structures was evaluated by capacitance voltage (*C-V*) measurements (Agilent E4980A LCR) conducted on metal-insulator-silicon (MIS) capacitors with evaporated Al top and backside electrodes.

4. Results and discussions

The obtained thicknesses of the (BO) layer, of the "stack layer" and of the (TO) layer in the sample shown in figure 2 before and after RTA are given in table 1 in comparison with the calculated nominal thicknesses by the known cycle numbers and GRC. The thicknesses of the layers are the results of the best fits using the measured VASE or MAIE data and the nominal thicknesses as model guesses. Our algorithm procedures allow testing of a different combination, even in Hilfiker ways [6], for best fits with appropriate guessed dielectric function dispersions of both the thin and the very thin oxides [7].

Table 1. Comparison of nominal and estimated by SE thicknesses of the TO layer, of the trapping stack and of the BO layer before and after annealing.

Layer	*d* (nm)	Nominal *d* (nm)	dielectric constant *k*
TO, as-grown	3.12 ± 0.08	3.0	7.3
TO, RTA	3.32 ± 0.05		7.7
(HfO_2):(Al_2O_3) stack as-grown	20.02 ± 0.08	16.5	averaged 10.0
(HfO_2):(Al_2O_3) stack RTA	17.36 ± 0.05		changeable; 15.3 in the middle
BO, as-grown	20.86 ± 0.08	20.0	7.1
BO, RTA	22.03 ± 0.08		8.1

As it is seen, a substantial thickness difference between the obtained value and the calculated nominal value only exists for the "stack layer". Some authors have commented on the discrepancies between the nominal and the measured thicknesses for thin multilayers. Our attention was fixed on determining the thickness of the "stack layer" and its blocks with sub-layers. As a complementary tool we used the depth profiling of the HfO_2 constituent as a relative volume fraction in the stack region. It is a q depth profile, as defined above. The uncommon character of the algorithms used here involves an uncommon estimation of the fit quality. The estimator is the resolution kernel R of the Newton-Kantorovich type iterations with Tihonov regularization [3, 4]. The shape of R is a Dirac delta function. At the ends of the investigated thickness, the values of R sharply increase as a "spike". The retrieved q depth profile of the annealed stack region of the sample in figure 2 is shown in figure 3.

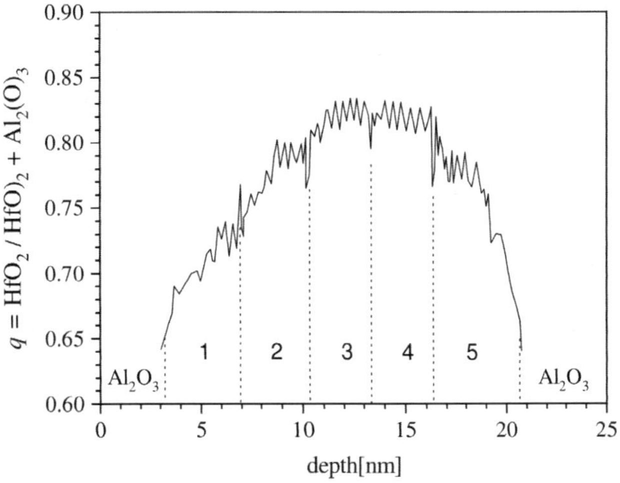

Figure 3. Retrieved HfO_2 relative volume fraction in depth of the stack region after annealing.

One can clearly see "spikes" in the errors at depths where the ends of the bi-layered blocks might be [4]. In every block, there are two sub-layers with the same topological trajectory in the profile. Moreover, there are oscillations of the profile errors, and what is essential, they are in that second sub-layer of each block where HfO_2 has to be. In the first sub-layers the oscillations are suppressed. Although the thicknesses of the first sub-layers are larger than the nominal Al_2O_3, we consider them as Al-rich HfO_2 or, more correctly, $(HfAlO)_x$ mixed oxides. In this sense, the second sub-layers are $(HfO_2)_{eff}$. Figure 4 compares the bi-layer blocks and their sub-layer thicknesses before $(AlHfO)_x$, $(HfO_2)_{eff}$ and after annealing, $(AlHfO)^I_x$, $(HfO_2)^I_{eff}$, respectively.

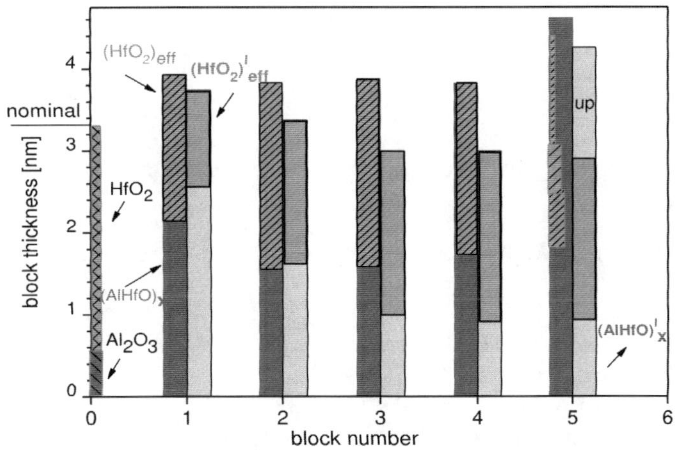

Figure 4. Estimated thicknesses of the blocks and sub-layers in the stack region before and after annealing. The left columns depict the results for as-grown samples; the right, the results for the annealed samples. Magenta represents HfO_2 content and green, Al_2O_3 one.

As it is seen, for the as-deposited stack, the three bi-layer blocks in the middle are very similar. As expected, the first and the last blocks differ from the middle ones as they are strongly affected by the neighboring TO and BO Al_2O_3 layers. The thinner $(HfO_2)_{eff}$ implies some growth inhibition in these numbers of cycles. The more complex structure of the last bi-layer is more likely due to some intermixing with the thick BO. Figure 4 shows that the O_2 annealing substantially alters the total thickness of the stack. The thicknesses of the blocks and sub-layers undergo serious reductions (see also table 1). The obtained thickness reduction varies from block to block. For the 1st block it is 0.19 nm, for the 2nd, 0.45 nm, for the 3rd, 0.88 nm, for the 4th, 0.83 nm and for the last one, 0.36 nm. Also, it is seen that the reduction of the $(AlHfO)_x$ sub-layers (except for the first block) is stronger, especially in the middle of the stack. These results imply that as a result of the annealing the Al_2O_3:HfO_2 stack densifies. Some phase transitions or out-diffusion of Al are also possible. By retrieving the HfO_2 relative volume fraction and the refractive index depth profiles of the stacks, we are able to calculate the profiles of the band gaps (E_g) and the values of the dielectric constant k (shown in table 1) of every block and sub-layer using the Clausius Mossotty relations. The values thus found fall in a range between the known values for very thin HfO_2 and Al_2O_3 layers. They serve as a complementary verification of the SE results and for the capability of the algorithms used.

The dielectric constants of the investigated stacks were also evaluated by standard capacitance-voltage measurements at a frequency of 1 MHz. The k values were calculated from the accumulation capacitance and the total SE thickness of the sample using the plain-parallel capacitor relation. The k obtained are effective ones, the complex multilayered dielectric being represented by a single insulator with permittivity k_{eff}. For the as-grown samples, k_{eff} is 10.3; for the annealed structures $k_{eff} = 10.04$. In order to compare these values with the data obtained by SE and listed in table 1, the effective permittivity of the dielectric structures were calculated using the relation for three serially connected capacitors: for the TO, the (HfO_2):(Al_2O_3) stack and the BO regions. The effective permittivities obtained from the SE thickness data are 8.2 and 10.0 for the as-grown and for the annealed sample, respectively. Therefore, the values of k_{eff} from both the optical and the electrical methods are generally in a good agreement.

5. Conclusions

The results presented demonstrate the capability of ellipsometry employed in combination with proper algorithms for VASE and MAIE data interpretation to investigate the physical properties of thin multilayers and very thin Al_2O_3:HfO_2 stacks. The depth profiling and its error peculiarities are complementary tools for determining the thickness of every layer in the investigated sample, and the thicknesses of the blocks and of their sub-layers. The thickness measured of our very thin stack is not equal to the nominal one. The values of the dielectric constants estimated by SE and an electrical method are in a good agreement.

Acknowledgements

The work was supported by Bulgarian National Science Fund under project KP-06-H37/32.

Reference

[1] Paskaleva A, Rommel M, Hutzler A, Spassov D and Bauer A J 2015 *ACS Appl. Mater. Interfaces* **7** 17032

[2] Spassov D, Paskaleva A, Krajewski T A, Guziewicz E, Luka G and Ivanov T 2018 *Phys. Stat. Solidi A-Applications and Materials Science* **215** 1700854

[3] Tonova D and Konova A 1996 *Surf. Sci.* **349** 221

[4] Tonova D and Konova A 2001 *Thin Solid Films* **397** 17

[5] Karmakov Y, Paskaleva A and Atanassova E 2012 *Appl. Surf. Sci.* **258** 4507

[6] Hilfiker J N, Pribil G K, Synowicki R, Martin A C and Hale J S 2019 *Surface & Coatings Technology* **357** 114–121

[7] Ming D, Bersch E, Consiglio S, Zhang T, Tyagi P, Clark R D, Leusink G J, Srivatsa A and Diebold A C 2009 *AIP Conf. Proc.* **1173** 104

Comparative study of the sidewall shape and proximity effect of the AR-N 7520 negative electron beam resist

I Kostic[1,4], K Vutova[2,5], E Koleva[2,3], A Bencurova[1], A Konecnikova[1] and R Andok[1]

[1]Institute of Informatics, Slovak Academy of Sciences, Dubravska cesta 9, 845 07 Bratislava, Slovak Republic
[2]Institute of Electronics, Bulgarian Academy of Sciences, 72, Tsarigradsko Chaussee, Sofia 1784, Bulgaria
[3]University of Chemical Technology and Metallurgy, 8, Kliment Ohridski Blvd., Sofia 1756, Bulgaria

E-mail: ivan.kostic@savba.sk, katia@van-computers.com

Abstract. This paper deals with the performance of the AR-N7520 (*Allresist*) negative electron beam resist (EB resist) which was selected as an etch mask for the fabrication of gratings on GaAs substrates. The developed resist sidewall shape is crucial for this purpose. The required near-to-vertical sidewall shape can be achieved by optimizing the electron beam lithography (EBL) process based on experimental investigations and computer simulations. The sidewall shape dependence on the EBL parameters (exposure dose, resist pattern, etc.) and the proximity effect are studied.

1. Introduction

Large-area gratings for optoelectronics require periodicity with a nanometer precision in order to be tuned for use at specific light wavelengths. The EBL allows the patterning and reproduction of gratings with nanoscale resolution and precision [1]. Among the many applications of EBL, the fabrication of gratings for semiconductor lasers with a specific wavelength has attracted particular attention. In view of optimizing the EBL process of producing such gratings, we performed AR-N 7520 negative EB resist profile measurements and profile simulations related to the fabrication of line gratings with a period of 820 nm on a GaAs substrate. The Novolak-based resist AR-N 7520 belongs to a family of promising high-resolution resists with high dry-etching durability [2-4]. This resist is investigated as a candidate for the fabrication of high-resolution and high-density gratings for optoelectronic applications.

2. Results and discussion

2.1 Experiment

The experiments were performed using the ZBA23 (*Raith*) EBL system with a variable-shape electron beam cross-section at an electron energy of 40 keV [5-7]. A thin film of the AR-N 7520 (*Allresist*)

[4] To whom any correspondence should be addressed.
[5] To whom any correspondence should be addressed.

Content from this work may be used under the terms of the Creative Commons Attribution 3.0 licence. Any further distribution of this work must maintain attribution to the author(s) and the title of the work, journal citation and DOI.
Published under licence by IOP Publishing Ltd

negative electron beam resist with a thickness of 400 nm was prepared by spin coating on a GaAs substrate. The development was conducted with the AR 300-47 developer for 60 seconds. The resist thickness measurements were carried out using a standard profilometer technique (Alphastep equipment). A high-resolution Quanta FEG 250 (*FEI*) scanning electron microscope was used for dimension measurements of the resist profiles at the magnification of 200 000×.

The tests resist profile consisted of a set of long single lines of linewidth 100 nm, 200 nm and 500 nm, and line gratings of *L/S* 100/700 nm, 200/600 nm and 300/500 nm, respectively (figure 1). The linewidths *L* were measured at heights *T* (at the resist thickness), ¾*T*, ½*T*, ¼*T* and 0*T* (at the substrate surface) (figure 2), for different exposure dose levels *Q* (from a minimal value up to a significantly higher one).

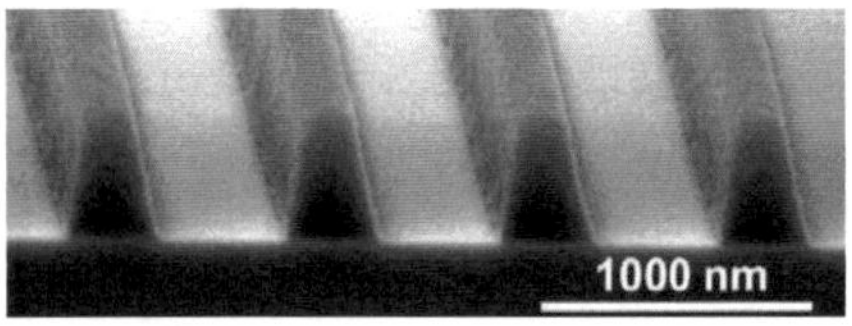

Figure 1. View of line grating profile exposed with 200×3000 nm^2 shots at exposure dose 175 μC/cm^2.

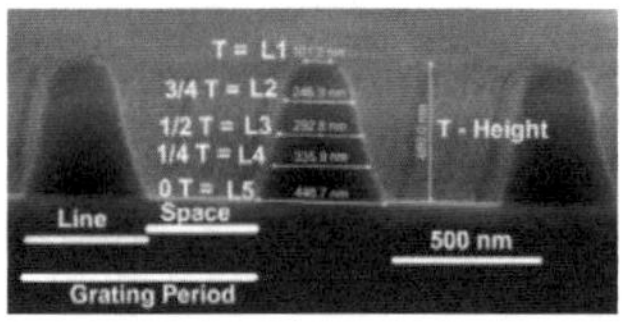

Figure 2. AR-N 7520 resist profiles and measurement dimension details.

Figures 3-5 show developed resist profiles for line gratings with linewidths 100 nm and 200 nm, and spacing 500 nm, 600 nm and 700 nm at various exposure doses for a 400-nm thin AR-N 7520 resist and 40-keV electron energy. The spacing and the linewidth are the limiting parameters in line grating fabrication, as demonstrated in figures 3 and 4. For the same exposure dose of 340 μC/cm^2, the profile of lines with a spacing of 500 nm is significantly rounded at the bottom (figure 3) in comparison to that with a spacing of 700 nm (figure 4). This fact is explained by the contribution of electrons to adjacent lines. Figures 4 and 5 demonstrates that for a 100-nm linewidth (figure 4), a nearly twice as high exposure dose is necessary to obtain the same linewidth as in case of 200 nm linewidth (figure 5) (in both cases the period is 800 nm).

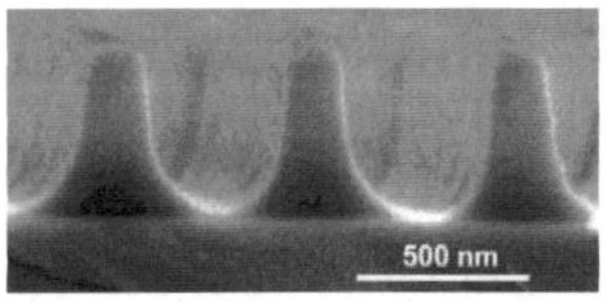

Figure 3. Profile of line grating with *L/S* 100/500 nm at exposure dose 340 μC/cm^2.

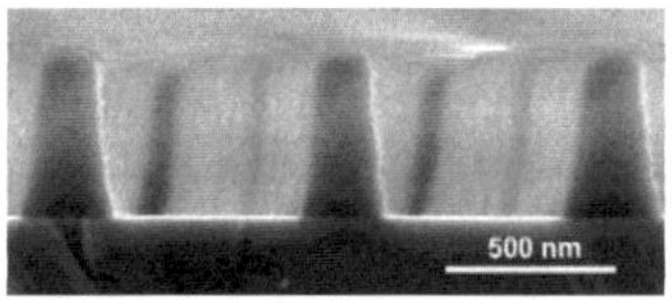

Figure 4. Profile of line grating with *L/S* 100/700 nm at exposure dose 340 μC/cm^2.

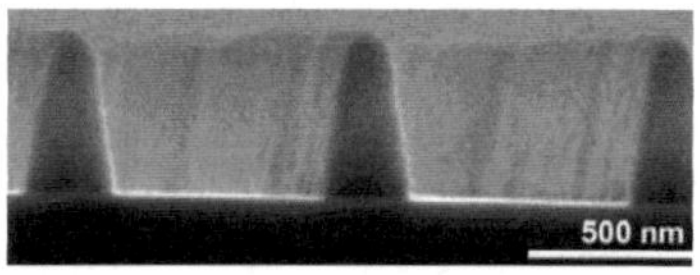

Figure 5. Profile of line grating with *L/S* 200/600 nm at exposure dose 150 μC/cm^2.

2.2 Models and simulation results

A statistical approach [8] was applied to estimate regression models describing the dependence of the profile geometry on the exposure dose based on experimentally obtained data (*2.1.*).

2.2.1 Regression model estimation and AR-N 7520 resist profile prediction. The linewidths dependences $Y_{LW\text{-}200}$ and $Y_{LW\text{-}500}$ for exposure of single lines of 200 nm and 500 nm as functions of the exposure dose (z_1) and the resist depth (z_2) are investigated based on the experimental data obtained. The ranges of the EBL process parameters considered in both cases are given in table 1.

Table 1. Process parameters for AR-N 7520 resist.

Linewidth	Parameter	Coded	$z_{min,\,i}$	$z_{max,\,i}$	Coding
$Y_{LW\text{-}200}$	Exposure dose z_1, μC/cm^2	x_1	200	760	$x_1 = (z_1 - 480)/280$
	Resist depth z_2, nm	x_2	0	475	$x_2 = (z_2 - 237.5)/237.5$
$Y_{LW\text{-}500}$	Exposure dose z_1, μC/cm^2	x_1	120	760	$x_1 = (z_1 - 440)/320$
	Resist depth z_2, nm	x_2	0	482	$x_2 = (z_2 - 241)/241$

The regression models are estimated for coded values of the parameters in the region [-1÷1] and are presented in table 2. The coded values x_i of the process parameters (table 1) can be calculated using their maximal ($z_{max,i}$) and minimal ($z_{min,i}$) values following the equation:

$$x_i = (2z_i - z_{max,i} - z_{min,i})/(z_{max,i} - z_{min,i}) \quad (1)$$

The accuracy is estimated by the root-mean-square errors (RSME) and the determination coefficients (square multiple correlation coefficients) R^2 and R_{adj}^2, which define the goodness of a fit of the estimated models (table 2).

Table 2. Regression models for the linewidths $Y_{LW\text{-}200}$ and $Y_{LW\text{-}500}$ of the developed AR-N 7520 resists.

Linewidth	Regression model	RSME	R^2, %	R_{adj}^2, %
$Y_{LW\text{-}200}$	$294.45+79.93x_1-28.20x_2-69.05x_1^2-6.61x_2^2+59.15x_1^3-14.15x_2^3-14.93x_1^2x_2+20.99x_1x_2^2$	6.37	99.41	99.34
$Y_{LW\text{-}500}$	$693.95+60.84x_1+356.89x_1^3-145.40x_2^3-151.30x_1^4-99.39x_2^4-370.29x_1^2x_2+279.70x_1x_2^2-177.87x_1^3x_2+302.01x_1^2x_2^2-78.13x_1x_2^3$	39.59	98.66	98.47

In figures 6-7, contour plots of the developed linewidth Y_{LW} as a function of the exposure dose (z_1) and the resist depth (z_2) are presented for single lines of 200 nm and 500 nm ($Y_{LW\text{-}200}$ and $Y_{LW\text{-}500}$).

Calculated and experimental (marked by red squares) developed AR-N 7520 resist profiles for single lines of 200 nm and 500 nm for different exposure doses are shown in figure 8. A very good agreement between the calculated and the experimental results can be seen.

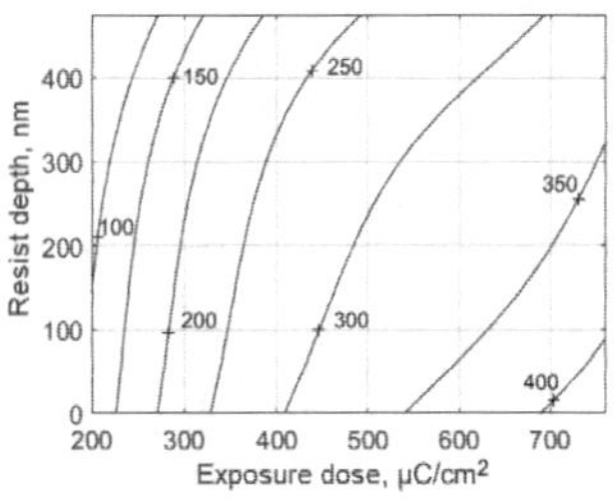

Figure 6. Contour plot of the developed linewidth $Y_{LW\text{-}200}$ as a function of the exposure dose (z_1) and the resist depth (z_2).

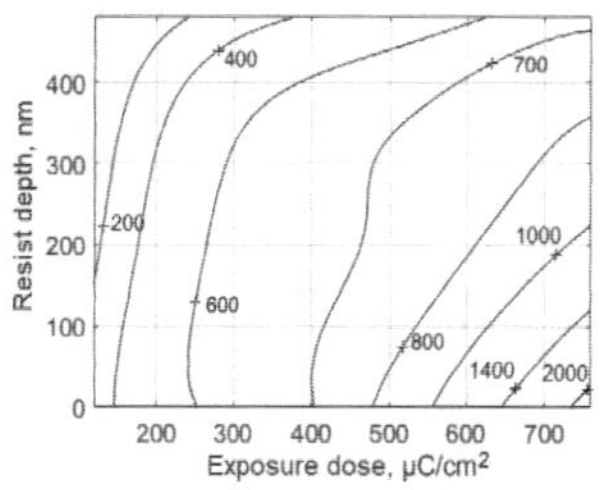

Figure 7. Contour plot of the developed linewidth $Y_{LW\text{-}500}$ as a function of the exposure dose (z_1) and the resist depth (z_2).

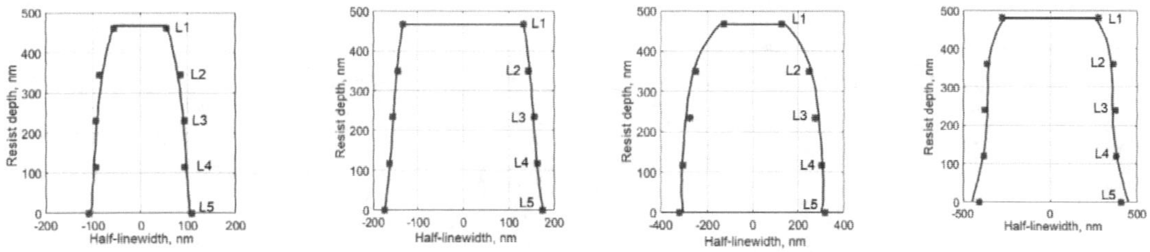

a) **b)** **c)** **d)**

Figure 8. Developed AR-N 7520 resist profiles for a single line of 200 nm and exposure doses: a) 280 μC/cm^2, b) 540 μC/cm^2, and for a single line of 500 nm and exposure doses: c) 260 μC/cm^2, d) 520 μC/cm^2. The experimental data are denoted by red squares.

2.2.2. Optimization criteria. As optimization criteria, the following parameters are proposed:

- the mean values of the linewidths of the developed resist $\overline{Y}_{LW}$ at different resist depth levels (*L*1 to *L*5) for a given value of the exposure dose;
- the sidewall angles α_1 defined between the half-linewidths at resist depths *L*5 and *L*2 (figure 9).

The parameters $\bar{Y}_{LW}$ and α_1 are calculated for the experimental data concerning single lines of 200 nm and 500 nm. The regression models are estimated for the proposed optimization criteria as functions of the coded values (x) of the exposure dose $Q(z)$. The results are presented in table 3. The regression models in the case of a 200 nm single line are estimated for 15 observations (without the first one at the exposure dose of 200 μC/cm^2). The relations between the coded (x) and the natural (z) values of the exposure dose are given by the following equations:

- for a 200 nm single line: $x = (z - 500)/260$ (15 observations);
- for a 500 nm single line: $x = (z - 440)/320$ (17 observations).

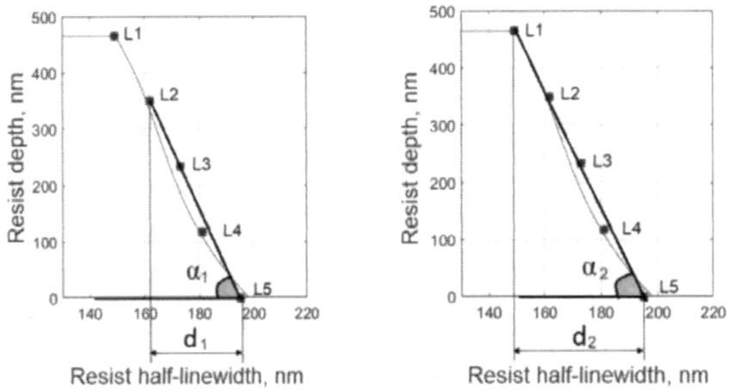

a) *L*5 – *L*2 resist levels. **b)** *L*5 – *L*1 resist levels.

Figure 9. Diagram and parameters for optimization criteria.

Table 3. Regression models for the optimization criteria as functions of the exposure dose z, μC/cm^2. The optimization criteria values based on the experimental data are denoted by red asterisks.

Parameter	Mean linewidth values $\bar{Y}_{LW}$	Profile sidewall angle α_1 for L5 and L2
Single line of 200 nm	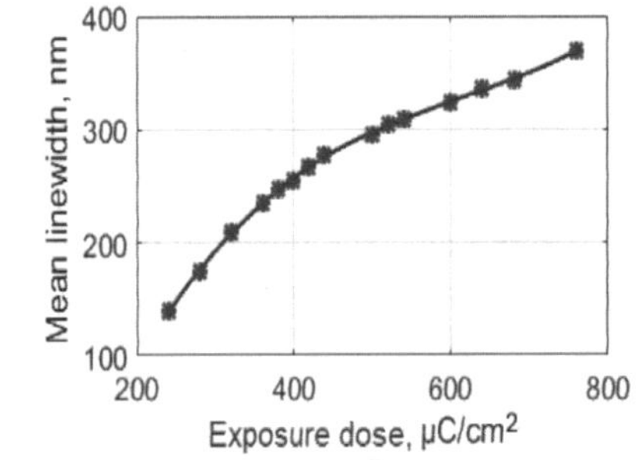 $\bar{Y}_{LW-200} = 297.24+83.00x-42.89x^2+32.08x^3$ *Determination coefficient:* $R^2 = 99.96$ %	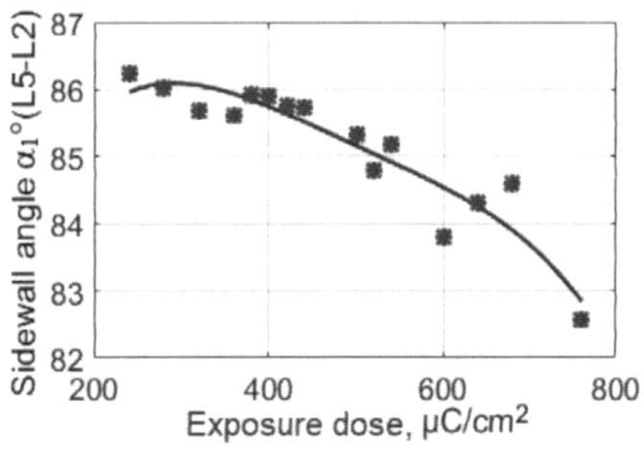 $\alpha_1 = 85.16-1.56x-0.76x^4$ *Determination coefficient:* $R^2 = 88.07$ %
Single line of 500 nm	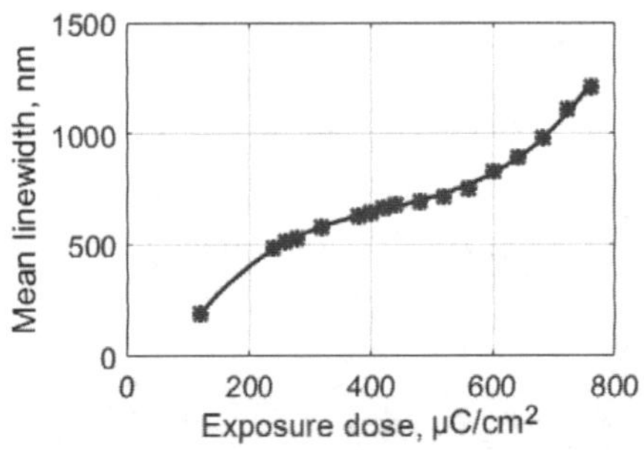 $\bar{Y}_{LW-500} = 669.79+208.19x+38.75x^2+308.99x^3$ *Determination coefficient:* $R^2 = 99.84$ %	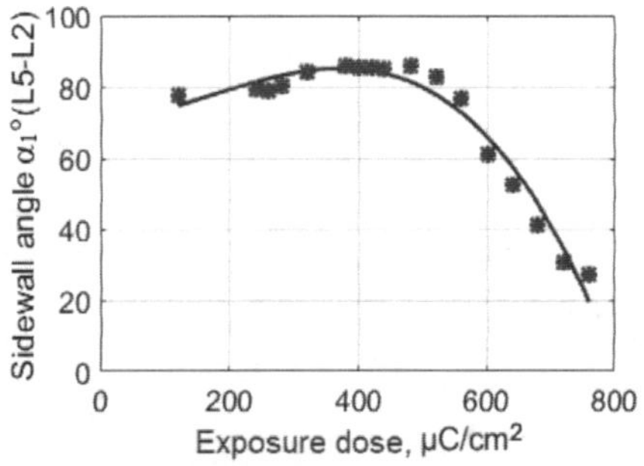 $\alpha_1 = 83.73-13.87x-36.69x^2-13.79x^3$ *Determination coefficient:* $R^2 = 96.12$ %

2.2.3. Optimization tasks. Using the estimated models obtained (table 3), different optimization tasks can be solved. Here are considered two single criterion tasks – optimization by the mean linewidth value $\bar{Y}_{LW}$ and by resist profile sidewall angle α_1 estimated at resist depths *L5* and *L2* (figure 9).

The optimal values for the mean linewidths ($\bar{Y}_{LW}$) and sidewall angles (α_1) for 200 nm and 500 nm single lines obtained at the corresponding optimal exposure dose values, and the calculated values of the other parameters (s_{LW}, d_1, d_2, α_2) are presented in table 4. The corresponding simulated optimal resist profiles obtained are demonstrated in figure 10.

Table 4. Optimization results – optimization by the mean linewidth value $\bar{Y}_{LW}$ and by the profile sidewall angle α_1.

Estimated parameters	Optimization task			
	$\bar{Y}_{LW-200}$= 200	$\bar{Y}_{LW-500}$= 500	$\alpha_{1,200}\to$ maximum	$\alpha_{1,500}\to$ maximum
Optimal exposure dose, μC/cm²	**308.48**	**253.22**	**291.35**	**371.17**
Mean linewidth value $\bar{Y}_{LW}$	**200.00**	**500.02**	186.43	623.72
Standard linewidth deviation s_{LW}	36.02	140.29	38.06	110.18
Half-linewidth difference d_1 at *L5* and *L2*	23.68	63.48	23.56	26.68
Profile sidewall angle α_1 at *L5* and *L2*	86.08	82.07	**86.10**	**85.15**
Half-linewidth difference d_2 at *L5* and *L1*	46.54	179.09	48.95	141.55
Profile sidewall angle α_2 at *L5* and *L1*	84.14	68.79	83.79	72.56

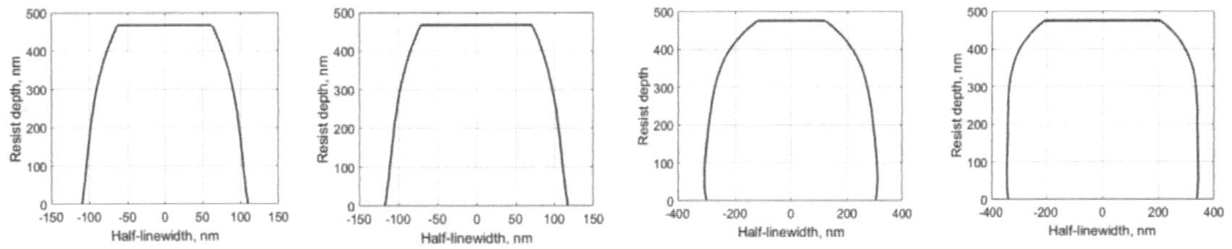

a) **b)** **c)** **d)**

Figure 10. Developed AR-N 7520 optimal resist profiles in the case of: a) 200 nm single line for exposure dose of 291.35 μC/cm², b) 200 nm single line for 308.48 μC/cm² exposure dose; c) 500 nm single line for 253.22 μC/cm² exposure dose; d) 500 nm single line at 371.17 μC/cm² exposure dose.

3. Conclusions

A study is presented on the negative EB resist AR-N 7520 for 40 keV electron energy. The influence of the pattern configurations and the exposure dose on the sidewall shape of the developed images in thin AR-N 7520 resist layers are studied experimentally and theoretically in order to achieve precise line gratings on a GaAs substrate. Models for prediction and optimization of the geometry of the developed resist profiles in the AR-N 7520 resist are estimated. Optimization criteria are proposed and optimization results concerning the sidewall shape are presented. The use of mathematical models, including certain characteristics of the object studied (resist profiles), makes it possible to determine more accurately the influence of lithographic process parameters (exposure doses, pattern dimensions, etc.) on the quality of the images developed in the investigated resist and to make recommendations on improving and optimizing the lithography process.

Acknowledgments

This work was supported by the Bulgarian Academy of Sciences and the Slovak Academy of Sciences (bilateral grant agreement between BAS and SAS) under a joint research project, by the Bulgarian National Science Fund under contract KP-06-N27/18, by the project SAS-BAS-21-04 of the Slovak Academy of Sciences, and by MESR and SAS under contract VEGA 2/0119/18.

References

[1] Zeitner UD, Oliva M, Fuchs F et al 2012 *Appl. Phys.* A **109** 789–796
[2] www.allresist.com
[3] Borzenko T, Fries P, Schmidt G, Molenkamp LW and Schirmer M 2009 *Microelectr. Engin.* **86** 726
[4] Lingkuan M, Xiaobin H, Jianfeng G, JunjieL, Yayi W and Jiang Y 2016 *Nanoscale Res Lett.* **11** 504
[5] Kley EB and Schnabel B 1995 *Proc. SPIE* **2640** *Microlithogr. and Metrol. in Micromachining* 71-80
[6] Andok R et al 2012 *Proc. 9th Int. Conf. ASDAM 2012 (Smolenice, SR)* (Piscataway: IEEE) 287–290
[7] Kostic I, Vutova K, Bencurova A, Ritomsky A and Andok R 2017 *Proc. 40th Int. Spring Seminar on Electronics Technology* (ISSE) *(Sofia, BG)* (Piscataway: IEEE) 8000969
[8] Kostic I, Vutova K, Koleva E and Bencurova A 2020 *J. Phys.: Conf. Ser.* **1492** 012015

Modification and activation of the surface of medical-grade PDMS after irradiation by ultrashort laser pulses

E Iordanova[1,3], G Yankov[1], N Stankova[2] and N Nedyalkov[2]

[1]Institute of Solid State Physics, Bulgarian Academy of Sciences, 72 Tsarigradsko Chaussee, 1784 Sofia, Bulgaria
[2]Institute of Electronics, Bulgarian Academy of Sciences, 72 Tsarigradsko Chaussee, 1784 Sofia, Bulgaria

E-mail: eiordanova@issp.bas.bg

Abstract. The present research aims to investigate the modification and activation effects on medical-grade polydimethylsiloxane (PDMS) polymer irradiated by ultrashort laser pulses. The motivation of the current study is based on the wide use of the PDMS material in medicine and, more specifically, in implantable neural interface applications. Systematic measurements are carried out to characterize the effect of the laser beam parameters on the optical absorption and the surface morphology with respect to the laser-treated zones. The PDMS polymer is modified by surface tracks after the femtosecond laser processing. The optical properties of the PDMS are investigated to prove the effective laser activation of the surface, which ensures further successful metallization of the modified tracks. It is seen that defective transformations occur in the entire laser-treated area when the number of pulses is increased. The preliminary observation indicates promising results regarding the implementation of such a laser-based method for micro- or nano-processing of optically transparent biopolymers for interface devices in bioengineering technologies, such as neural implants and interface applications.

1. Introduction

Due to its exceptional properties, the synthetic elastomer polydimethylsiloxane (PDMS) is the preferred biocompatible polymer in bioengineered interface technologies, such as long-term neural implants (bio-microelectrode arrays) in soft tissues. Among these properties are, e.g., high biocompatibility and biostability, high non-toxicity, mechanical flexibility and chemical stability, optical transparency from the UV to the near IR spectral ranges. The PDMS has thus become an excellent choice for fundamental studies in the fields of biology, chemistry, physics, medicine, and applications in neural implants and interface devices. The diversity of applications of medical grade PDMS requires an insight into surface modification in view of changing its morphology and chemical composition to functionalize the surface in the manner desired. The surface modification methods include, e.g., plasma treatment, chemical vapor deposition, layer-by-layer deposition, protein absorption, etc. [1-5].

An alternative method for modifications of PDMS, which has become the most preferred one, is laser treatment. The use of pulsed laser irradiation offers a new approach to the processing and selective surface functionalization of polymers. The laser treatment could be further applied to facilitate selective metallization of the areas activated by laser irradiation [6-9]. It is important to mention that treating the

[3] To whom any correspondence should be addressed.

Content from this work may be used under the terms of the Creative Commons Attribution 3.0 licence. Any further distribution of this work must maintain attribution to the author(s) and the title of the work, journal citation and DOI.
Published under licence by IOP Publishing Ltd

PDMS polymer using additives absorbing laser radiation is limited by the stringent requirements of non-toxicity. However, with the requirements for extremely high-quality and specificity of the properties needed in miniature volumes as used in implantology, laser processing is the preferred method for direct surface modification. It allows one to increase of the number of tracks and electrodes on a unit area (miniaturization of the device) while further increasing the resolution. The laser energy delivered to the polymer is sufficient to break the Si-O bond and cause chemical activation of the surface. Also, laser ablation can be used to form openings, craters or channels that possess a specific micro or nanostructured morphology. All these processes contribute to simplifying the subsequent functionalization of the treated area, because of, e.g., a higher adhesion of the noble metal (Pt) ions in the electroless plating, or a surface wettability change (hydrophobic surface to hydrophilic or vice versa). This is why, laser modification of the PDMS elastomer and other biopolymers, such as polymethyl methacrylate (PMMA), polymethyl acrylate (PMA), polystyrene (PS), polycarbonate (PC), etc., is currently the subject of intensive practical and fundamental research due to wide possible applications as microfluidic channels, electro-optical components, waveguides, storage devices, thin coatings in photolithography and biomedicine, neural-computer interfaces, lab-on-chip devices, dielectric elastomer actuators.

The present study deals with the effects of irradiating medical-grade polydimethylsiloxane (PDMS) elastomer by femtosecond UV laser pulses. Results of the surface modification and activation after laser irradiation are presented concerning the PDMS surface micromachining with the purpose of (i) fabricating tracks, (ii) activating the surface without altering its bulk properties, (iii) functionalizing successfully the tracks by electroless metallization with platinum (Pt). All the procedures enumerated together would allow the PDMS to be further used in specific applications, such as multielectrode arrays (MEAs) neural implantable interface devices.

2. Experimental configuration and settings

2.1. Materials and methods

The biopolymer sheets used in the study were medical-grade polydimethylsiloxane (PDMS) elastomer MED 4860, NUSil. The size of the samples was 1×2 cm with a thickness between 80 μm and 200 μm. The PDMS elastomer is optically transparent in the UV-Vis-NIR spectral ranges.

The laser-processed areas were characterized and analyzed by using various analytical techniques, including scanning electron microscopy (SEM) to image the surface morphology, optical spectrometry to analyze the transmission spectra in the UV-Vis-NIR range, and optical microscopy to observe the visible permanent surface modifications. Once the laser treatment was performed, the samples were metalized with a noble metal (Pt) by using an electroless deposition technique (autocatalytic bath). The electroless plating process is based on a hydrazine hydrate reducer at an optimal concentration determined empirically. Detailed information on the metallization procedure can be found in our previous work [10].

2.2. Setup configuration

The experimental setup consists of a femtosecond laser system (Spectra-Physics) and the corresponding periphery, including optical and optomechanical elements, see figure 1. The femtosecond laser system comprises several units – an amplifier (Spitfire Ace), a pumping laser (Empower), an oscillator (Mai Tai), and an optical parametric amplifier (TOPAS). The Ti: Sapphire regenerative amplifier provides a pulse width of ~35 fs, a pulse energy of ~6 mJ, and operates at a 1-kHz repetition rate. The main wavelength of the laser system is 800 nm, with the TOPAS unit providing a wavelength range from 240 nm to 2600 nm. The femtosecond laser is further equipped with a high-precision planar XY stage (Aerotech, Two-Axis Motor Stage), a power-meter (Ophyr, Nova II), and optical and optomechanical elements. An autocorrelator system (GRENOUILLE model 8-20-USB) is used to control and monitor the main output parameters of the fs laser. The laser beam distribution is analyzed by a laser-beam profiler (Spiricon, Inc. LBA-PC Laser Beam Analyzer). The PDMS samples are positioned on the XY planar stage to ensure smooth and precise translation of the sample through the laser beam focus. The laser beam is focused via a lens with a focal distance of 20 cm. The laser beam radius is determined to be 30 μm. The size of the beam is measured and controlled by the laser beam profiler.

Figure 1. Photographs of the experimental setup configuration presenting: a) the femtosecond laser system and b) the main periphery (e.g., autocorrelation unit, planar XY stage, power meter) as well as indications of the laser beam direction and the sample position.

3. Results

The medical grade PDMS elastomer investigated was modified by lasers ablation in air by the UV harmonic of 266 nm of the femtosecond Ti:Sapphire laser. The PDMS samples were irradiated and modified under a variety of laser conditions, including different laser fluence, number of pulses, and wavelengths. The number of applied laser pulses was varied from 100 to 10 000 in view of finding the optimal conditions with respect to the number of laser pulses. The threshold laser fluence achieving ablation and activating surface modification was found to be 0.18 J/cm^2.

3.1. Optical properties

The PDMS is optically transparent from the UV up to the NIR region of the optical spectrum. Laser treatment of such materials is possible due to the incubation effect, which increases the optical absorption because of the accumulation of defects and transformations below the surface as the number of pulses increases. Consequently, changes in the relief and browning of the native material occur. The moving speed s of the XY planar stage is related to the number of pulses N needed to achieve ablation in a unit area a (beam spot diameter) and to the laser repetition rate f by $N = af/s$. Thus, it is important to determine how the number of consecutive laser pulses (the incubation process) affects the optical parameters of the native material. This is why we performed optical transmittance measurements after subjecting a stationary sample to a rising number of laser pulses and a fixed laser fluence. Figure 2 shows the optical transmission of native (untreated) and of laser-treated PDMS samples at pulse numbers ranging from 300 to 10 000 at a wavelength at 266 nm and a laser fluence of 0.18 J/cm^2. It is seen that increasing the number of laser pulses results in the incubation occurring in the entire laser-treated area. This leads to an increase of the optical absorption and to a saturation of the incubation and, as a result, the laser ablation process becomes efficient. In fact, the optical microscope measurements demonstrated that the ablation depth increased from 4 μm to 20 μm after applying number of pulses increasing from 300 to 10 000. It was found that visible defects appear due to the incubation process after laser irradiation for an initial number of pulses higher than 300 pulses.

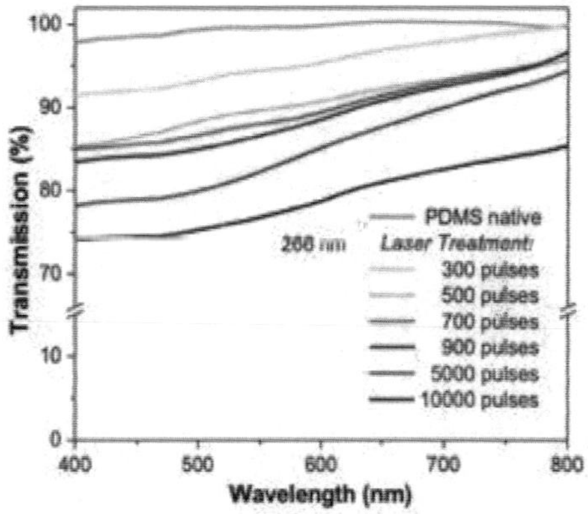

Figure 2. Optical transmission of the laser processed tracks at wavelength of 266 nm and different number of pulses.

As a result, the optical absorption of the laser-modified zones starts to increase. A further increase of the number of pulses leads to a significant decrease of the optical transmission, as shown in figure 2.

3.2. Morphology properties

The changes in the samples morphology were followed by scanning electron microscopy. Figure 3 presents tracks at different magnifications ((3a) at 20 μm and (3b) at 2 μm). It is seen for a number of laser pulses exceeding 1000, the surface morphology is significantly modified compared to the non-treated surface due to the efficient laser ablation. The tracks exhibit symmetric and wide-open shapes. No debris are observed. The image with higher magnification of the track interior (see figure 3b) shows a fine micro-structured relief resulting in activation of the surface. The SEM images prove that high-quality tracks can be formed by UV fs-laser treatment; the procedure facilitates the following process of functionalization by electroless metallization with Pt in view of application as multielectrode arrays of implantable interfaces.

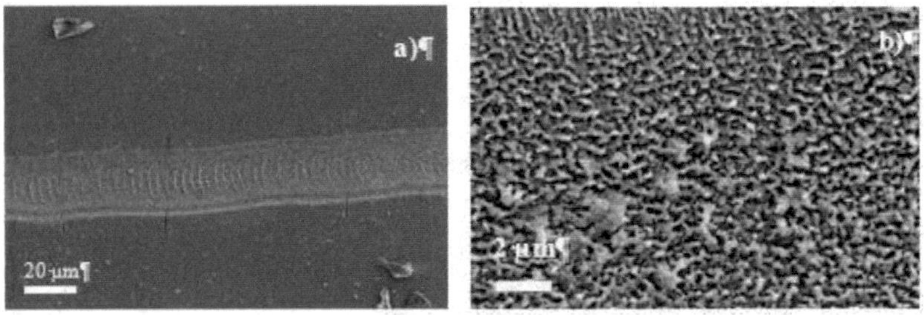

Figure 3. Scanning electron microscope images of PDMS MED 4860 tracks at different magnification ((**3a**) at 20 μm and (**3b**) at 2 μm), wavelength of 266 nm, pulse duration of 35 fs, laser fluence of 0.18 J/cm^2 and number of pulses 5 000.

3.3. *Metallization*

Once the samples were (i) modified successfully by laser irradiation under the selected optimal working conditions and (ii) analyzed by optical transmission and SEM, they were further subjected to the final step of the study, namely, metallization. The laser treatment was followed by a successful electroless deposition of Pt in the tracks without sensitization by Sn and chemical activation by Pd, which are usually preceding this process. Selective metallization was thus performed avoiding the application of masks or external templates. We also found that the time interval between the laser treatment and the metallization is not a critical process parameter, as suggested in [11]. Figure 4 presents tracks on the PDMS surface produced by the fs-laser processing at a wavelength of 266 nm and 5 000 pulses, before (figure 4 a) and after (figure 4 b) metallization demonstrating that a good-quality uniform metal coating deposition was achieved. The DC resistance measured was between 0.2 Ω/mm and 5.0 Ω/mm.

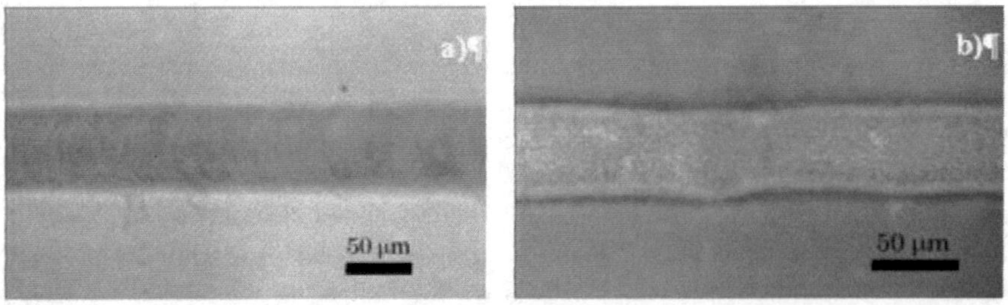

Figure 4. Optical microscope images of tracks produced at 266 nm and 5 000 laser pulses before metallization (**4a**) showing browning and ablation of the track after laser irradiation, and after metallization (**4b**)).

4. Conclusions

The modification and activation of the surface of a medical grade PDMS elastomer were studied after irradiation by femtosecond laser pulses at the wavelength of 266 nm. Despite the low optical absorption of the native PDMS in the UV range, successful laser treatment was accomplished due to the defects induced below the polymer surface. The optical absorption thus increased and saturation caused by the incubation process was observed as the number of pulses was increased. This process significantly enhanced the laser ablation efficiency and the optical transmission started to decrease significantly after irradiation with a number laser pulses exceeding 300, since the number of defects visible under the sample surface increased sufficiently because of the incubation – a process typically occurring during laser processing of optically transparent media, such as the PDMS elastomer. The ablation of the PDMS-elastomer started after 300 pulses and occurred efficiently after reaching incubation saturation at 5 000 pulses with the ablation depth increasing from 4 μm to 20 μm. The results obtained demonstrate the achievement of this study main aim, namely, uniform metallization with Pt of the laser-treated tracks by electroless plating. This study will be extended by a future investigation concerning the structural properties changes of the PDMS processed by femtosecond laser pulses in the VIS and NIR spectral ranges.

Acknowledgments

The authors acknowledge the financial support of the Bulgarian National Science Fund under project No. KP-06-H38/5, 2019, and of the bilateral joint research projects "Laser assisted processing of materials and characterization for top technology applications" and "Processing and characterizations of materials by ultrafast laser pulses" between the Bulgarian Academy of Sciences and the Romanian Academy of Sciences.

References

[1] Raj M K and Chakraborty S 2020 *Applied polymer* **137** (27) 48958
[2] Zhou J, Khodakov D A, Ellis A V and Voelcker N H 2011 *Electrophoresis, Microfluidics and Miniaturization* **33** (1) 89–104
[3] Vladkova T G, Keranov I L, Dineff P D, Youroukov S Y, Avramova I A, Krasteva N and Altankov G P 2005 *Nuclear Instrum. Methods Phys. Res.* B **236** 552–562
[4] Hillborg H, Ankner J F, Gedde U W, Smith G D, Yasuda H K and Wikstrom K 2000 *Polymer* **41** 6851–6863
[5] Donzel C, Geissler M, Bernard A, Wolf H, Michel B, Hilborn J and Delamarche E 2001 *Adv. Mater.* **13** 1164–1167
[6] Stankova N E, Atanasov P A, Nedyalkov N N, Kolev K, Valova E and Armyanov S 2019 Laser processing of biopolymers for development of medical and high-tech devices *Materials for Biomedical Engineering* ed A-M Holban and A M Grumezescu (Elsevier) chapter 15 pp 487–526
[7] Dicara C, Robert T, Kolev K, Dupas-Bruzek C and Laude L D 2003 *Proc. SPIE* **5147** 255–265
[8] Laude L D, Kolev K, Dicara C and Dupas-Bruzek C 2003 *Proc. SPIE* **4977** 578–586
[9] Laude L D, Cochrane C, Dicara C, Dupas-Bruzek C and Kolev K 2003 *Nuclear Instrum. Methods Phys. Res.* B **208** 314–319
[10] Stankova N *et al.* 2021 *Polymers* **13** (17) 3004
[11] Dupas-Bruzek C, Drean P and Derozier D 2009 *J. Appl. Phys.* **106** 074913

AUTHOR INDEX

Aceti, D M 217
Aceti, D 222, 227
Aleksandrov, L 260
Alipieva, E 118
Anchev, A 28
Andok, R 271
Andreeva, A 217
Andreeva, R 1
Angelov, O 47
Angelova, L 217, 222, 227, 238
Angelova, R 98
Anguelov, V 138
Atanasova, A 34
Atanasova, G 87
Atanasova, M 28
Avdeev, G 47
Avramova, I 7, 18, 23
Babij, M 123
Bahri, Hocine 128
bahri, Hocine 133
Bankova, A 12
Belina, E 82
Bencurova, A 271
Beshkova, M 18, 23
Bezdushnyi, R 233
Blagoev, B S 7, 23
Blagoev, B 98
Boev, V 103
Brazhnikov, D 118
Buchvarov, I 217, 222, 227
Carette, X 227
Chaikovsky, A 171
Chamati, H 103
Chitanov, V 76
Cholakova, T 76
Closset, R 123
Daskalova, A 217, 222, 227
Daskalova, Al 238
Dechev, D 54, 60, 233
Deleva, A 171
Delibatov, D A 7
Deminskyi, P 18
Dikov, Hr 161
Dikovska, A O 42
Dikovska, A 87, 238, 244
Dilova, T 87, 238
Dimitrova, T L 82
Dimov, D 47
Dobrev, D 12
Donchev, V 255
Donkov, N 66, 71
Dreischuh, T 143, 149, 171, 180
Dubrava, T 71
Dudin, S 71
Dunchev, V 28
Dyankov, G 82
Evgenieva, Ts 138, 143, 149
Filipov, E 217, 222, 227
Galeva, H 212
Genkov, K 47
Genova, J 103
Georgiev, N 156
Georgiev, S 255
Georgieva, B 23, 47, 92, 123
Georgieva, V 92
Gergova, R 161
Ghelev, Ch 98, 123
Goltsev, A 71
Grancharov, G 161
Grigorov, I 149
Grochowska, K 244
Groudeva, V 98
Gurdev, L 138, 143, 149
Hocini, Abdesselam 128, 133
Hristeva, N 76
Hristova-Vasileva, T 34
Hsu, C-W 18
Iliev, M 98
Ingebrandt, Sven 128
Iordanova, E 277
Iordanova, S 156, 166
Ivanov, I 108
Ivanov, K 47
Ivanov, N 54, 233
Ivanov, Tz 249
Ivanov, V 156, 191
Kaisheva, D 28
Kakanakov, R 76
Karaivanova, D 47
Karashanova, D 42, 47, 222
Karmakov, Y 266
Katrova, V 34
Khedrouche, Djamel 133
Kirilov, K 255
Kisov, H 82
Kiss'ovski, Zh 108, 196

Kolaklieva, L 76
Kolchev, St 76
Kolev, G 206
Kolev, N 149
Kolev, S 47, 118, 123
Kolev, St 156, 166, 191
Koleva, E 200, 206, 271
Koleva, L 200, 206
Koleva, M E 42
Koleva, M 238
Konecnikova, A 271
Kostic, I 271
Kotlarski, G 28, 54
Koutzarova, T 118, 123
Kovacheva, D 98, 123
Krezhov, K 123
Lazarov, J 92
Lazarova, S 156
Mankov, V 82
Marinov, S 108
Marinova, V 222
Markov, M 113
Mehandzhiev, V B 7
Mehandzhiev, V 23
Milanova, M 255
Milenov, T 47
Mincheva, R 227
Mitov, M 12
Nedkov, I 98
Nedyalkov, N N 42
Nedyalkov, N 87, 238, 244, 260, 277
Nikov, Ro 238, 244, 260
Nikov, Ru 87, 238, 244
Nikova, T 260
Ormanova, M 28, 54, 60, 233
Pachauri, Vivek 128
Pashinski, Ch 76
Paskaleva, A 249, 266
Paunska, Ts 156, 191
Pedersen, H 18
Penchev, S 180
Pencheva, V 180
Pescherenkov, V 171
Peshev, Z 171
Petrov, P 28, 54, 60, 233
Petrova, Z 206
Popkirov, G 161
Rabadzhiyska, S 28, 60
Rafailov, P M 7
Safonov, V 66, 71
Salah, Hocine Ben 128,133
Savov, P 149
Sendova-Vassileva, M 161
Seremak, T 66
Shtepliuk, I 18
Slavkova, Z 103
Slavov, L 98
Sofronov, Y 212
Sotelo, L 217
Spassov, D 249, 266
Stanchev, T 249
Stankova, N 238, 277
Starbov, N 92
Starbova, K 92
Stefanov, P 92
Stefanova, V 185
Steller, J 66
Stoyanov, B 28
Stoychev, D 1
Strijkova, V 23, 34
Sveshtarov, P 7
Tarnev, Kh 191
Taskova, E 118
Terziyska, P 7, 23, 47, 60, 161, 255
Todorov, G 212
Todorov, R 34
Todorov, V 28
Toncheva, E 143
Tran, L-M 123
Trifonov, A 217, 222, 227
Trushnikov, Dm 206
Tsonev, P 191
Tsonevska, Ts 200
Tzaneva, B 12
Uzunov, T 212
Vachkov, V 108, 196
Valcheva, E 47
Valkov, S 28, 54, 60, 233
Vasilev, V 166
Vassileva, V 113, 185, 200
Vertruyen, B 123
Videkov, V 12
Vladkov, E 166
Vulkova, L 171
Vutova, K 113, 185, 200, 271
Yakimova, R 18, 23
Yakovin, S 71
Yancheva, D 103
Yankov, G 277
Zaleski, A 123
Zegaar, Imane 128, 133
Zykova, A 66, 71